Sketching User Experiences

Sketching User Experiences

getting the design right and the right design

Bill Buxton

AMSTERDAM • BOSTON • HEIDELBERG • LONDON
NEW YORK • OXFORD • PARIS • SAN DIEGO
SAN FRANCISCO • SINGAPORE • SYDNEY • TOKYO

Morgan Kaufmann is an imprint of Elsevier

Publisher: Diane Cerra
Publishing Services Manager: George Morrison
Senior Project Manager: Brandy Lilly
Editorial Assistant: Asma Palmeiro
Copyeditor: Adrienne Rebello
Indexer: Distributech Scientific Indexing
Interior printer: Transcontinental Interglobe
Cover printer: Transcontinental Interglobe

Book Design: Henry Hong-Yiu Cheung
Information Graphics: Henry Hong-Yiu Cheung
Typography Consultant: Archetype
Illustrations: Adam Wood

Morgan Kaufmann Publishers is an imprint of Elsevier.
500 Sansome Street, Suite 400, San Francisco, CA 94111

This book is printed on acid-free paper.

Library of Congress Cataloging-in-Publication Data

Buxton, William.
Sketching user experience : getting the design right and the right design / Bill Buxton.
p. cm.
Includes bibliographical references and index.
ISBN-13: 978-0-12-374037-3 (pbk. : alk. paper)
ISBN-10: 0-12-374037-1 (pbk. : alk. paper) 1. Design, Industrial. I. Title.
TS171.B89 2007
658.5'752--dc22
2006036416

ISBN 13: 978-0-12-374037-3
ISBN10: 0-12-374037-1

For information on all Morgan Kaufmann publications, visit our Web site at www.mkp.com or www.books.elsevier.com

Printed in Canada
08 09 10 11 5 4 3 2

To the love of my life, Lizzie Russ.

Contents

Preface

If Ernest Hemingway, James Mitchener, Neil Simon, Frank Lloyd Wright, and Pablo Picasso could not get it right the first time, what makes you think that you will?
— Paul Heckel

This is a book about design. Mainly, it is about the design of appliances, structures, buildings, signs, and yes, computers, that exist in both the physical and behavioral sense. That is, there may be something concrete that you can touch, see, and hear. But there is also something that you can actively experience: something that involves dynamics or time; something with behavior that is usually the result of software running on an embedded microprocessor; and something whose design needs to be grounded in the nature of that experience.

The underlying premise of the book is that there are techniques and processes whereby we can put experience front and centre in design. My belief is that the basis for doing so lies in extending the traditional practice of sketching.

So why should we care about any of this?

Hardly a day goes by that we don't see an announcement for some new product or technology that is going to make our lives easier, solve some or all of our problems, or simply make the world a better place.

However, the reality is that few of these products survive, much less deliver on their typically over-hyped promise. But are we learning from these expensive mistakes? Very little, in my opinion. Rather than rethink the underlying process that brings these products to market, the more common strategy seems to be the old shotgun method; that is, keep blasting away in the hope that one of the pellets will eventually hit the bull's eye.

Now if this is a problem, and I believe that it is, it shows every indication of getting worse before it gets better. The pundits, such as Weiser (1991) and Dourish (2001), as well as those contributing to Denning & Metcalfe (1997) and Denning (2001), tell us that we are in the midst of a significant transition in the very nature of the products that are going to emerge in the future. Others, such as Forty (1986) and Borgmann (1987) would say that this transition began a long time ago. Both are accurate. The important point to recognize is that whenever it started, the change has reached a tipping point (Gladwell 2000), where we ignore it at our peril if we are in the business of creating new products.

By virtue of their embedded microprocessors, wireless capabilities, identity tagging, and networking, these products are going to be even more difficult to get right than those that we have produced (too often unsuccessfully) in the past. For those of us coming from computer science, these new products are going to be less and less like a repackaging of the basic PC graphical user interface. For industrial designers, they are no longer going to be the mainly passive entities that we have dealt with in the past. (The old chestnut problem of the flashing "12:00" on the VCR is going to look like child's play compared to what is coming.) For architects, buildings are going to become increasingly active, and reactive, having behaviours that contribute as much to their personalities as do the shapes, materials, and structures that have defined their identity in the past.

And then there is the business side. These new products are going to present a raft of new challenges to the product manager. Finally, company executives are going to have to acquire a better understanding of the pivotal role of design in achieving their business objectives, as well as their own responsibilities in providing the appropriate leadership and environment where innovation can thrive, as opposed to just survive.

On the one hand, the change that confronts us is rooted in the increasingly rich range of behaviours that are associated with the products we are being asked to create. These products will be interactive to an unprecedented degree. Furthermore, the breadth of their form and usage will be orders of magnitude wider than what we have seen with PCs, VCRs, and microwave ovens. Some will be worn or carried. Others will be embedded in the buildings that constitute our homes, schools, businesses, and cars. In ways that we are only starting to even imagine, much less understand, they will reshape who does what, where, when, why, how, with whom, for how much, and for how long.

On the other hand, as suggested by this last sentence, the extended behaviours of these products will be matched, and exceeded, by the expanded range of human behaviour and experience that they enable, encourage, and provoke—both positive and negative.

Some academics, such as Hummels, Djajadiningrat, and Overbeeke (2001), go so far as to say that what we are creating is less a product than a "context for experience." Another way of saying this is that it is not the physical entity or what is in the box (the material product) that is the true outcome of the design. Rather, it is the behavioural, experiential, and emotional responses that come about as a result of its existence and its use in the real world. Though this may always have been the case, this way of describing things reflects a transition to a different way of thinking, a transition of viewpoint that I characterize as a shift from object-centred to experience-centred.

And it is not just academics touting the experience-centric line for both products and services. It is also reflected in the titles that we find in the business sections of airport bookstores, such as *Priceless: Turning Ordinary Products into Extraordinary Experiences* (LaSalle & Britton 2003), *Building Great Customer Experiences* (Shaw & Ivens 2002) and *The Experience Economy* (Pine & Gilmore 1999). However, my favorite way of hitting this particular nail on the head comes from a designer friend, Michael Kasprow, the creative director of Trapeze, a design firm in Toronto:

Content is content ... Context is KING.

New labels tend to carry with them the risk of being reduced to a trendy change in language, rather than any significant change in substance. It is one thing to talk about experience design; embracing it in one's practice is quite another. Expanding the sphere and responsibilities of design to include such experiential concerns carries with it a very real burden—a whole new level of complexity, especially if we factor in the broad range of emerging technologies that are involved. It really requires a rather different mind-set and range of concerns than those that traditionally have driven the practice of design and engineering.

For example, think about the introduction of texting (more properly called Short Messaging Service, or SMS) into mobile phones. The traditional object-centred approach would view SMS as the design of a protocol to enable text messages to be sent between phones, and then its implementation in hardware and software (along with the associated model for billing for the service). Yet that description does not even begin to accurately characterize the real nature of SMS. This is far more accurately reflected by activities such as voting for your favourite performer in American Idol, or flirting with someone across the floor in a dance club. That is SMS, and I don't believe that you will find anyone involved in its design who would claim that they anticipated, understood, much less considered any of that when they were designing the feature.

This SMS example leads us to yet another dimension in which these emerging products are becoming more complex: increasingly, the technologies that we design are not islands—that is, they are not free-standing or complete in their own right (to the extent that they ever were, but more on that later). Rather, they are social entities. As with people, they have different properties and capacities when viewed as a collective, within a social, and physical context, than they have when they are viewed in isolation, independent of location or context. For example, just as I behave differently when I am alone than I do when with others (among other things, I talk with them, but hopefully not to myself), so it will be with our devices. When they approach other devices, or possibly people, they will become social animals. Just like you and me, their behaviour will vary, depending on whom they are with, in the same way you and I behave differently with family than we do with strangers, business colleagues, or alone.

Success in this emerging world is going to depend on significant change in how we work. Nevertheless, I believe that this change can respect the best of the traditions of the past—that is, it involves change that builds on, rather than replaces, existing skills and practice. It is change that must recognize, acknowledge, and respect the importance and interdependence of the different design, engineering, management, and business disciplines involved. Each is essential, but no single one is sufficient.

As it is with people, so it is with technology. Industry also must learn to reconcile these interdependencies with the idiosyncratic properties—and demands—of the new technologies and types of products that it is trying to bring to market. And to really succeed, these products must be reconciled to the needs and values of the individuals, societies, and cultures to which they are being targeted.

However, today's reality is that in this equation, the value of design is too often being questioned, and the contributions of the designer are being seen as an expensive luxury. Similarly, in software products, we are seeing the notion of user interface design disappearing as a professional description, too often being replaced by usability engineering, something that is ever more remote from something an industrial designer, for example, would recognize as design.

Against this backdrop is the compelling observation that there may never have been a time when design was more important, and the specific skills of the designer more essential. And yet, with far too few exceptions (such as Sharp, Rogers & Preece 2007 and Moggridge 2006) design as it is currently taught and practiced is better suited for how things have been in the past, rather than meeting the demands for what is coming in the future.

The psychologist Jean Piaget has defined intelligence as the ability of an organism to adapt to a change. The substantial—and largely technology-induced—changes affecting us now are a clear challenge for the design professions to adapt their skills to the redesign of their own practice. This is not only the intelligent response, it is essential if design is to fulfill its potential role in shaping our collective future.

> Technology isn't destiny, no matter how inexorable its evolution may seem; the way its capabilities are used is as much a matter of cultural choice and historical accident as politics is, or fashion. (Waldrop 2001; p.469)

This book is based on the premise that design is a distinct discipline. It involves unique skills that are critical to the molding of these emerging technologies into a form that serves our society, and reflects its values. Far from being a luxury, informed design is essential from the technical, economic, and cultural perspective.

A second key premise is that although design is essential, it is not sufficient. Design is just one—albeit an important one—of the components requisite to the development of successful, appropriate, and responsible products.

But what is the role of the designer? How does design fit in among all the other components of the process? For example, how does design fit in with engineering, marketing, or the corporate plan and executive objectives? If I am an architect, industrial designer, environmental graphic designer, or software developer, what is my role? What skills do I need to cultivate? How can they best be deployed? If I am an educator, what should I be teaching so as to prepare my students for what is coming, rather than how things have been done in the past? If I am an investor, businessperson, or manager who aspires to bring new products to market, how do I staff my teams? What kind of process should I put in place?

These are the types of questions that motivated me to write this book. For over thirteen years, but especially in the past four or five, I have had the pleasure and privilege to work with some of the world's most outstanding designers, from almost all disciplines. I was a willing student, and they were generous teachers. Having come from a background in the arts (music) and technology (computer science), as well as some experience as an executive of a mid-sized company, I have been lucky. I was in an unusual position to see things from a unique and privileged perspective. My hope is that what I have written here is respectful of those who were my teachers, and worthy of the confidence that they placed in me.

Ultimately, this book is about product design, with an emphasis on products that have dynamic behaviour due to the incorporation of embedded digital technology. It tries to address the topic while looking in a few different directions: outward, to other parts of the overall organization that is trying develop the product; further out, to the users and even culture within which the product is destined; and inward, to the staff, techniques, and methods of the design team itself.

My approach is built largely around case studies and examples, supplemented by a discussion of the underlying issues. Hopefully, along the way, I will shed light on some of the key questions that might help us innovate effectively in this ever more technologically complex environment. What are some of the core skills that one should expect in a modern design team? What should be taught in design programs, and how? What are some of the issues of managing a modern design team compared to engineering?

This is a start. It is a rough sketch. The best that I can hope is that its timeliness will in some way compensate for its broad strokes.

After all, isn't that what sketching is all about?

Audience

To have the intended impact, this book must address multiple audiences. No matter how well one group performs, it is unlikely that the overall job will be done successfully unless the rest of the organization is working in concert. For this to happen, everyone has to be singing from the same song sheet. In order to help bring this about, I have tried to speak to the following key groups:

User interface designers: people who have primarily a software and/or psychology background, and who have traditionally dealt with things like icons, navigation, menus, search, etc.

Industrial designers: people who largely studied at a design school, and whose specialty is product design.

Related design professionals: people who have studied architecture, environmental graphic design, illustration, film making, etc.; that is, those who make up part of the ever-richer mosaic of the new experience design team.

Software engineers: people who are hard-core computer scientists, who have the responsibility to build shipping code that is robust, maintainable and meets specification.

Usability engineers: people who test and evaluate products during their development in order to ensure that they are fit for human consumption and that there are no unexpected negative surprises for the user.

Product managers: people who typically come from marketing or an MBA type of background, who have to perform like the conductor of an orchestra of disparate instruments.

Executives: the people at the top of organizations who ultimately own responsibility for providing both leadership as well as a physical, intellectual, and organizational ecology in which design and innovation can thrive.

That is a start, at least. I know that I have left off some, but the list is already daunting, if not foolhardy. And, just to make things more complicated, my hope is to write something that speaks to each of these groups, regardless of how you slice the pie among the following categories:

Student
Teacher
Practitioner
Researcher

So although this is not a textbook, it should be of use to teachers and students. It is not a recipe book that you can throw at a product development team and say, "Go and implement this." But it should help companies understand how to improve their performance.

I have tried to provide examples and ideas that will help all readers in the practice of their day-to-day profession. But what is foremost in my mind is to paint a larger, holistic picture. My overarching goal is to help this diverse cast of characters better understand their role in the larger intertwined performance that constitutes our companies, schools, and practices. In this I have attempted to find a balance between going into sufficient depth in any topic so as to have relevance to the specialist, while still sustaining comprehensibility and interest for others.

Overview

The rest of this book is structured in two parts.

Part I lays the foundation. It talks about the current state of design, as well as much of the underlying belief system that drives my thoughts on experience design. Much of the focus is on software product design. Initially, this may seem remote, or not too relevant to an industrial designer, for example. However, I think it is important for someone from the traditional design professions to understand the state of software design.

There are two reasons for this. First, it will probably remove any lingering delusions that the software industry will come in with some magic bullet and solve all our current problems. The second reason is that there is nothing even vaguely resembling a "design process" in software, at least not in the way that an industrial designer would understand that term.

I realize that this is both a contentious and provocative statement, and in making it I am not trying to denigrate software developers. Rather, the exercise that I am trying to bring about is for them to look more deeply into the skills and practice of design professionals, and compare, contrast, and understand these with their own. Neither design nor engineering is sufficient for the task at hand, and both are essential. What is required is a new relationship involving an adaptation of both skill sets that reflects the demands of the new design challenges.

Part II is about methods. It moves to a more pragmatic realm. It is primarily made up of a number of case studies and examples. If we bump up a level, it is about techniques for bringing design thinking to the design of interactive products and the experiences that they engender. It is also a bit of a history lesson. It is intended to provide some exercises and examples that one can work through in order to help build a base literacy around the problems of designing in this new space. These include problems of process. I have chosen many examples from the prior art. Despite repeatedly talking about "new" problems and "new" design challenges, there is simply not a broad awareness of what has already been done. This is good news. It means that we don't have to start from scratch, and by building our awareness of the literature, we can proceed from a much better position. To emphasize that what I am advocating is within our reach, I have balanced the examples of past masters with several from young students of today.

We end with a coda. It tries to synthesize some of what we have seen. It is both a summary as well as an essay around the implications of what we have discussed. This section will be of aid not only to the student and practitioner, but to the educator, the manager, companies developing new products, and even governmental policymakers.

Finally, in addition to the book itself, there is a web page containing supplementary material (www.mkp.com/sketching). Perhaps the most important of this is the collection of video clips that are referenced throughout the text. Given that we are dealing with experiences that have a strong temporal component, these really help bring much of the material to life.

Overall, my goals are probably overly ambitious. But after all, aren't designers supposed to be dreamers?

Leadership = CEO

Stewardship = COO

Resource Management = CFO

Technology = CTO

Design = ?

The issue of executive responsibility is especially important to me. This stems partially from my experience as an executive, and partially by analyzing the structure of companies that I would describe as successful repeat offenders—that is, companies that have grown their business through consistent innovation and design. So let's get to the point by asking a few questions:

Who makes design decisions in your company?
Do they know that they are making design decisions?

If you are the CEO of the company:

Is design leadership an executive level position?
Do you have a Chief Design Officer reporting to the president?

If the answer to the last two questions is no, remember that your actions speak louder than your words. In this case, the likely message that you are telegraphing to all of your employees is that you are not serious about design or innovation, and you are also sending an implicit message that they need not be either.

Why employ designers, researchers and other creative people, if you are not going to set them up to succeed?

As an executive, of course you have to have creative and innovative ideas. But perhaps at the top of the list should be ones around (a) how important innovation is to the future of your company, (b) the role of design in this, (c) a recognition that innovation cannot be ghettoized in the research or design departments, since it is an overall cultural issue, and (d) an awareness of the inevitable and dire consequences of ignoring the previous three points. (Buxton 2005; p. 53)

Acknowledgements

I am seriously indebted to a number of people who helped me develop my understanding of design, and provided suggestions and materials for this book. These people are far more than colleagues—they are friends, and partners in crime.

To begin, there are a few people with whom I have worked intimately over the years, who have had a huge influence on my thinking and experience in general, as well as in terms of this specific project. In particular, I want to acknowledge my colleagues and friends Ron Baecker, George Fitzmaurice, Gord Kurtenbach, Abigail Sellen, and K.C. Smith.

As I was getting started, some of my friends from the design community were instrumental in giving me encouragement, confidence, and materials. To this end, I especially have to acknowledge the contributions of Alistair Hamilton of Symbol Technologies and Michael Sagan of Trek Bicycles.

Along the way, a number of people have contributed insights, materials, suggestions, and in general, much of which brought this book to life. These include, listed in alphabetical order, Mattias Arvola, Linköpings Universitet; Tim Bajarin, Creative Strategies; Richard Banks, Microsoft Research Cambridge; Larry Barbera, Kaleidoscope; David Biers, University of Dayton; Ron Bird, 2 Birds Design, UK; Andreas Butz, University of Munich; Dawn Chmielewski, San Jose Mercury News; Hugh Dubberly, Dubberly Design Office; Herbert Enns, University of Manitoba; Lars Erik Holmquist, Viktoria Institute; Adi Ignatius, Time Magazine; Hiroshi Ishii, MIT Media Lab; Scott Jenson, Google; Steve Kaneko, Microsoft; Dave Kasik, Boeing; Ianus Keller, TU Delft; Butler Lampson, Microsoft Research; Joseph LaViola, Brown University; Don Lindsay, Microsoft; Gene Lynch, Design Technologies; Alex Manu, Ontario College of Art and Design; Roger Martin, Rotman School of Business, University of Toronto; Bruce Mau, Bruce Mau Design; Tyler Millard, Toronto; Dan Olsen, Bringham Young University; Mike Roberts, Brook Stevens Design; John Sören Pettersson, Karlstad University; Frank Steinicke, Westfälische Wilhelm Universität of Münster; Dan Rosenfeld, Microsoft; Bob Spence, Imperial College; Louise St. Pierre, School of Design, University of Washington; Ilene Solomon, Institute without Boundaries; Pieter Jan Stappers, TU Delft; Erik Stolterman, Indiana University; Alex Tam, Toronto; John Tang, IBM Research; Lucia Terrenghi, University of Munich; Richard Turnnidge, IDEO; Frank Tyneski, Research in Motion, Smart Design, New York City; Karon Weber, Yahoo!; and Shumin Zhai, IBM Research.

I also want to acknowledge the huge contribution of my friends, Richard Harper, Caroline

Hummels, Jonas Löwgren, Steve Pozgaj, and Ken Wood. After a thorough reading of an early draft, Steve and Richard made a huge contribution to improving the content, the tightness of my prose, and flow of my story. Jonas, Caroline, and Ken made a similar, really significant contribution through their detailed comments on later drafts. Then Richard went through the whole thing again, helping me restructure things and bring the project to a close. I could not have done it without them.

Likewise, I want to thank my friend, partner in the mountains, and colleague, Saul Greenberg of the University of Calgary, who fed me examples and provided professional guidance, moral support, and welcome diversions throughout this project.

In transforming my manuscript into this book, I want to thank my cycling partner and designer, Henry Hong-Yiu Cheung and my son and illustrator, Adam Wood. They did a wonderful job of making the book itself reflect the message that it contained. From Morgan Kaufmann, I want to thank my editor, Diane Cerra for her unflagging support, Asma Palmeiro, the associate editor working with us, Brandy Lilly for her great project management, and Marisa Crawford who helped with the not inconsiderable task of securing permissions.

Furthermore, I would be remiss if I did not acknowledge the help and support that I have gotten from my colleagues at Microsoft, and the support of Microsoft Research, which gave me the opportunity to finish this project.

Perhaps most of all, I have to single out Azam Khan. He was both my biggest cheerleader and toughest taskmaster. His help, prodding, editorial suggestions and detailed proofreading—not to mention his friendship—have been as much appreciated as they were valuable.

Collectively, all these people lend proof to my conjecture that we are far less smart individually than we are collectively. This sharing of knowledge and insight is fundamental to intellectual and cultural growth.

Finally, those mentioned being good teachers, friends, and colleagues does not make me a good student. Consequently, any mistakes, distortions, misinterpretations, stupidities, or other generally goofy things in this manuscript are the result of my own shortcomings, not theirs. God knows they tried. And so did I. It is just that we are all human. That we are all human, by the way, is the point of the whole exercise. We need to keep that in the forefront, all the time—but especially in design.

Author's Note

In some ways, compromise can be defined as annoying everyone equally. So, if you are annoyed by some of the decisions that I have taken in writing this book, take solace in the fact that I have done my best to make sure that you aren't alone.

On the one hand, I have tried to write in an informal and approachable style. Simply put, a large part of the audience that I am trying to address, and think critical to address, are not interested in reading a deep academic tome. That is just how it is, and I am a pragmatist. On the other hand, I have worked very hard to ensure that there is solid scholarship behind what I write. The challenge is to figure out how to balance these two things.

For one group of readers, my casual style may be off-putting. For the other, my embedding of citations in the text will be unfamiliar, and disrupt the flow of their reading. But despite some strong suggestions to the contrary, I have chosen to keep the references in. Besides just wanting to remind the reader that this is not all coming off the top of my head, I believe that after very few pages, the references will fade into the background for those readers who want to ignore them, while being invaluable for those who don't. As I said, it is a compromise.

Note: Photos without credit are the work of the author.

Finally, throughout the book the reader will encounter relatively brief sections that are broken out from the main text. These fall into three categories: extended quotes or thoughts of others, points of my own that I want to emphasize, and sidebars (that is, side stories that relate to or amplify the adjacent text). I have used the following conventions to distinguish these:

- Quotes and thoughts of others are indented and in small type

- Points of my own are indented and in bold

- Sidebars are hi-lighted with yellow background

I hope this helps. Thanks for your patience.

Part I:
Design as Dreamcatcher

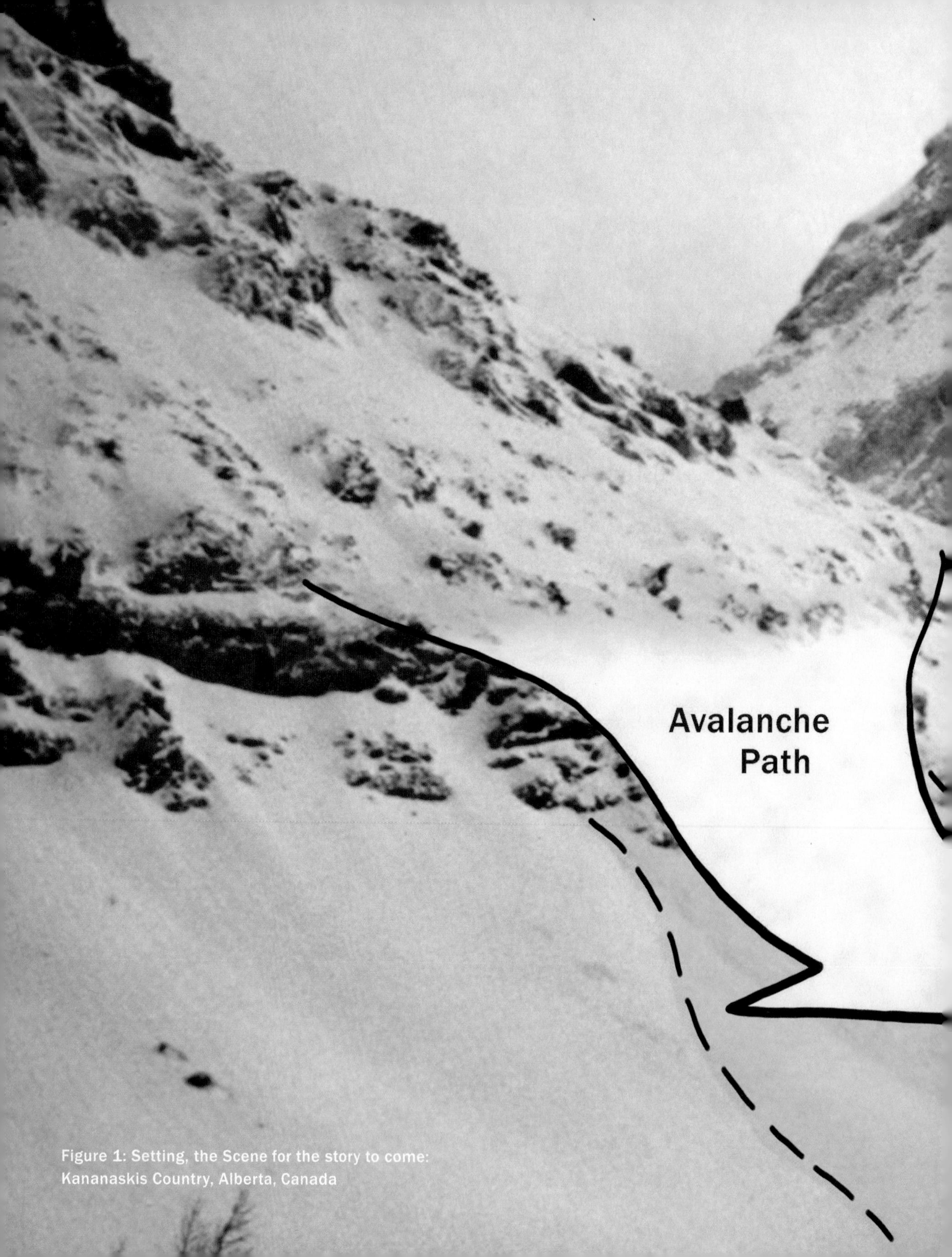

Avalanche Path

Figure 1: Setting, the Scene for the story to come: Kananaskis Country, Alberta, Canada

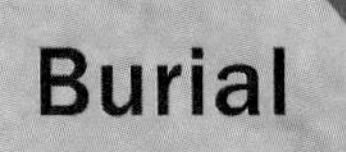
Saul's Ski Path
Burial

Figure 2: The protagonist, Saul Greenberg.
On Pigeon Spire (3156m) in the Bugaboos in Canada, Summer of 2006

Design for the Wild

No Risk is the Highest Risk.
—Aaron Wildavsky

I want to start with a story.

I have a good friend named Saul Greenberg. He is a professor at the University of Calgary, and one of the world's leading researchers in the area of human–computer interaction.

It is through that profession that we met. We collaborated on a book (Baecker, Grudin, Buxton & Greenberg 1995), and met at conferences. But it was our mutual passion for mountains and the outdoors that cemented our relationship.

I have been told on occasion that I am extremely competitive. Confessing this will hopefully give all the more weight to my public acknowledgment that Saul is a far more experienced mountaineer than I am. I love the time that we have spent in the mountains, not just because I love mountains, or because he is great company. Saul is also a great and generous teacher and someone whom I could, and frequently do, trust with my life.

A couple of years ago, Saul almost lost his life. He was skiing with his wife, Judy, and three other friends. They were, essentially, in their backyard, behind Canmore Alberta. It was spring, conditions were great, and they were in terrain that they knew extremely well. And yet, Saul was caught in an avalanche and buried under almost two metres of snow.

In many ways, this book is about why Saul is alive today and why I still have the pleasure of climbing and skiing with him—frequently in terrain not unlike where he was caught.

So, given how lethal avalanches are, why is Saul still alive? The simple answer is that Judy dug him out before he suffocated.

However, if you have any experience with such things, you will know that nothing is that simple. For example, why weren't Judy and the others also swept up by the slide?

Let me answer that question. To minimize the likelihood of this happening, the normal procedure whenever traversing avalanche terrain is to spread out. If the risk is perceived to demand it, you go one-by-one, and in either case, you always try to have lookouts. These people remain in a safe position, spotting the location of the person(s) doing the traverse. That way, there is someone who knows where the victims were last seen in the event that something happens.

Although the avalanche risk on this particular day was considerable, they had been skiing terrain far more severe, so they deemed the lesser precaution of spreading out as adequate for this slope. As it turned out, this was a bad call. They had not spread out far enough so two others in the party also got caught. One was buried up to her shoulders. The other, Shane, ended up on

the surface. Saul was the only one completely buried.

The people caught were the inner three of the five. Judy, at the front, had traversed the slope safely, and was playing the role of lookout from lower down. The last person in the party, Steve, had held back, and was spotting from above.

Therefore, when the slide occurred, Steve and Judy were well positioned, organizationally as well as physically, to do what was required. Normally, if there are enough people, one of the lookouts will stay in a safe spot. This is a safety precaution in the (all too frequent) event that a second slide catches the would-be rescuers. In this case, because they were the only two not caught, and given their assessment of the risk, both lookouts went to the aid of those caught.

Steve, who was higher, checked up on Shane (who was okay), and then immediately went to his wife. He freed her arms, and made sure her head was above the snow. He then went down to where Saul was buried. (This is a form of triage that you do, making sure the most visible people are clear just to the point that they are safe, and then go for the longer or deeper burials.)

Judy went directly to the spot where she had last seen Saul. What she did not do is immediately start digging, since the chance of her finding him based on that would be almost nil. Among other things, the slide would have likely carried him from where she last saw him. Furthermore, even if she guessed approximately where he might be, she would not be able to dig around looking for him. No matter how powdery soft the snow was before the avalanche, the heat generated by friction during the slide would melt it, and when it stopped, it would freeze to a consistency most resembling concrete. It is difficult enough to dig one hole—digging holes as a search strategy is an exercise in futility, almost certainly with death as a consequence.

In order to pinpoint Saul's location, Judy used her avalanche transceiver. In computerese, this is a wireless collaborative PDA with a multimedia (audio/visual) user interface, such as the one shown in Figure 4. Using this, she walked a particular pattern on the snow, employing the loudness of a ping (determined by the strength of a signal from Saul's transceiver) to guide her closer and closer to a spot above where he was buried.

When the transceiver search has indicated the likely burial point, normally the next step is to pinpoint the exact location using an avalanche probe. Illustrated in Figure 5, these look like three- to four- metre long skinny tent poles. You push them down into the snow, in a regular pattern, in the area indicated by your beacon. When the victim is felt you then start to dig.

In this case, Judy took a calculated risk, and skipped the probe step. She started to dig immediately at the location indicated by her transceiver. This was when Steve arrived. Having verified that she had not confirmed the exact location with her probe, he asked if he should do so. She said no, stating that her visual tracking followed by the transceiver search left her reasonably confident that she was in the right spot. However, she had to dig so deep that this confidence was starting to waver just before she got to Saul.

So let's focus on Saul. What did he do once he realized his situation? The first thing he tried was to ski out of it. This is generally difficult; avalanches can travel at up to 200 km/hr. (For most of us, 40 km/hr is skiing really fast.) In this case, the combination of the conditions and Saul's technique enabled him to ski down with the slide. The problem was at the base of the slope, where it immediately rose to a knoll, thereby creating a kind of trough. This is what we call a

Figure 3: Spreading Out in Avalanche Terrain

When traveling in terrain where there is avalanche risk, you spread out so that if there is an avalanche, you minimize the number of people caught in it. Someone always watches from a safe position (in this case it is the photographer) in order to spot where people are, in the event of a problem. Photo: Pat Morrow

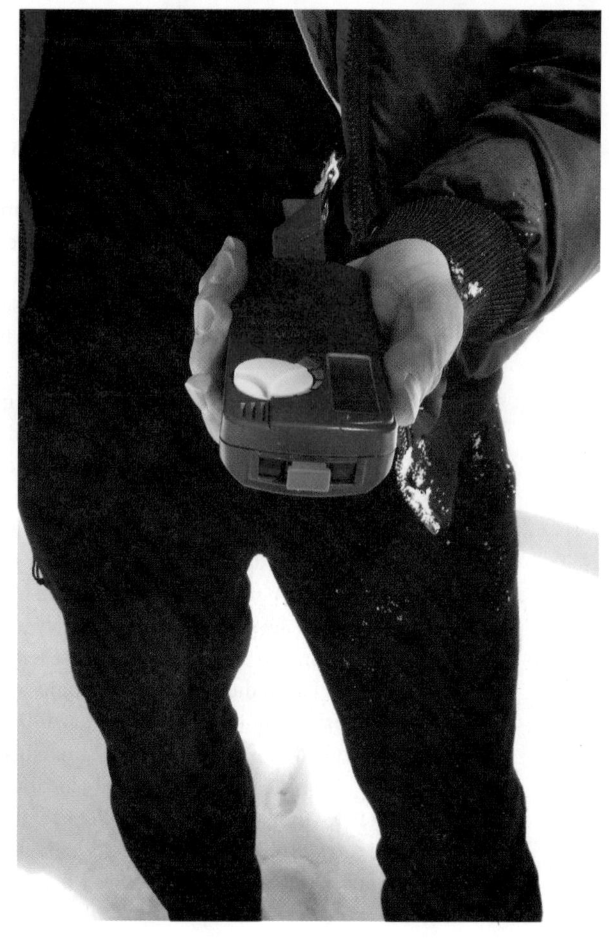

Figure 4: An Avalanche Transceiver

A transceiver is worn under your outer layer of clothing in order to minimize the risk of it being torn off by the force of an avalanche. The harness is a fundamental part of the design. The device works in one of two modes: transmit or receive. The default is that they are always in transmit mode. That is, normally, all of them should be transmitting a signal that can be picked up as an audible "ping" by a receiver. You want this as a default if you are buried, because the last thing that you want when caught by an avalanche is to be fumbling around with your transceiver. Snapping the waist strap into the device, which you have to do to wear it (left photo), automatically sets the device in the correct mode. That same connector has a fast-release mechanism. If someone is buried, the survivors undo the snap, and they are automatically in receive or search mode (right photo). Note that the device still is secured to the wearer by the yellow strap, in the event that the rescuers are hit by a secondary avalanche. All controls of the transceiver can be operated while wearing gloves (although I am not doing so in these photos).

Photos: Liz Russ

Figure 5: An Avalanche Probe

The probe is like a long thin tent pole. It is light, and collapses for easy portability (left photo). It has a cable running through the centre of the poles that enables it to be assembled very quickly. When assembled (right photo), it is pushed down into the avalanche debris, enabling the rescuer to probe for the victim.

Figure 6: An Avalanche Shovel

An avalanche shovel appears to be just a conventional shovel. However, it has a few conflicting constraints on its design. First, it must be compact and light, since you need to carry it in your pack. On the other hand, it needs to have a blade strong enough to penetrate the very hard consolidated snow encountered in avalanche debris, and be large enough to let you dig quickly and efficiently.

feature trap. It not only trapped Saul at the bottom of the slope, it also provided a natural basin to collect the avalanching snow. (It was also why Judy felt comfortable skipping the probe step. She had been on the top of the knoll, and knew that since Saul was at the bottom when buried, he was not going to be carried further downhill.)

As the river of snow started to slow down and cover him, Saul did the most important thing in terms of staying alive: he cupped his hand over his nose, and mouth, making sure that he kept an air space within which he could breathe.

Once buried, he went very Zen, and purged every effort to struggle or rescue himself. Struggle and fight is what you do on top of the snow. Underneath, you must wait. And wait. And have faith in your partners, their training, and everyone's gear.

In Saul's case, this trust was well placed. Judy found him. The total time from the slide to rescue was about 10 minutes. That was good—after 20 minutes, he probably would have been dead.

What Saved Saul?

Is Saul happy to be alive? Without a doubt, yes. Is he lucky to be alive? I don't think so.

Sure, there was some luck involved in what happened, both good and bad. But the final outcome was far more than luck. What saved him began well before they ever set out that day. It was a combination of on-the-spot problem-solving and performance, along with a combination of training, procedure, and equipment. Furthermore, I would argue that the answer transcends any single one of these factors. Even though each part was essential, on both the human and the technological level, neither was sufficient on its own. The tools shown in the preceding photos, for example, would have been useless had they not existed in a larger ecosystem. What is especially significant is that the system worked despite the fact that they did not exactly follow the prescribed procedure; it was sufficiently robust to be adapted to the local circumstances by appropriately skilled practitioners.

As it was in the story, so it is for almost all the tools that we might design. Consequently, any design methodology that does not take full account of the relevant ecosystem can have serious consequences.

As technology becomes more and more pervasive, it is finding itself in increasingly diverse and specialized contexts. The technologies that we design do not, and never will, exist in a vacuum. In any meaningful sense, they only have meaning, or relevance, in a social and physical context. And increasingly, that social context is with respect to other devices as well as people. (For a trivial example of what I mean by this, just think where Saul would be right now if he had been the only one with a transceiver. Not to put too fine a point on it, the answer would be the same no matter how well that transceiver worked.) As much as people, technologies need to be thought of as social beings, and in a social context.

Wild Cognition

Let's return to Saul. What fills me with wonder in this story is that none of the participants had ever been in that situation before. They had no indication in advance that this would happen, when, how, or to whom. And yet, they executed brilliantly, even given the added strain that the victim was the spouse of one of the rescuers.

In short, it is a wonderful and extreme example of a self-organizing system, and what Hutchins (1995) has called cognition in the wild.

Hutchins is a psychologist and a professor at the University of California in San Diego. He is passionate about two things: cognitive science and open-ocean sailboat racing. His book, *Cognition in the Wild,* is a study that combines these two passions. It is a detailed look at how navigators on naval vessels do their work. It is an analysis of how cognition is distributed among the group, and in particular, how it is tightly coupled to, and embedded in, the physical and cultural ecology within which the activity takes place.

Stated differently, Hutchins argues that the cognitive activity is embodied within the location of the activity and the tools used, as well as the navigators themselves.

By tracing the history of navigation and of the associated tools, Hutchins shows how different contexts demand different solutions. He makes one especially insightful observation. He argues that tools often don't so much enhance our ability to do a particular task, such as trigonometry (something that is fundamental to navigation); rather, more often than not they recast the problem in a different representation that simply side-steps the need for that task to be done.

That is, rather than being seen as a better calculator, or some prosthesis to enhance our speed or accuracy at doing a particular calculation, the tool is often better understood as a notational or representational device. We can think about this in the perspective of those old adages that we learned in Philosophy 101:

> Notation is a tool of thought.
>
> A problem properly represented is largely solved.

What Hutchins reinforces is that notational or representational systems are not restricted to things that we draw or write. Rather, physical devices can have the same impact on the representation of a problem—such as navigation—as the zero and decimal had in facilitating our ability to do multiplication and division, compared with doing them using Roman numerals (and keep in mind that multiplication and division were difficult in ancient times, even if you were an expert in Roman numeral calculations).

However, the representational power of the tool is meaningful only within the larger social and physical context within which it is situated.

One example of this—one that sticks to Hutchins' theme of navigation—was introduced to me by my friend, the industrial designer Alistair Hamilton. If we want to talk about design in the wild, then this is about as wild as it gets.

Let's go to the arctic.

Figure 7: Canadian Inuit in Kayak in Arctic Waters
Photo: W. McKinlay / National Archives of Canada

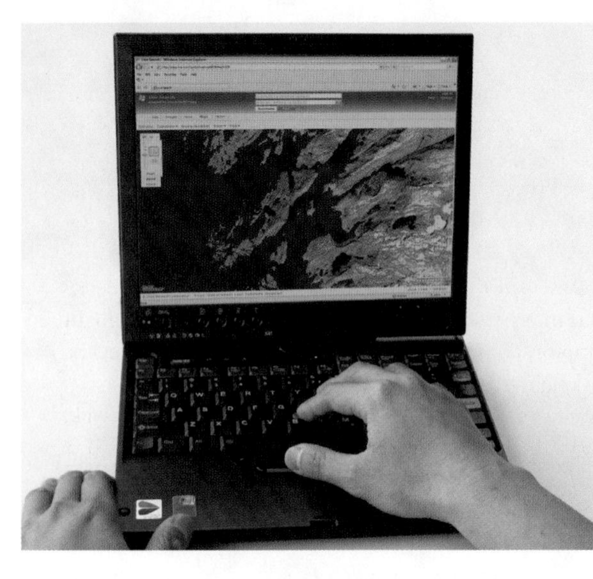

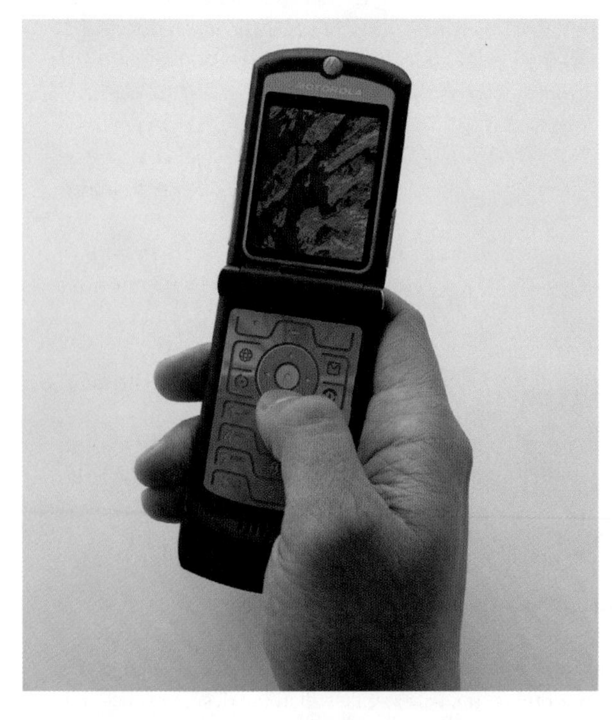

Figure 8: Navigating the Coast of Greenland

Two navigational aids for paddling along the coast of Greenland are shown. The first is the map as it would appear on your computer, the second is what would appear on your mobile phone.

Figure 9: A Difference That You Can Feel

These are 3D wooden maps carved by the Ammassalik of east Greenland. The larger one shows the coastline, including fjords, mountains, and places where one can portage and land a kayak. The thinner lower map represents a sequence of offshore islands. Such maps can be used inside mittens, thereby keeping the hands warm; they float if they fall in the water; they will withstand a 10 metre drop test; and there is no battery to go dead at a crucial moment. Credit: Greenland National Museum and Archives, Nuuk. See also woodward and Lewis, 1998, p.167-169.

Imagine that you were kayaking along the coast of Greenland, and needed a chart to find your way. You might have a paper chart, but you will probably have trouble unfolding it with your mittens on, and in any case, it will probably get soaked in the process and become unreadable. From the urban perspective, an alternative solution might be to go to your PC and use a mapping program on the Internet. If you did so, you might see something like what is shown in Figure 8. However, there is a minor problem here, too. You don't have your PC with you in the arctic, much less in your kayak. We all know about Internet-enabled cell phones and PDAs—they might provide another alternative. Why not jump on the Internet using your cell phone, and get the map that way? If you were successful, you might even get something like the figure.

But here is the problem. You probably can't get cellular service where you are in your kayak. And even if you can, your battery is probably dead because it is so cold. Or, your phone won't work because it is wet. Even if your mobile phone does work, and you have service, you probably can't operate it because you can't do so without taking your mittens off, and it is too cold to do so.

Now let's look at a third approach, one that the Inuit have used. It is the solution shown in Figure 9. This shows two tactile maps of the coastline, carved out of wood. They can be carried inside your mittens, so your hands stay warm. They have infinite battery life, and can be read, even in the six months of the year that it is dark. And, if they are accidentally dropped into the water, they float. What you and I might see as a stick, for the Inuit can be an elegant design solution that is appropriate for their particular environment. These are a wonderful example of "design for the wild," and like the avalanche beacons that helped save Saul, one that could mean the difference between life and death.

What this example reinforces is my thesis that in order to design a tool, we must make our best efforts to understand the larger social and physical context within which it is intended to function.

Hutchins refers to such situated activities as "in the wild" in order to distinguish their real-world embodiment from some abstract laboratory manifestation that is as idealized as it is unrealistic.

I call the process that expressly takes these types of considerations into account "design for the wild." To do this effectively, we ideally need to be able to experience our designs in the wild during the early stages of the process. Failing that, we have to do the next best thing, whatever that might be. Implicit in this is the following belief of mine:

The only way to engineer the future tomorrow is to have lived in it yesterday.

To adequately take the social and physical context into account in pursuing a design, we must experience some manifestation of it in those contexts (the wild) while still in the design cycle—the earlier the better.

I realize this is perhaps superficial "hand waving." But please, read on and see how I explain my case. The key point at this stage is this: I believe that we are at an important juncture in the history of design, where taking "the wild" into consideration will be fundamental to success.

Moving to a Solution

Architecture, industrial design, and environmental graphic design are all in a period of transition. This is a result of microelectronics and telecommunications technologies being incorporated into their designs. Likewise, the nature of software applications is being transformed by virtue of their being embedded into devices that have broken away from the anchor of the stationary PC, where their user interface and user experience is as much a part of the physical device and context as the software.

These trends create great opportunities, but also significant challenges as to how we develop products. Our approaches to designing even today's software and technology-based products are already broken. Without a change in approach, these problems will be multiplied many times over if we try to apply today's inadequate process to the products of tomorrow. The stakes, from both a cultural and economic perspective, are high. In the larger sense, the results of all of this will largely be determined by the design decisions that we make in the coming years. How I approach this challenge personally has been shaped largely by one of my favourite historians of technology, Melvin Kransberg, and in particular, by his first law:

> Technology is neither good nor bad; nor is it neutral. (Kransberg 1986)

What this says is that whenever we introduce a product into the market and our society, it will have an impact—positive or negative. I have a corollary to Kransberg's First Law. It is:

> **Without informed design, technology is more likely to be bad than good.**

Once acknowledged, Kransberg's "law"—and my corollary—imply that before introducing any new technology, we must make our best effort to inform its design such that the balance is more weighted on the positive than the negative. This is simply responsible design.

Yet, there is currently a general lack of informed design, and I believe our traditional methodologies are not up to the task of changing this situation. However, understanding how to take the larger ecological, contextual, and experiential aspects of "the wild" into account is a good start. Developing such a process may well provide the means to break out of the status quo. It is toward achieving this end that this book is devoted.

But is the status quo so bad?

I guess that depends on your perspective. Too much of user interface and interaction design, for example, are still rooted in the conventional notions of the PC, browsers, the Web, and the graphical user interface. I view this in terms of what I call The Rip Van Winkle Hypothesis. (For those unfamiliar with the story, it appeared in 1819–1820, in a serialized book called, *The Sketchbook of Geoffry Crayon, Gent.* written by the American author Washington Irving. It is one of the first and best-known American short stories. It is about a man who goes to sleep sometime before 1775 and wakes up 20 years later, having missed the entire American Revolutionary War.) So, in honour of Rip, my hypothesis goes as follows:

If Rip Van Winkle woke up today, having fallen asleep in 1984, already possessing a driver's license, and knowing how to use a Xerox Star or Macintosh computer, he would be just as able to use today's personal computers as drive today's cars.

Sure, the computers and cars would be faster, smoother, and more refined than when he last used them, but the essential conceptual model and operating principles would be as he remembered them.

Some might say, "So what? PCs and conventional models of computation are obsolete anyhow, replaced by things such as the cell phone. Consequently this is a nonissue." To them I would reply, "Have you looked at a modern cell phone? It is becoming more like a PC every day, with all the problems and repeating all the mistakes." I would like to move beyond the constraints of the conventional PC model as much as anyone. I just don't believe that a simple change in form factor is going to bring that about.

Conversely, others will argue that the stability of the GUI design is a good thing, that it reflects the power and benefits of consistency, and (in terms of computers) the significance of the design. Now don't get me wrong. I think that the graphical user interface was a great idea. But I am also equally certain that the originators (such as Johnson, Roberts, Verplank, Smith, Irby, Beard & Mackey 1989) would agree that regardless of how good it was, it was not the final word. Simply stated, user interface design in 1982 was nowhere near as mature (and therefore worthy of standardization) as the design of the automobile.

In terms of stifling innovation, good ideas are far more dangerous than bad ones. They take hold, assume momentum, and therefore result in inertia. Consequently, they are hard to displace, even when they are well past their prime.

It is, nevertheless, not simply the lack of innovation in user interface design that concerns me; it is more the state of the process of developing new interactive products.

iBook G4

Case Study: Apple, Design, and Business

You can't depend on your eyes when your imagination is out of focus.
—Mark Twain

On December 20, 1996, Apple Computer announced their intention to purchase NeXT Computer. The deal closed on February 4, 1997, at a price of $427 million. With this acquisition, Apple got at least two things. First, they got a technology that could serve as the foundation for their next generation operating system. Second, they got Steve Jobs, who was thereby repatriated with the company that he had cofounded.

At the time, Apple was in trouble. Its market share was falling, and its share price was hovering around a 12-year low. There was little to convince investors or customers that the company possessed any of the innovative vision that had characterized its glory days. Things came to a head when the second-quarter results came in. Apple had lost $708 million. On July 9, CEO Gil Amelio, who had engineered the NeXT purchase, was gone, and Jobs' role as a special advisor to the executives and the board took on a new importance.

On September 16, Jobs was named interim CEO. The question on the minds of the press, the market, competitors, and customers was this: Can Jobs bring back some of the magic that had created Apple's success in the first place?

Looking back with today's eyes, we know that the answer was yes. This was certainly not clear at the time, but like so many things, it is not the answer that is important. Rather, it is the path followed to get there. Therein lies the purpose of telling this story. That, and a desire to use a real-world example to illustrate the codependence and intertwining relationship among business objectives and management, industrial design, software, marketing, luck, and skill.

On his second day on the job—not the second month or second week, but the second day—Jobs held a meeting at Apple's Cupertino headquarters. It was with six of the top analysts and journalists that covered the company (Chmielewski 2004). His purpose was simple: to explain to them how he was going to turn Apple around. As described to me by one of the analysts present at the meeting:

> He specifically emphasized getting back to meeting the needs of their core customers and said that Apple had lost ground in the market because they were trying to be everything to everybody instead of focusing on the real needs of their customers. He also pointed out that Apple had broken new ground with the original Mac OS and hardware designs and that he would now make industrial design a key part of Apple's strategy going forward. (Tim Bajarin, personal communication)

Figure 10: The Original iMac
Announced in May 1998, and shipped in August of that year. Steve Jobs had said that industrial design was key to his strategy for turning Apple around, and this was the first of many products that turned those words into reality.

The analysts were intrigued, but not completely sold. However, knowing Jobs, they were perhaps more willing to give him the benefit of the doubt than they may have been with most others who might say the same thing. Tim Bajarin describes his impressions on leaving the meeting like this:

> On the way out of the meeting, I remember us discussing the idea of design as a key issue for saving Apple. I can't remember the reaction of the others, but I know that my first impression was that Apple had so many problems that I could not see how industrial design needed to be a key part of his strategy to save Apple.
>
> I also was concerned that given Apple's serious financial condition, whatever he was going to do needed to be rock solid and make an impact quickly. But, I also remember telling the people I was with that you can never under estimate Steve Jobs and that if anybody can save Apple, it would be Jobs. (Tim Bajarin, personal communication)

It didn't take long for Jobs to begin to deliver on his strategy. His first real salvo was fired on May 7, 1998, when the first iMac was announced. From a design and business perspective, this machine, illustrated in Figure 10, had almost as much impact as did the original Macintosh in 1984. Like the original Mac, the iMac integrated the computer and the monitor into a single highly styled package. But the most distinctive aspect of the design language of the iMac was the translucent "Bondi Blue" plastic of the case.

This machine began to ship in August 1998, and was an immediate success in the marketplace. Apple was hot again, and it was primarily the industrial design of this machine that made it so. There is little doubt that it was the key catalyst to

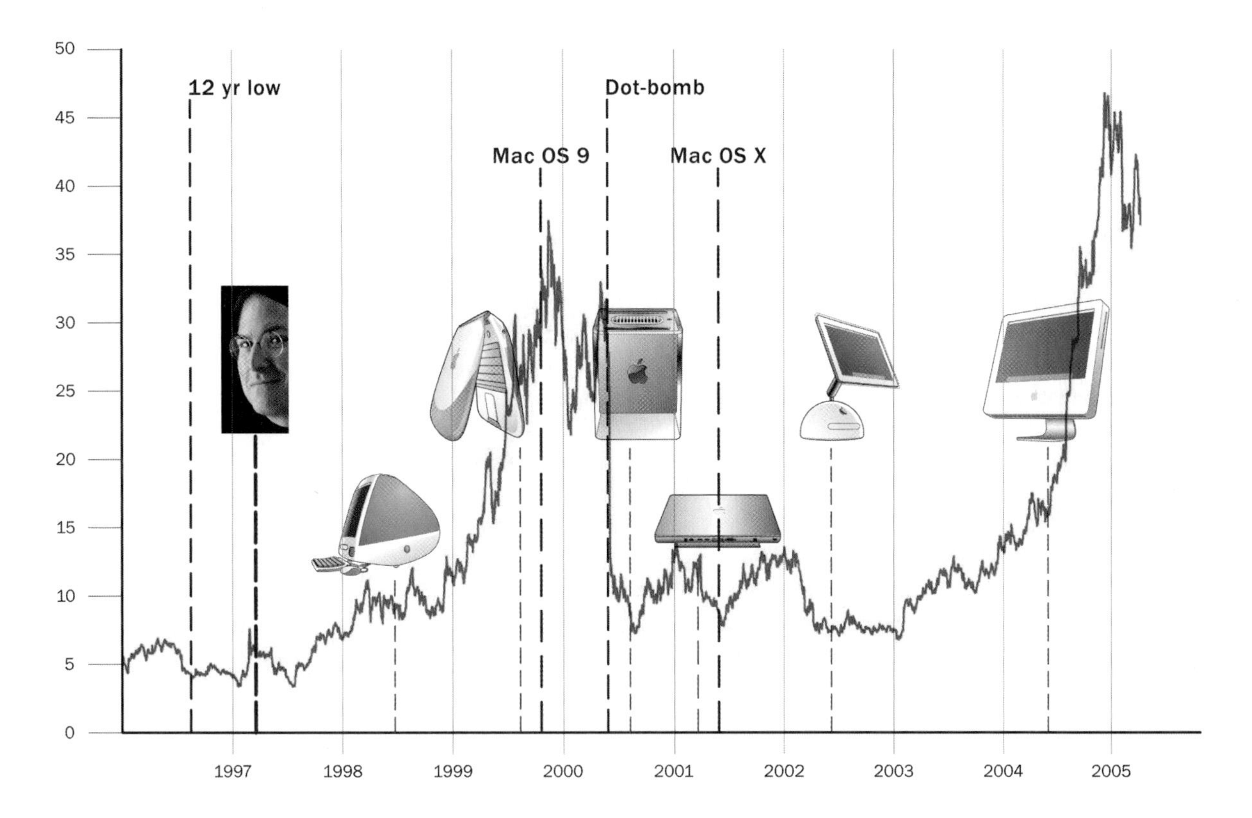

Figure 11: iMac Product Timeline
Key product introductions and events relating to the launch of the iMac are shown along a timeline, along with a plot of the daily close price of Apple's stock from July 1, 1996 to June 13, 2005 adjusted for dividends and splits. Note that the iPod needs to be taken into account in the rise in the share price from 2003 on.
Stock information source: Yahoo! Finance

the ensuing turnaround in Apple's fortunes. But it was just the start.

In early 1999 the iMac was followed up by versions that expanded the palette of available colours to include blueberry, strawberry, lime, tangerine, and grape. These preserved the translucency of the original iMac, and helped consolidate the design language that was to characterize the product family. These were machines that looked like nothing else in the market and made a clear statement that Apple was different.

Here are eight points that I think are particularly relevant in the story of the iMac—equally relevant to the business manager, industrial designer, teacher, or user interface designer:

1. Design saved Apple. This is redundant, but I think that it is important enough to reiterate, especially in light of what follows. One way that I have tried to capture this is in the graph shown in Figure 11. It plots the company's share price (adjusted for dividends and splits), along with some key product launches, from the time of Steve Jobs' return until mid 2005. The first thing to note is that the share price did not start to climb until the launch of the initial iMac. At the other end of the timeline, one has to remember that I have shown just the iMac, not the iPod, so not all the later activity can be attributed to the computer side of Apple's business (but more on the iPod later).

2. The design innovation was done with the existing team. The core of the team that designed the products that revitalized Apple, from the original iMac to the iPod, were already at Apple before Jobs returned. That is to say, the company was largely saved by existing talent. For example, Jonathan Ive, VP of Industrial Design at Apple, and who has been the lead designer on all these products, joined Apple back in 1992, when John Sculley was CEO. He was there when Michael Spindler had that position, and he was there through Gil Amelio's tenure as well. That is, one of the prime talents that helped save the company was in its employ through the full period of its slide to near oblivion!

3. Executive vision was critical to success. The lesson from the previous point is that it does not matter if you already have the talent to save your company among your current employees. If you do not have the vision, will, and power at the highest level, then that talent is almost certain to remain as wasted as it is frustrated.

4. Momentum was sustained and rapid. Despite its success, the iMac alone did not save the company. It simply revived it. What saved Apple was that the company repeatedly did to the iMac what the iMac did to its predecessors. As illustrated in Figure 11 (which shows only about half of the new computer products introduced during the period covered), the innovation was constant and rapid, and the design language of the products kept changing and developing. It ranged from the candy-coloured translucency of the original iMac, to the minimalist form of the Power Mac G4 Cube, to the iMac G4, to the iPod-inspired iMac G5.

5. There were failures. Steve Jobs is the prime example that demonstrates my thesis: Your failures are all but forgotten as long as you also have great success. Set aside the fact that Jobs got removed from Apple in 1984, and that NeXT was a failure (although, selling the company for $427 million is my kind of failure). Even after he came back to Apple, during this period where design reigned supreme, Jobs took some serious missteps.

One misstep was actually part of his initial success, the first iMac. The Achilles heel of the product was its hockey-puck shaped mouse, illustrated in Figure 12. The key problem with the mouse was its uniform circular shape. It looked beautiful, and was in keeping with the rest of the computer. However, the regularity of its circular shape provided no affordances, or tactile cues, that let you know its orientation when you grabbed it. Hence, as likely or not, when you moved your hand one way, the cursor on the screen went another. The design was rapidly replaced, and there was no long-term negative impact on either the product or the company.

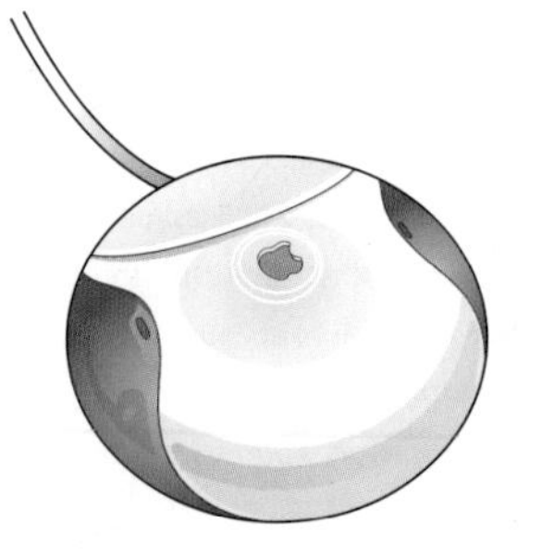

Figure 12: The Hockey-Puck Mouse for the Original iMac

The original mouse for the first iMac was in the form of a hockey puck with its edges rounded. It looked great but that didn't carry across into its usability.

This was not the case with his next major failure of design, the Power Mac G4 Cube, which was announced in July of 2000.

The G4 Cube, illustrated in Figure 13, might be the most beautiful computer ever built. To steal a word from Jobs, it was "insanely" stunning. Its styling was something that you would expect to see inside of Architectural Digest, or in New York's Museum of Modern Art (in whose collection it actually does exist—see Antonelli 2003). The only problem was, it was too beautiful for the real world of everyday use. If you put it on your desk, by contrast, your desk looked hopelessly messy. And as soon as you connected wires to it (and there were lots of them), it looked like hell, because of the contrast between its clean symmetrical form and the chaos inherent in the multiple cables. But the problems were worse than that. The early versions had material problems that caused hairline cracks to appear,

Figure 13: The Power Mac G4 Cube
One of the most beautiful computers ever produced. However, it is also one that was more successful as a piece of sculpture than as a viable design for the home or workplace. It was withdrawn, but not before its disappointing performance in the marketplace brought about a steep fall in the company's share price.

which marred its surface. And the same convection cooling system that made it so quiet also led to frequent overheating, since the flat surface on the top, where the hot air escaped, also afforded a convenient place to put things like papers or books, thereby blocking the hot air escape vent.

Coupled with issues around price and performance, the problems with this product were not so easily glossed over as those with the iMac mouse. As Figure 11 clearly shows, its launch was followed by a significant collapse in the share price, a drop that was to last until 2004. (However, it would be unfair to blame the drop in Apple's share price just on the G4 Cube, since its launch was just four months after the March 10, 2000 bursting of the dot-com bubble.)

6. The failures were key to success. I believe that if Jobs had played it safe, and not risked periodic failure, he never would have succeeded the way that he did. My rule here is this: In the long term, safe is far more dangerous than risk. That is not to say that we should be reckless or make uninformed decisions. To the contrary. Risk can be mitigated by having the right tools, the right training and technique, the right fitness, and the right partners. But it can't be eliminated. Things will not always go right, and you have to factor that into your plan—not as an error, but as a valuable (and expensive) "learning experience." Jobs and Apple had to fail (but not always) in order to succeed.

7. The design that led to success was largely in the realm of styling, bordering on the superficial. What I find most interesting about the use of design in Apple's computers since 1998 is that virtually all the changes were, as I describe it, on the front side of the glass. That is to say, the impact would have been essentially identical even if there had been no changes in the software system and applications that ran behind the glass of the monitor. They had to do with external appearance rather than internal look, feel, or behaviour. They certainly didn't redefine computing or how we interact with computers. Rather, they simply redefined what the computer looked like—the styling—which is important but nevertheless, superficial. Be clear, I don't mean this in a pejorative sense. The style of these machines gave them character that clearly resonated with people, and helped reshape their perception of what a computer might be for. But underlying these systems was the old familiar graphical user interface (GUI), with perhaps a bit of an updating in graphical style.

8. There was almost no interaction between industrial design and user interface design. Personal computers and GUIs were essentially mature technologies at the end of their fundamental innovation cycle. Hence, the industrial designers knew that if there was a mouse, keyboard, and display, then the user interface (UI) could be supported. Likewise, the UI designers knew that their systems could, and must, run equally well on any of the company's platforms, regardless of industrial design.

Not only could the industrial design and interface design teams work independently, Jobs actively discouraged communication or collaboration between them—with one exception. In order to provide the illusion of hardware/software integration, the industrial design team designed the default desktop pictures for each revision of the iMac (however, as of Mac OS X, the UI team "owned" the desktop picture).

In many ways, this represents a success story. At least as far as it goes. But how far is that? How much of the preceding can, and should, be adopted or emulated by other companies?

I think that the story speaks very favourably in terms of the role that design can play in affecting a company's fortunes. I think that it also emphasizes the importance of the role of executive management and vision in creating an environment where design can succeed. In fact, the challenges we have demand a very different approach to product design management than that outlined in the last two points. Briefly stated, it is going to be ever harder to separate the software aspects of the user interface from the physical aspects—that is, the part done by user interface designers that normally would be done by industrial design. There cannot just be an appearance of collaboration. Holistic design, which truly integrates both of these aspects of the design, must be there in fact.

Take 2: The Apple iPod

In order to explore this issue further, let's extend our Apple case study a bit further by looking at the evolution of the iPod.

Actually, I want to discuss the iPod for another reason as well. Sometime around February 2005 I got a call from a friend who is lead designer at a company whose products I love and use with delight. To place the chronology of the call in context, this was right after "iPod Christmas." That is, the Christmas when it seems that every store in North America was sold out of iPods (I bought four).

Figure 14: The First Generation iPod

The first iPod was introduced in October, 2001. If you got an iPod in the past year or two, a nice exercise that will help reinforce some of what we will discuss later is to compare your iPod to this one. Notice that there are significant differences, yet they are both clearly iPods.

Photo: Apple Computer Inc.

Here is the gist of what he said.

> Bill, I am at a loss. I just got out of a meeting with the president of the company [significantly, not a designer or anyone with a design background]. He said that what they wanted design to do was to come up with an iPod, and I don't even know where to begin in terms of responding. You have to help me.

Now of course, the request was not to make an iPod per se. Rather, it was to make a product that was to their traditional product line what the iPod was to Apple's. And, of course, dazzled by the success of the iPod, the president also wanted his "iPod" to perform as well in the marketplace as Apple's did for them.

I understood my friend's problem. But I wasn't that surprised by what he recounted, and suspect that a variation of his president's request was heard in more than a few companies around that time. So how to respond? The one thing that my friend couldn't say was no. That would be an undesirable (and unnecessary) career-limiting move.

But more importantly, the president's request was both reasonable and the right thing to ask for. Why shouldn't executives want to have their company create breakthrough products that generate great returns? By the same token, why wouldn't any designer worth their salt leap at the chance to work on such a project? Look at Jonathan Ive. His life went from frustration bordering on misery to every designer's dream when Jobs gave him and his team the chance to do the iMac and the iPod.

But none of this helps the immediate situation between my designer friend

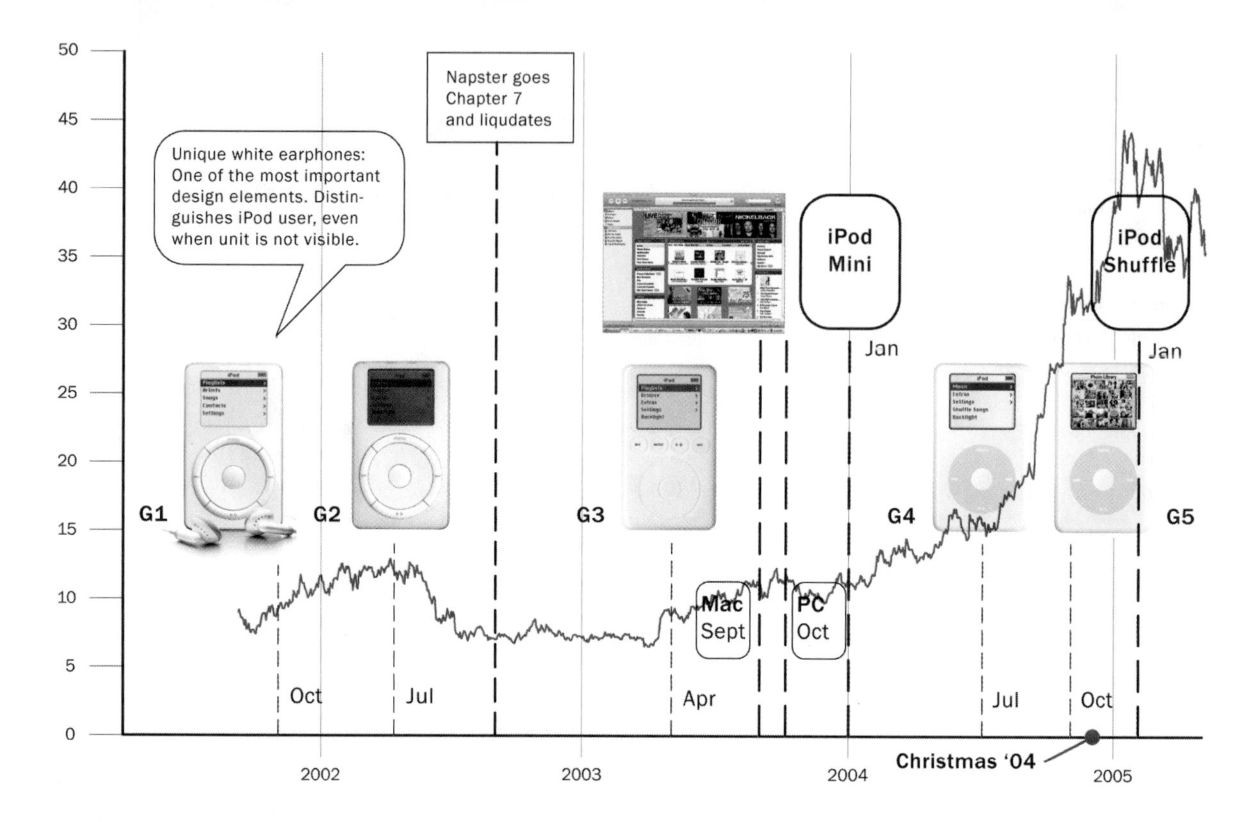

Figure 15: What it Took to Make the iPod an "Instant" Success

Key product introductions and events in the life of the iPod are shown along a timeline. Superimposed is a plot of the daily close price of Apple's stock from September 4, 2001 to June 13, 2005 adjusted for dividends and splits.

Stock information source: Yahoo! Finance

and the company president. The objective was (ideally, at least) desirable to both, but something was fundamentally broken, as it is in most companies that I have had any contact with. The problem as I saw it was a total disconnect on the president's side in terms of his having any understanding of what it actually took to create "iPod the physical MP3 player" much less "iPod the phenomenon." And, to be fair, there was probably some disconnect on the design team's side too, around some of the issues and concerns of the company's executive management.

My advice to my friend was to arrange a conversation with the president and try and come to a common understanding of what his request implied to the business. And if the iPod was going to be the model, then the conversation needed to be framed around an analysis of the history of its development, in the broader sense. Without that, I just didn't see how anyone could make any informed or meaningful decisions.

What follows is an expanded version of what I gave him to help support that conversation.

To begin with, an operational understanding of the iPod story hinges on appreciating the difference between when the iPod was introduced versus when it reached its tipping point (Gladwell 2000)—that is, distinguishing when it was launched from when it became a major public phenomenon. The former was in October 2001. The latter—which is likely where the executive's knowledge began—was in the second half of 2004.

A lot happened in the interim that was significant to an understanding of the eventual success of the product. In order to try and provide a quick snapshot of

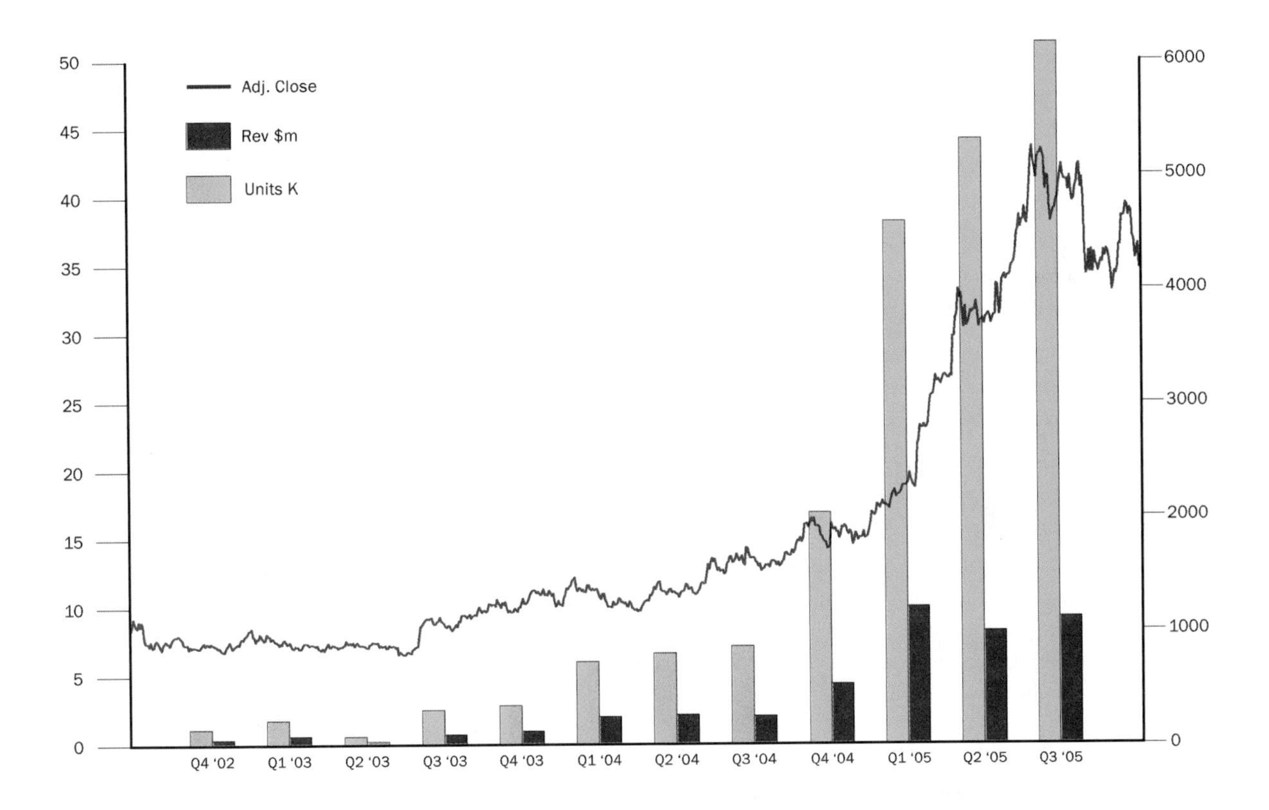

Figure 16: Apple Share Price and iPod Revenue and Volume

Apple's adjusted closing share price from Q4 2002 to Q3 2005 (the red line) is plotted over top of iPod unit sales and revenue for the same quarters.

Share price: Yahoo! Revenue and sales volume data: Apple Computer financial statements.

this, I created an early version of the graphic seen in Figure 15, and then sent it to my friend.

Later, in order to get a better sense of the performance of the iPod, I created a second graphic that shows, quarter-by-quarter, the sales volume and revenue generated by the iPod, with Apple's share price superimposed (see Figure 16).

Let me highlight what I think are some of the key points to note about the life of the iPod, most of which are in the figures.

1. It took three years for the iPod to become an "overnight success." In fact, it took even longer, since the time line starts when the product was launched, not when the project began.

2. The iPod was not the first product in this space. The Sony TPS-L2 Walkman was introduced in 1979. The first portable MP-3 player, the Eiger Labs MPMan F10, was introduced in the United States in the summer of 1998, and was soon followed by the better-known Diamond Rio PMP300. The first portable MP-3 player to use a hard disk, rather than flash memory, was developed by Compaq Research in 1998 (three years before the release of the iPod!). The Personal Jukebox PJB-100 had a 6 GB disk, and was commercialized by a Korean company called HanGo Electronics Co. It was first shown in the United States at COMDEX in November 1999. Other hard–disk-based devices also preceded the iPod, including the Nomad, released by Creative Labs in 2000.

3. Apple was its own strongest competitor. Even though they were not first to offer a disk-based MP-3 player, once they entered the market, their pace of introducing new products and services left little space for competitors to grab market share.

4. Apple arguably had the best-designed product, but that is a relative thing. The iPod gets a lot of press due to its design and especially its usability. It is important to recognize that this is a relative thing. Despite the positive press, even the current model has serious usability problems. It is just that the competition does too. There has always been, and likely always will be, significant room for improvement.

5. Style and fashion are really important. This is obvious to people from consumer products or haute couture. But it is not so well appreciated in the high-tech sector. What is especially worth noting is how the appeal of styling can be used to overshadow a product's weaknesses. For example, each generation of the iPod had its design problems, but these were more than compensated for by the iPod's strengths as a fashion item. The designer Jonas Löwgren calls this the iPod's "jewelry aspect." The thing to bear in mind in leaning too hard on this aspect of design is that fashion can be very fickle, and can cut both ways.

6. It took four generations of the basic iPod before it "tipped." Evidence to support Point 4 can be seen in the fact that Apple itself made repeated and significant improvements to their product. And by this, I do not mean just adding bigger disks or adding PC compatibility. Their changes affected the most iconic aspects of the iPod's design, namely the user interface and the scroll wheel. (See the accompanying sidebar, *Revision While Maintaining the Basic Design Language.*) The first and fourth generation iPods are clearly of the same family, but by comparison to today's model, the first one feels almost clunky and coarse—a very different sensation than what it provoked when it first came out.

7. The success of the iPod depended on a much larger ecosystem. There were a number of key steps in establishing the product and the brand. The saturation ad campaign featuring the black silhouettes on pastel backgrounds was just one example of how the creativity of the marketing has to match that of the product. This campaign was so distinguishable from anything else that you could put a silhouette of Bill Gates holding an X-Box on that background and it would still say "iPod," not Microsoft.

Likewise, a critical part of Apple's success was in their parallel initiative with iTunes and the associated music store. Again, however, the success here is not just in recognizing the need for the software or its design (which I think has real problems). The hard part here was what it took to feed the software with music. That is, someone had to convince the music companies, who were more than a little reluctant, to let Apple (who they did not know or trust) sell their music in the associated iTunes Music Store.

8. Jobs turned the Gillette model on its head. At one level, we can say that the key to Apple's success lay in recognizing the potential of leveraging the player (the iPod) and the music sales and distribution mechanism (iTunes). But the creativity of the design of the business model goes far deeper.

Gillette sells razors at a loss in order to make money on the blades. Xerox made a major part of its income from the paper and toner consumed by their copiers. And videogame platforms and cell phones are sold at a loss in order to realize the upside from game sales and network charges, respectively. Well, Apple has managed to do just the opposite. They built their business around making their margin on the "razor" (the player) and accessories, and then selling the "blades" (the music) at the minimum that they could. The music not only helped drive revenue from iPods, its pricing also reflected the reality of the market, where until recently, music downloads had been free.

9. Growth in revenue does not keep pace with growth in sales. The previous point notwithstanding, Figure 16 shows that revenue does not continue to grow proportional to sales volume. This is normal, since as the product matures, there is downward pressure on price and margin, and much of the volume comes from the introduction of lower-priced units. We can expect this trend to continue, with the eventual necessity to introduce new products or new sources of revenue in order to sustain growth. To grow the installed base, for example, in January 2005 the inexpensive iPod Shuffle was introduced. This exploited the desirability value of the brand. It was then followed in September by the more up-market (and stunning) iPod Nano. Then in October, the video-enabled iPod came onto the market. Through the introduction of video, the company was able to sustain the demand for high-power, high-storage, and consequently, high-margin devices, and to do so with a product that complemented rather than competed with the Nano. The next step in jumping up the food chain was the January 2007 introduction of the iPhone.

10. There was some (a lot?) of luck involved in Apple's success. A great deal of what is represented in Figure 15 came about through hard work, good management, execution, and creative thinking. But not all of it. For example, it was not until after the second-generation iPod had been released that Napster was finally forced to close its doors. Apple might have hoped, or even guessed, that this would happen. But they could not be sure. They were lucky. But then, as the Roman philosopher Seneca said:

> Luck is what happens when preparation meets opportunity.

11. Even Steve Jobs had no idea how successful the iPod would be. This is a variation on the previous point. I am certain that Jobs hopes that every product that he brings out, including the Power Mac G4 Cube, will have this kind of success. And in planning, he anticipates what will need to be done in the event that it does. But by the same token, he knows that batting 100% is really unlikely. He hopes for it. He anticipates it. But he doesn't expect it or take it for granted.

12. Le bon Dieu est dans le détail (Gustave Flaubert) / God is in the details (Ludwig Mies van der Rohe). I want to pick just one detail of the iPod design that illustrates the power and importance of simple details that generally get overshadowed by the much-discussed aspects of the design, such as the smooth corners, white face with chrome back, white colour, and such. Remind yourself that under normal conditions, the iPod is not visible. It is in one's pocket or bag. So now look at Figure 17 and tell me what kind of personal music device Lance Armstrong listens to. Then tell me how you know. In the meantime, guess what percentage of the people watching the 2005 Tour de France could answer that question. I suspect that more people know what device he listens to than know what kind of bike he rides. (It is a Trek Madone, by the way, a wonderful piece of carbon fiber—but more on that in a later chapter.)

13. Holistic design not only requires an ecosystem, it also feeds one. I made the point earlier that the success of the iPod required a holistic approach to the design, from the product itself, through the marketing, sales, and business model. There is another side to this coin. When all of these components are firing together, there are real tangible benefits that can result. I am only going to point out two. First, I am starting to think that Apple might make close to as much money (or even more) selling iPod accessories as it does selling iPods themselves. Second, the investment in design sells the design. Look at how marketing can leverage great design, as outlined in the sidebar, *Whose Budget Should Pay for Design?*

14. From the design and management perspective, the iPod is a different class of product than the iMac. Of course an MP-3 player is different than a PC. My point is that the iPod signals the way toward a type of information appliance that is going to dominate the technology-based product space in the future, much in the same way that the PC dominated the past. Furthermore, I would suggest that we will not, and should not, see what we saw with the iMac. There should not be a separation between the software aspects done by user interface design, and the hardware aspects done by industrial design.

To understand the significance of this last point, just look again at the iPod. Note how integral the physical scroll wheel is to the device's user interface. Yet, a huge part of the interface lies in the embedded software. In terms of the overall experience, it is impossible to decouple the hardware and software components. They function as a unified whole.

I am on dangerous ground here. The reason is that despite what I have just said, the basic idea for the iPod scroll wheel was done before any of the user interface group knew about the project, much less were part of the team.

But I would argue that that is merely a testament to the quality of Jonathan Ive and his team, and the fact that over the years Apple has assimilated into the entire culture a strong feeling for user interface design. To use another cliché, unless you have the skill and experience of Steve Jobs and Jonathan Ive, my recommendation is: Kids, don't try this at home. You will probably get hurt.

I am being generous here, largely because it is hard to argue with the iPod's success. However, I believe that some of the design flaws in the device's user interface could—and likely would—have been caught if the user interface designers had been involved earlier in the project.

Regardless of any shortcomings, it is clear that the iPod was an overwhelming business success. Less obvious, but perhaps more important is the following observation:

> **I am hard pressed to think of one part of Apple that wasn't critical to the success of the iPod.**

The company had to be firing on all cylinders, with all parts going in more or less the same direction. Sure, there are some superstars among the protagonists. But despite that, and in keeping with a theme that runs throughout this book:

> **Everyone is essential, but no person or group is sufficient on his or her own.**

Figure 17: Lance Armstrong Spinning at the Tour de Georgia, April 2005
This photograph prompts me to ask two questions. First, what kind of personal music player is he listening to? Second, how do you know, since it is not visible? The white earphone cable broadcasts the answer as loudly as his shirt announces that he is riding a Trek bike, sponsored by Nike, and on the Discovery team.
Photo: Trek Bicycles

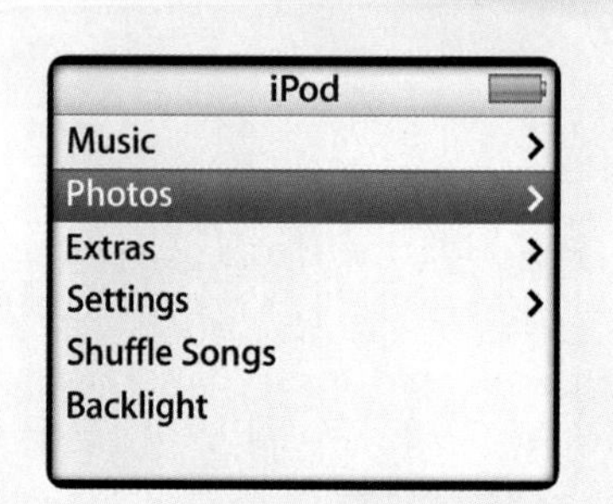
iPod
Music
Photos
Extras
Settings
Shuffle Songs
Backlight

Revision While Maintaining the Basic Design Language

The signature element of the iPod's design and user interface is the scroll wheel. Yet, for usability and cost reasons, it has gone through four distinct generations in the short life of the product. What is interesting is that the industrial design (ID) team, led by Jonathan Ive, has been able to accomplish this, and still preserve the essence of the design language.

Generation 1 (October 2001)

The first iPod had a mechanical scroll wheel. That is, the wheel physically rotated when manipulated by the finger. The buttons were also physical buttons that you pushed, and they were laid out around the circumference of the scroll wheel.

Generation 2 (July, 2002)

In the second-generation device, the mechanical scroll wheel was replaced by a touch-sensitive device. That is, the wheel did not actually turn. Rather, it just sensed the touch of your finger as it rotated around its surface. The buttons, however, were mechanical, and were the same as on the first generation device.

Generation 3: (April, 2003)

The third-generation iPod kept the touch-sensitive scroll wheel of the previous version. However, it replaced the mechanical buttons with touch-sensitive ones. Furthermore, it repositioned the buttons from around the scroll-wheel, where they had been in the previous two generations, and placed them in a horizontal row between the LCD screen and the scroll wheel.

Generation 4 (July, 2004)

The fourth generation introduced what was called the click wheel. It got rid of the row of touch-sensitive buttons, and repatriated them down to the scroll wheel. However, rather than place them around the wheel, as in the first two generations, they were placed under the wheel. This was possible because the wheel was not only touch sensitive, it had some mechanical "give" under the North, South, East. and West points, where you could feel and hear a click when pressure was applied.

JANUARY 14, 2002

www.time.com AOL Keyword: TIME

AFGHANISTAN: DEADLY HUNT ■ INDIA & PAKISTAN: WAR DANCE

TIME

FLAT-OUT COOL!

Steve Jobs thinks he has seen the future—again. Apple's new iMac is an all-in-one hub for music, pictures and movies. It's elegant and affordable. But will millions of PC users get it?

January 14, 2002

Whose Budget Should Pay for Design?

Getting your product on the cover of a magazine is like an ad that doesn't look like an ad—one that probably has more value with skeptical readers than an outright ad in the same place, particularly in a credible publication like the ones shown here. So how much is such exposure worth? Let's take Time Magazine as an example. For the U.S. edition, the list price for a full-page colour ad is $234,000. The same ad on the back cover costs $315,900. The reason for the difference is that the back cover provides more exposure. Now if we could buy an ad for the front cover—which we cannot—we would expect it to be more expensive than the back, since it gives yet even more exposure. As an estimate, let's assume that percentage-wise, the front cover would be as much more expensive than the back, as the back cover is to a full-page ad on the inside. By this reckoning, the value of the front-cover product placement would be around $426,465. I think that this is a reasonable guestimate. Although ad space is often discounted, these prices are for the United States only, and the cover generally appears worldwide.

Now multiply that number by the number of magazine covers your design makes. Then ask yourself if perhaps marketing, rather than product development, should be paying for great design!

One final aside. Note that the likelihood of an Apple product making it onto the cover is significantly increased if the company's PR department makes Jobs available to the right reporters and photographers at the right time. Here we see PR and the CEO fulfilling yet two more key roles in the heterogeneous intertwined team that is required to exploit the full potential of the investment in design.

February 2, 2004

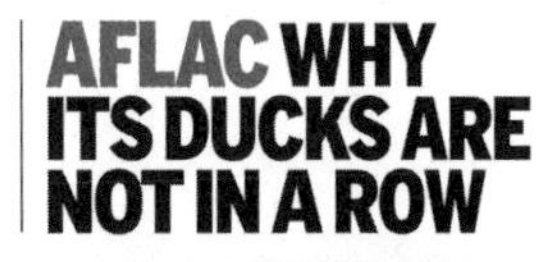

INVESTING THE BEST-PERFORMING BOND FUNDS
HIGH TECH THE INDUSTRY'S MOST HATED COMPANY
AFLAC WHY ITS DUCKS ARE NOT IN A ROW
The McGraw·Hill Companies
BusinessWeek
FEBRUARY 2, 2004
www.businessweek.com
SHOW TIME!
Moving beyond computers, Steve Jobs has become a real force in music and movies. But rivals loom, from Microsoft to Wal-Mart. Can he stay on top this time? BY PETER BURROWS

July 26, 2004

Figure 18: Bossy the Cow

Eventually the milk gives out. Not all at once, but gradually, as she ages. And, at the same time, she gets more expensive to keep.

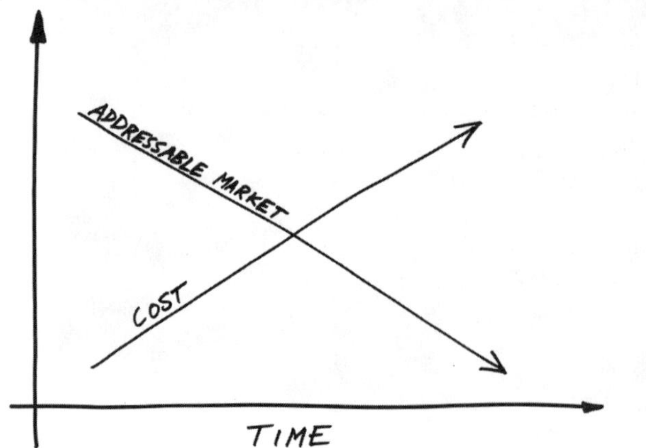

Figure 19: Relationship of Development Cost and Addressable Market

As a software product reaches maturity, the cost to bring each new release to market increases while the size of the addressable market decreases.

The Bossy Rule

You can't milk that cow forever.
— Farmer Brown

This book is about design. It should not be surprising, therefore, that one of its underlying premises is a basic belief that companies need to make new products. This statement may seem banal, but nevertheless, I feel that it is important to say explicitly, since so many companies are not very good at it.

Sure, in order to get started, companies generally need to have at least an initial product offering. If the company survives, then that first product likely is reasonably good, or it fills a need. Depending on the type of company, the product may be a physical object, such as a computer; a service, such as a new type of health insurance; or software, such as a word processor. But there must be something to sell.

That gets the company started. That and a lot of sales, marketing, manufacturing, distribution, and so on. Now let's assume that the product and the company are successful. The product sells well and the market grows. With that growth comes expectations on the part of both customers and investors. What the investors want is for the growth to go on forever, and ideally, at ever higher rates. But generally speaking, no matter how good that initial product was, the company will not be able to sustain itself, much less sustain its growth, if it continues to sell that initial package unchanged. As an initial product, it was probably not perfect, and even if it was pretty good, others will eat into the company's business through competitive pricing, features, or both.

So, to stay ahead of the game, what most technology companies do is put out a new release of the product every year or so. Hence, the initial launch of Bossy 1.0 is typically followed up a year later by Bossy 2.0, then 3.0, and so on. This dynamic is true whether the product is a word processor or a computer. In our discussion, let us describe such subsequent releases as "n+1" rather than "new" products, since they are essentially the same product with additional features, bug fixes, and such.

For example, I would argue that the various versions of the iMac were n+1 products, but the iPod was a new product. Likewise, Microsoft's Word 2003 is an n+1 product, whereas Windows Movie Maker 1.0 was a new product.

For many companies, especially software companies, the corporate strategy is to build a business around the development of n+1 products. However, my observation is that this is not sustainable in the long term. As a rule, companies will eventually have to release truly new products, or see their business shrink to the point where it is likely no longer of interest to investors, and perhaps not even viable.

To understand why, let's narrow our discussion and focus on software companies.

With software products, each n+1 release must contain sufficient incremental improvement to contribute to two objectives. First, it should help motivate those who have not already purchased the product to do so. Second, and perhaps more important, it should convince those that have bought previous versions to buy the upgrade. Let's say that that the new release has crossed the *Value Threshold* if it meets these two objectives.

The problem with relying on n+1 products is that the cost of achieving an improvement that is greater than or equal to this Value Threshold increases with each release. Furthermore, my experience suggests that the higher the release number, n, the higher the jump in incremental cost will be.

Why should this be so?

First, as systems go through successive releases, they grow in complexity. This in itself negatively affects their malleability, and increases the cost of adding value or rewriting (refactoring) the code. However, it is not just the number of lines of code or additional features that cause this. As they approach maturity, usually around the 9.0 release, the legacy of the initial underlying architecture, technologies, and paradigms creates a straightjacket that severely affects the cost of change.

Second, as the product matures, the size of the installed base starts to reach critical mass. At this point, the developers are constrained by the user community in terms of what they can or cannot change. Briefly, users have made an investment in learning the product, which has given rise to certain expectations and hard-won skills. Hence, any significant changes that affect the user interface or the workflow need to take this into account. Just imagine if you could quadruple the size of the addressable market for typewriters if you just switched the "b" and the "e" key on the keyboard. The new users might love you. On the other hand, existing users will likely hate you. They had already developed the skills to cope with the old way. The "improvement" may do nothing for them other than cause them a headache due to having to relearn something that they were already managing perfectly well. Virtually all products have analogous aspects of their design that could be changed to improve things, but the cost is potentially high.

Third, as products get more mature, the low hanging fruit has all been picked, both in terms of features and customers. The product more or less works as intended, and the markets are approaching saturation. For most users, the current version is "good enough." Changes increasingly tend toward tweaks and tuning rather than major improvements. Or, they are directed at ever more specialized functions needed by an ever smaller and smaller segment of the market. Consequently there is less to motivate most customers to upgrade. Just think about what version you are using of your current word processor and you have a concrete example of the dynamic that I am talking about.

Fourth, as a product reaches maturity, the number of features have grown (so-called "feature bloat") to the point where those that you do add represent an ever-smaller part of the overall application (despite their increased cost) and are harder to find, even for the smaller number of users who may need or want them.

In summary, the cumulative effect is that, as the product reaches late maturity, development costs for the next release increase at precisely the same time that the size of the addressable market diminishes. This is illustrated in Figure 19, and in case you need reminding, this is not a good thing—except for your competitors.

The consequence to the company is that software sales alone are no longer sufficient to cover the growing development costs. Some strategies that companies adopt when they reach this stage is to rely more and more on annual support contracts, or services, rather than software sales for revenue. However, even this is generally not sustainable. Costs will still rise, and support contracts, at best, will be challenged to cover both R&D costs and deliver the financial returns required to keep the company interesting to investors. At best, the switch to a reliance on support revenue will just delay by a few release cycles the inevitable collision of technology costs and economic viability.

The net effect is that the law of diminishing returns kicks in for the developer, the investor, and the prospective customer. In general, n+1 products alone cannot sustain a company in the long term.

For this, companies need new products, and these have to come from somewhere. On the hardware side, we have already explored some of this space in our discussion of the Apple iPod. And, from this discussion about new products, perhaps it is even clearer why investors have shown such enthusiasm for Apple as a result of their success with it.

The problem, as we shall soon see, is that besides their first one, software companies do not have a good track record in developing new products. Simply stated, the history of the industry suggests that they are hopeless at the task. And this affects all of us, since software is such an important component of the types of products that are in demand today. This is why understanding the dynamics of the software industry is so important, even if you are not from that world.

All of which leads me to a personal story.

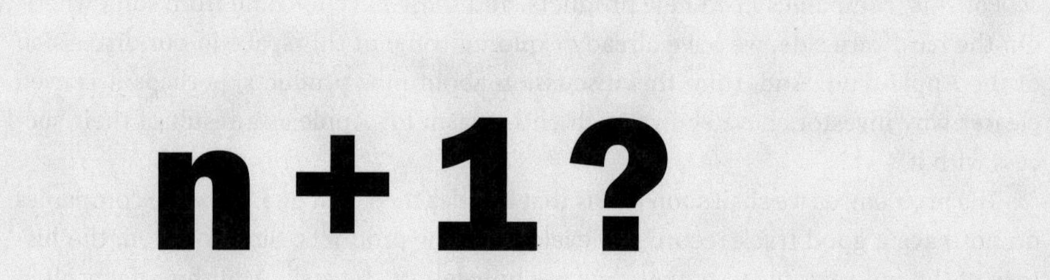
n + 1 ?

In case some of you might believe that what I am describing might apply only to software, let me dissuade you of that impression. First, look again at Figure 16. Despite the phenomenal success of the iPod, the writing is already on the wall. Revenue growth is not keeping pace with sales. At the same time, other companies are playing a more significant role in terms of fighting for market share. To sustain their growth, Apple will repeatedly have to introduce something new that will be relative to the iPod as the iPod was relative to the iMac. Hence the introduction of the iPhone for example.

Now what about developing successive versions of hardware? Do the costs go up on successive n+1 versions in the same way as they tend to do with software? Let's look at an example, the mobile phone. A good indication comes from a story on page 22 of the September 6, 2005 edition of the Financial Times. What it said is that it took fifty times as many people to do the R&D to develop Japan's third-generation CDMA cell phone technology compared to their second-generation standard, PDC. Fifty times! According to the article, Japanese companies lost Y35bn on combined revenues of Y1,000bn ($9.2bn US) in 2004, despite having 70% market penetration.

The cautionary lesson here is that the escalation of costs in providing the additional functionality perceived as necessary to sell such n+1 technologies can have a huge negative impact on an industry, regardless of whether it is hardware or software. It is also worth noting that this is generally associated with a push to add more and more functionality into the product. With mobile phones, for example, squeezing such additional functionality into such a device while maintaining acceptable usability and performance puts ever more pressure on design.

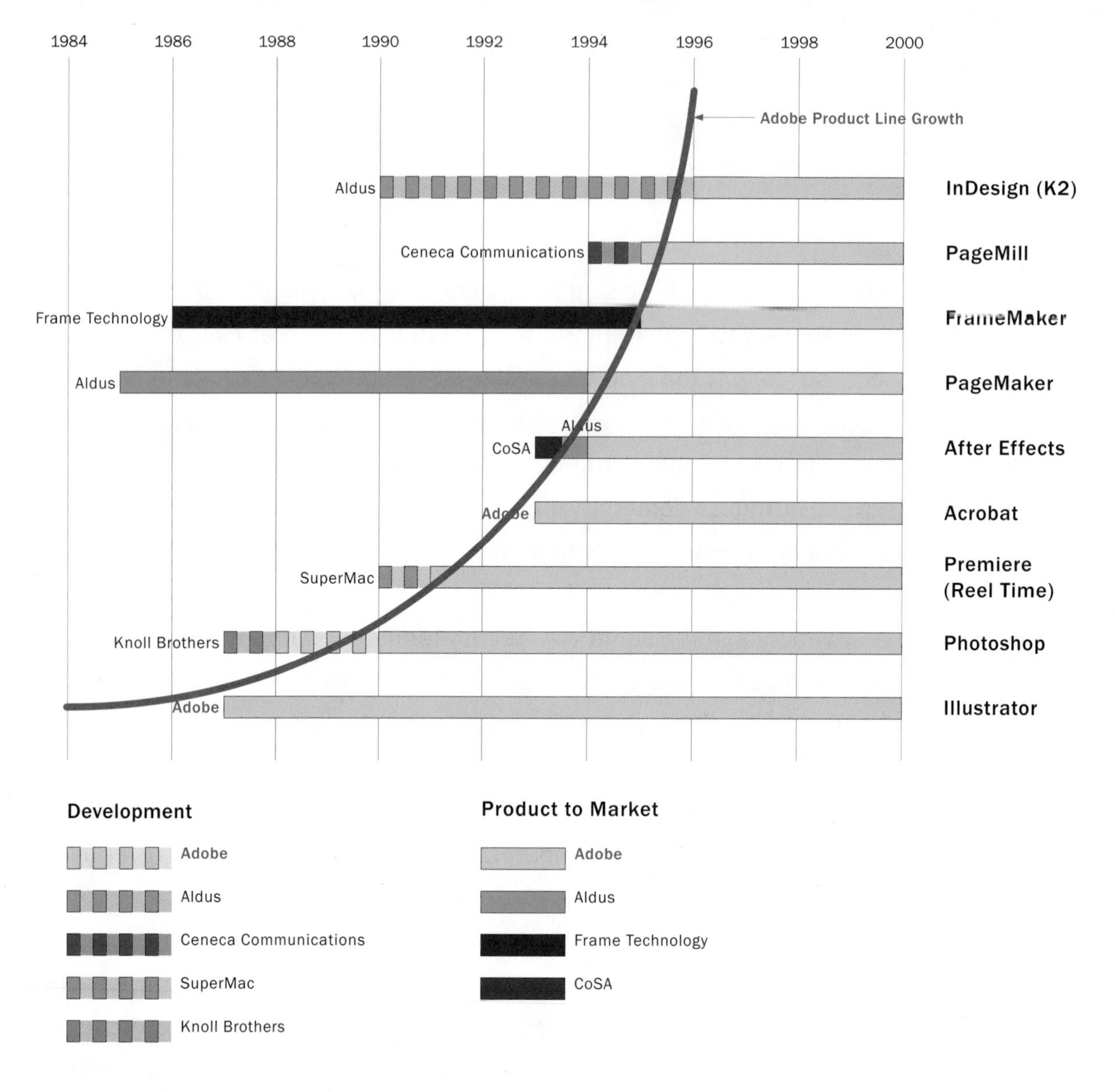

Figure 20: Genesis of Adobe Desktop Applications

Of the applications listed here, only two were developed in-house by Adobe. The rest were the result of mergers or acquisitions. Data: Pfiffner, 2003.

A Snapshot of Today

The future is no longer a synonym for "progress."
—Bruce Sterling

I had six years of increasing frustration. Why? Because the software company that I worked for, and of which I was an executive, was a total failure at the in-house development of new (as opposed to n+1) products. As Chief Scientist, I wasn't even directly responsible for product design. But that was little consolation. I still felt that we should be able to create new products.

Early in 2002, however, something significant happened. It suddenly occurred to me that our company was not alone in this situation. Rather, as far as I could make out, virtually every other software company was pretty much in the same boat. After establishing their initial product, they were about as bad at it as we were. When new products did come from in-house development, they were generally the result of some bandit "skunk-works" rather than some formal sanctioned process (not a comforting thought if you are a shareholder or an employee). Across the board, the norm was that most new products came into being through mergers or acquisitions. Let me illustrate this with a case study.

Adobe is one of the largest software companies in the world. It is also one of the older ones. Consequently, it is probably fair to call it very successful, and to use it as an example. Figure 20 shows the genesis of a number of Adobe desktop applications. What is of interest to me is that only two of them, Illustrator and Acrobat, were developed in-house. Illustrator was their first desktop application, *so after that, they have developed only one new desktop product in house!*

Following the pattern that I outlined earlier, the rest are all the result of mergers and acquisitions. This includes the most successful of these products, *Photoshop*. It also includes the full suite of Macromedia products, which are not shown but are now part of Adobe's product line.

The page creation product *In Design* warrants special comment. Although much of its development was undertaken by Adobe, the core technology came from Aldus, who had been working on it since 1990. Getting access to it was one of the key motivations for Adobe's 1994 acquisition of Aldus.

As an aside directed at anyone who still needs to be convinced about the need for executive leadership and vision, the official history of Adobe (Pfiffner 2003) points out that the acquisition of Photoshop met with strong resistance from within the product group (despite the product being demonstrable at the time). They viewed Photoshop as falling outside of the company's core competence and technology. From their perspective, Adobe was in the Postscript business (Postscript being the core technology around which Adobe was founded, and which underlies Illustrator). They didn't view Adobe as a 2D graphics company. Since it didn't employ Postscript, Photoshop was considered a bad fit. Luckily for Adobe and its shareholders, this view was not shared by John Warnock, the company's founder. If not for his intervention, Adobe would never have taken on the product. But they did, and the rest, as they say, is history.

There is a further argument that had it not been for Warnock, one of Adobe's other run-away successes, Acrobat, would never have seen the light of day either, or stayed on the market long enough to take off. Without enlightened executive vision and leadership around these two products, the history of Adobe may well have been much different, regardless of how creative the ideas may have been lower down in the organization.

But is there really a problem with basing one's growth on mergers and acquisitions? As I have already said, Adobe is one of the largest software companies. Its success suggests that this strategy has worked for them.

To better understand things, we need to dig a bit deeper. Here is where Adobe has done a service by commissioning the Pfiffner book on its history. It makes something clear that will be verified by anyone who has been through the process: merging corporate cultures, technologies, and geographically separated facilities comes at a very high cost in terms of people, efficiency, and money. And as a general strategy for new product development, this approach works only if there is a company available with the desired technology, if they are willing to let you have it, if you have the resources to acquire it, and if you have the skills and resources to negotiate the deal. Those are a lot of "ifs."

Sometimes all these factors do line up, and acquiring a technology is absolutely the right strategy. For example, companies can often get a product to market sooner, and therefore gain considerable competitive advantage, when the product is acquired rather than built in-house. And, at least as far as the product is concerned, there is often less left to the imagination, and therefore less risk with acquisitions. One hopefully knows what one is buying. These benefits may even mean that it is worth paying a premium for an acquired company.

Certainly any responsible manager should constantly be vigilant about the "not invented here syndrome" and always evaluate "buy" rather than "build" in any technology or product considerations.

But can you rely upon mergers and acquisitions as your sole strategy for growth? Not in any company that I want to manage, work for, or invest in. It is just far too limiting.

What happens when all the ducks do not line up? What are your options when there is no company or technology to buy, license, or merge with? What do you do when the company is there, but you do not have the skills or the resources to do a deal? Or what happens if the available technology is available only from a company whose geographic location or corporate culture is such that a merger would simply be a recipe for disaster? And what of the business literature, such as Porter (1987), that suggests that more often than not, in the long term, acquisitions fail to increase the value of the purchasing company?

Mergers and acquisitions cannot be the only alternative.

Ideally, one of the options that we would like to have open to us is to have a reliable way to develop new products in-house, within our own corporate culture, tailored to our own strategic plan, in a managed (rather than bandit) process, and where we can take into account the technologies employed in the rest of our product offerings. However, at least for the software industry, the track record says that this option is simply not viable today.

One underlying objective of this book is to help change this situation.

Furthermore, as the software and hardware components of products become ever more integrated, this is going to be increasingly important.

My underlying approach in what follows will be to put forward a holistic approach to experience-based design. Along the way, I will show how the weaknesses of software product development can be complemented by the strengths of traditional product design, and likewise, how the weaknesses of traditional product design can be complemented by the very real strengths of software developers. But my strongest argument is for the need for an explicit and distinct design process, integrated into the larger organization, supported by appropriate executive leadership.

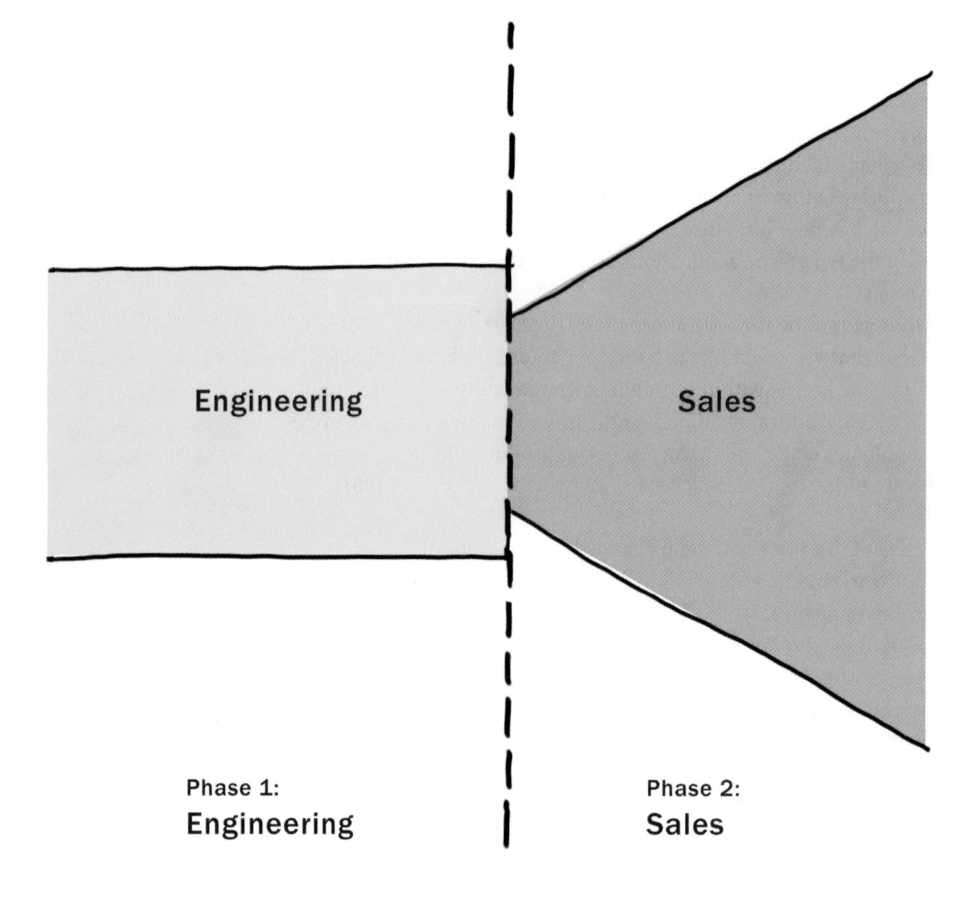

Figure 21: The Status Quo in Software Product Development
Projects get a green light right at the start, and go directly to engineering where they are built. The next phase is when they ship—usually late, with bugs, over budget, and missing functionality.

Figure 22: A Preproduction Clay Model
Before making a commitment to put a new model into production, automotive executives insist on having a very clear idea of what they would be investing in. Hence, the approval process includes evaluating a full-size clay model of the car that to you or me would be indistinguishable from the real thing. This is in marked contrast to the vague state in which most software products are when they get the green light.
Credit: General Motors

The Role of Design

Our inventions are likely to be pretty toys, which distract our attention from serious things. They are but improved means to an unimproved end.
—Henry David Thoreau

My belief is that one of the most significant reasons for the failure of organizations to develop new software products in-house is the absence of anything that a design professional would recognize as an explicit design process (Buxton 2003). Based on my experience, here is how things tend to work today.

Someone decides that the company needs a new product that will do "X." An engineering team is then assigned to start building it. If you happen to be a shareholder of the company, the most frightening aspect at this early stage is that a projected revenue number for this new product is already placed in the company's financial forecast, despite the fact that at this point nobody knows for sure if it can be built, how it will be built, how much it will cost to build, or when it will realistically be shipped (if ever).

Nevertheless, the product goes straight into engineering. If it ever emerges, it is almost certainly seriously late and over budget (Brooks 1975).

The only good thing about this approach is that one will never be accused of not conforming to the initial design. The bad news is that this is because there is no initial design worthy of the term. As illustrated in Figure 21, the project is given a green light, is committed to product, and has engineering resources allocated to it, right from day one. The fact that I am not exaggerating is reflected in some of the books that you will find on project management and product development. In the chapter entitled, "What Is the Design Phase?" in a book modestly titled, *The Definitive Guide to Project Management*, one can find the telling (and for me, frightening) words:

> ... on software projects, for example, the design and build phase are synonymous. (Nokes et al. 2003; p. 157)

Such is the status quo. In it, there is nothing that compares to preproduction in film-making, or the design phase that is a standard step in the development of a new automobile.

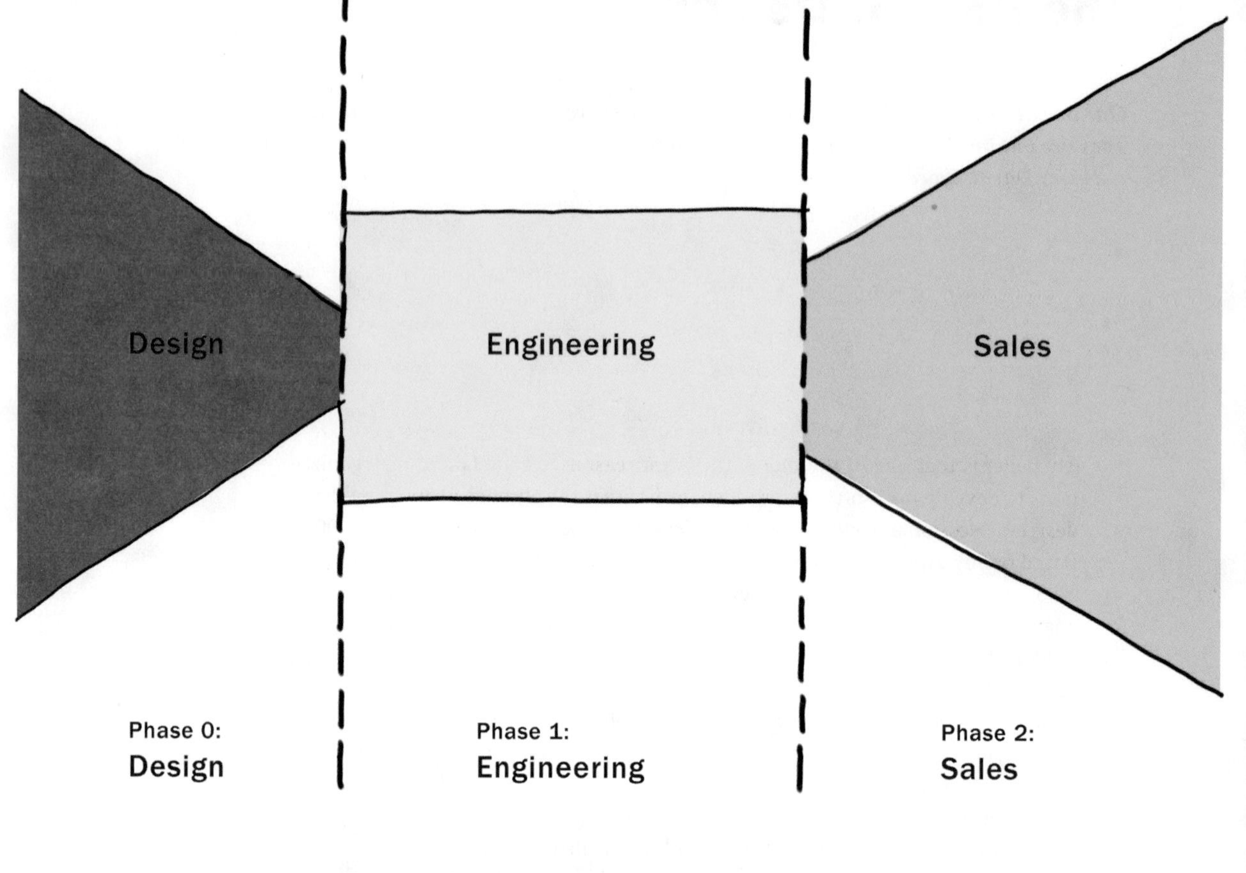

Figure 23: Inserting an Explicit Design Process Prior to Green Light
The process is represented as a funnel, since the number of concepts to emerge is always anticipated to be fewer than the number that enter. Design is a process of elimination and filtering as well as generation.

The culmination of the design phase for a new car, for example, involves the construction of a full-size clay model, such as is illustrated in Figure 22. These models can take more than a month to build and cost over a quarter of a million dollars.

When they are done, they are indistinguishable from the real thing (I still have to ask, "Is this real or a model?"). Despite the cost of doing so, it is only by bringing the concept to this level of fidelity in terms of its form, business plan, and engineering plan that a project can enter the green-light stage. And, despite being visually indistinguishable from the real thing, if the project receives a green light, there is still typically a year of engineering before the project can go into production.

As we shall discuss later, preproduction in feature films has a comparable set of hoops that one has to jump through before a project can be considered for green-lighting. Is this up-front process expensive? Absolutely. But it costs nothing compared to the costs that will likely be incurred if you don't make that investment!

There is a lesson here for the software industry—and a warning. The heart of the lesson is that there needs to be a distinct design phase followed by a clear green-light process before proceeding to engineering. This is illustrated in Figure 23. Now the warning. Since it varies so much from the status quo, my claim is that this design process must not be undertaken or owned by product engineering. Despite statements to the contrary, such as the quote from Nokes et al. earlier, design and engineering are different. They employ different processes. They require different skills, different styles of management, and the people suited for one are typically not suited for the other. Let me summarize this last point as follows:

> **It is just as inappropriate to have an engineer manage the design process as it is to have a designer who graduated from art college be responsible for the product's engineering details.**

In this I am not casting aspersions on either the design or engineering professions. To the contrary. I have the highest regard for both, and each is essential to the overall process. By the same token, however, neither is sufficient. The latter point is too often neglected in our engineering-dominated practice. I am just trying to shed some light on how I believe that these two critical skill sets can best complement each other.

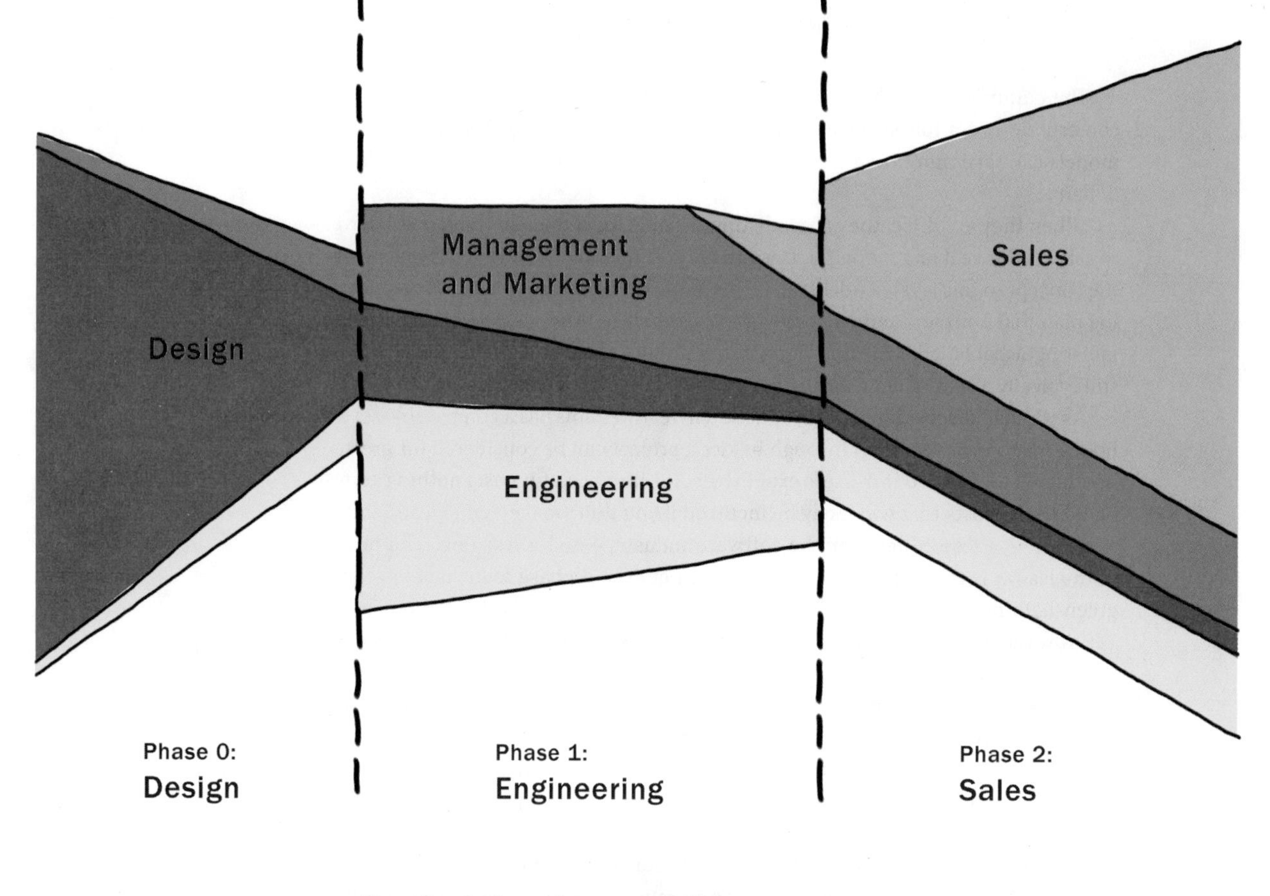

Figure 24: No Silos

A representation of the proposed ongoing responsibilities of the various teams throughout the overall process. Of central importance for our purposes is the notion that "design" includes the design of the business and engineering plans, as well as the product itself. In this way it is no different than filmmaking.

A Sketch of the Process

The things we have to learn before we do them, we learn by doing them.
—Aristotle

My focus in all of this is largely on design, but it is not my intention to present you with a deep treatise on some idealized design process. Frankly, I would doubt the value of such a thing, even if I thought one could be written. However, that doesn't mean that there is nothing to talk about. There are some general comments and concepts that I do think warrant discussion. Think of them, and the way in which I approach them, as analogous to a sketch. There is no pretense here to provide a final rendering of the design process.

In the simplest terms, the difference between Figure 21 and Figure 23 illustrates the basic argument that I am making: the need to insert a design process at the front end of product development. The primary assumption in advocating doing this is that the cost and time lost due to this additional stage will be significantly less than the cost and time lost due to the poor planning and overruns that will inevitably result if it is not included.

My perspective is that the bulk of our industry is organized around two all-too-common myths:

1. That we know what we want at the start of a project, and
2. That we know enough to start building it.

Like the sirens who tried to lure Ulysses to destruction, these myths lead us to the false assumption that we can adopt a process that will take us along a straight path from intention to implementation. Yes, *if* we get it right, the path is optimal. But since there are always too many unknowns actually to do so, the fastest and most efficient path is never a straight line. Furthermore, my experience suggests that embarking on the straight-line path, and then having to deal with the inevitable consequences, is the path with the highest risk.

At best, it is a route to mediocre products that are late, over budget, compromised in function, and that underperform financially. At worst, it leads to product initiatives that are cancelled, or fail miserably in the marketplace. And with it, design, such as it exists, typically is limited to *styling* and *usability*.

Here is the reality, at least as I see it. Even if we think we know at the outset what the product needs to be, in nearly all cases its definition changes, or is significantly refined as we progress. This may be because of insights that we gain along the way, or because the market has changed in the interim, or both. Hence, our process must be designed to accommodate such changes as the norm, and be tailored to uncover them as early as possible, which is when learning, mistakes, and fixes are the least expensive.

A friend, the sociologist Gale Moore, introduced me to some useful vocabulary from Donald Schön (1983). It sheds some light on how to counter these two myths. Schön talks about the importance of distinguishing between two aspects of design: *problem solving* and *problem setting*. The former can be characterized by the question, "How do we build this?" The latter addresses the question, "What is the right thing to build?" By assuming that the product specifications or design brief define the problem adequately enough to start building the product, as per our first myth, problem setting falls off the design agenda. The folly of this type of assumption, however, is obvious to almost anyone who has actually built a product. What they will tell you is that they really didn't know enough to start until they had finished.

As Schön was one of the first to point out, product development demands attention to both problem setting and solving, and this is fundamental to the design process. Unfortunately, industry has not paid sufficient attention to his teaching.

An underlying observation that helps position the respective roles of product design and engineering is this: Even if you do a brilliant job of building what you originally set out to build, if it is the wrong product, it still constitutes a failure. Likewise, you also fail if you build the right product the wrong way. Stated another way, we must adopt an approach that inherently aspires to *get the right design* as well as *get the design right*. The former, which is one of the prime objectives of the up-front design phase, is the part that is too often absent in today's practice. Hence, it occupies a significant part of our discussion. Our premise is that we must get both right. Doing so requires an adequate investment in *both* design and engineering. But as we shall see, it requires far more than that, and may well involve every arm of the company, which is why executive understanding of these issues is so important (Buxton 2005, 2006).

What is critical to note in all of this is that breaking the product lifecycle into three basic phases, as shown in Figure 23, does not mean that the process should be viewed as three silos, roughly partitioned among design, engineering, and sales. Quite the opposite. As Figure 24 tries to capture, the business, technical, and creative elements must be active in all three phases. The three phases have more to do with what the primary focus is, and who has responsibility for that phase, than with who is involved.

We can return to film making for an example. Here, the equivalent to the design phase is called preproduction. Most film buffs will know this term and may even be familiar with some of its components, such as story development and the assembly of the key players. But it actually goes much further than that. Typically, you consider a film for green-lighting only upon the presentation of a complete package (Brouwer & Wright 1991). The package includes not only script, director, and key cast members, it also specifies the budget, release date, marketing plan, and a detailed production plan. In other words, before committing the resources to shoot the film, there must be an extremely detailed creative, production (technical), and business plan in place. It is an evaluation of all three criteria—creative, technical, and business—that determines if the film gets made. Although this represents a significant up-front investment, with no assurance that the film will be produced, the costs and competitive nature of the industry are such that it is generally unthinkable to commit to something where these things are not nailed down in advance.

There is room still for creativity and innovation during production. However, it could be argued that it is precisely the up-front planning in the preproduction phase that provides this freedom, and lets the director and producer manage it, while staying within the framework of what was approved.

Likewise, the "design phase," as I define it, does not imply just the design of the actual product. Rather, as in film, it also includes the design of the engineering process, the design of the marketing plan, and the refinement of the overall business model.

Note that green-lighting a project is not a signal that the designers are now off the project. Think of the ongoing relationship and division of responsibilities between design and engineering by comparing it to the role of the architect and structural engineer in designing a building. The architect has prime responsibility for the design. However, the architect clearly must work with the structural engineer in order to ensure that the design can be built. The more technologically innovative the design, the more likely this is true. Furthermore, during the design phase, the expertise of the structural engineer often can suggest new possibilities for the architect, thereby contributing to a much-improved result. On the other hand, during construction, the architect is still involved in the project, even though the primary responsibility now lies with engineering.

Returning to our comparison with film, if we consider the director as the head of the creative department of a film, a useful exercise or test on any software product would be to ask, "Who is the equivalent of the director? Do they have comparable power and responsibility? If not, why not? If not, why do we believe that the integrity of the design will be maintained?" Likewise, an analogous set of questions needs to be asked in order to ensure that the integrity of engineering and business concerns are sustained through the design process. And throughout, one must keep in mind that everyone on the film probably works at the pleasure of the producer, who ultimately has financial responsibility.

The equivalent of the producer, in the context of product development, is the person or team that has financial and overall business ownership. However, though it is important to recognize that the producer owns the film, it is equally important to recognize that having hired the director and agreed on the overall objectives and guidelines, undue interference with the film's direction by the producers will almost guarantee that it fails. Yes, checks and balances are necessary. But there must also be a recognition that people are in their place because they are good at it. Generally, the director shouldn't be producing nor the producers directing. Talking and working together? Absolutely; it is a team. But each must be able to play their own position.

Let me end this section by addressing a comment that usually comes up at this stage:

> This is all well and good, and in the ideal world this might be fine. But in the real world, we have to meet deadlines and work under limited budgets. We simply don't have the time or money to add this kind of process. Shooting for the ideal is just not realistic. "Good enough" is all we have time for.

My standard response to such comments nearly always takes the form of the following two counter-questions:

> **How is it that we can never afford to do proper planning and design, yet we always seem to be able to afford to pay the cost of products being late as well as the cost of fixing all of the bugs that inevitably result from inadequate design, planning, and testing?**
>
> **How can this be when the cost of design and planning is nothing compared to the cost of being late to market or having a defective product?**

It takes very strong and brave management to admit that we don't know what we are doing at the start, and therefore need to accommodate that in our process. But the reality is, if we factor out luck and a few rare exceptions, it is always faster, cheaper, and leads to a better product if you take the extra step of incorporating an explicit up-front design phase rather than going directly for the status quo shortcut.

Design for the Wild - Sketching Experience
Author: Bill Buxton
Publisher: Morgan Kaufmann - Diane D. Cerra
Project Plan
Designer: Henry Hong-Yiu Cheung
Draft v.2
10.25.2006

Designer Workdays | Publisher Workdays | Author Workdays

OCTOBER — Week 1: S 8, M 9, T 10, W 11, Th 12, F 13, S 14 — Week 2: S 15, M 16, T 17, W 18, Th 19, F 20, S 21 — Week 3: S 22, M 23, T 24, W 25, Th 26, F 27, S 28 — S 29, M 30

	TASKS	DURATIONS		
		Start Date	End Date	Number of Business Days
1.0	**PROJECT START-UP**			
1.1	Sketch Concepts			7
1.2	Sketch Layout Templates	11-Oct-06		7
1.3	Review		19-Oct-06	7
1.4	Project Plan Development	19-Oct-06		1
1.5	Contract Development	11-Oct-06	25-Oct-06	10
2.0	**SCHEMATIC DESIGN**			
2.1	Refine Overall Book Concept		27-Oct-06	6
2.2	Font Selection			6
2.3	Cover Concept Design			11
2.4	Develop Layout Templates	20-Oct-06		16
2.5	Image and Content Text Integration I			16
2.6	Copyediting (Publisher)			16
2.7	Index Development (Publisher)		13-Nov-06	16
2.8	Develop Book Specification Options			10
2.9	Coordinate w/ Publisher and Printer	26-Oct-06		10
2.10	100% Schematic Design Review	13-Nov-06		1
2.11	Identify and Commission Illustrations	29-Oct-06	27-Nov-06	20
3.0	**DESIGN DEVELOPMENT**			
3.1	Refine and Finalize Cover Design			4
3.2	Refine and Finalize Layout Templates			9
3.3	Image and Content Text Integration II		27-Nov-06	9
3.4	Image Refining - Color Correction / Formatting / Resolution	14-Nov-06		9
3.5	Assemble Image Source List - (Publisher)			19
3.6	Procure Image Copyrights - (Publisher)		11-Dec-06	19
3.7	Consolidate and Integrate Illustrations from Artists			4
3.8	Design Index			5
3.9	Finalize Book Specification	20-Nov-06	27-Nov-06	5
3.10	Coordinate w/ Publisher and Printer			5
3.11	100% Design Development Review	27-Nov-06		1
4.0	**PRE-PRODUCTION**			
4.1	Final Design Files for Typesetting	27-Nov-06	01-Dec-06	5
4.2	Final Review (Publisher / Designer / Author)	04-Dec-06	08-Dec-06	5
4.3	Indexing (Publisher)			5
4.4	Prepare and Organize Final Files for Printer		15-Dec-06	10
4.5	Finalize Index Design	11-Dec-06		5
5.0	**PRODUCTION**			
5.1	On-Press Review	08-Jan-07	12-Jan-07	5
5.2	Printing		02-Mar-07	40
6.0	**DELIVERY**			
6.1	Shipping	05-Mar-07	01-Apr-07	10
6.2	Book Launch	01-Apr-07		1

The Ying and the Yang of Design

There is another side of the design profession that is generally not too visible. It is the organized professional who knows how to plan, and understands deadlines, when to dream, and when to get down to it. Let me give you a concrete example. When I spoke with Hong-Yiu Cheung about doing the design of this book, the first thing that we did is sit down for an afternoon and go through almost every design book in my library and his, discussing what we liked, and what we thought might fit this title. Within twenty-four hours Hong-Yiu produced his first deliverable. It was not a package of sample page designs. Rather, it was the spreadsheet, above. It worked backwards from when we

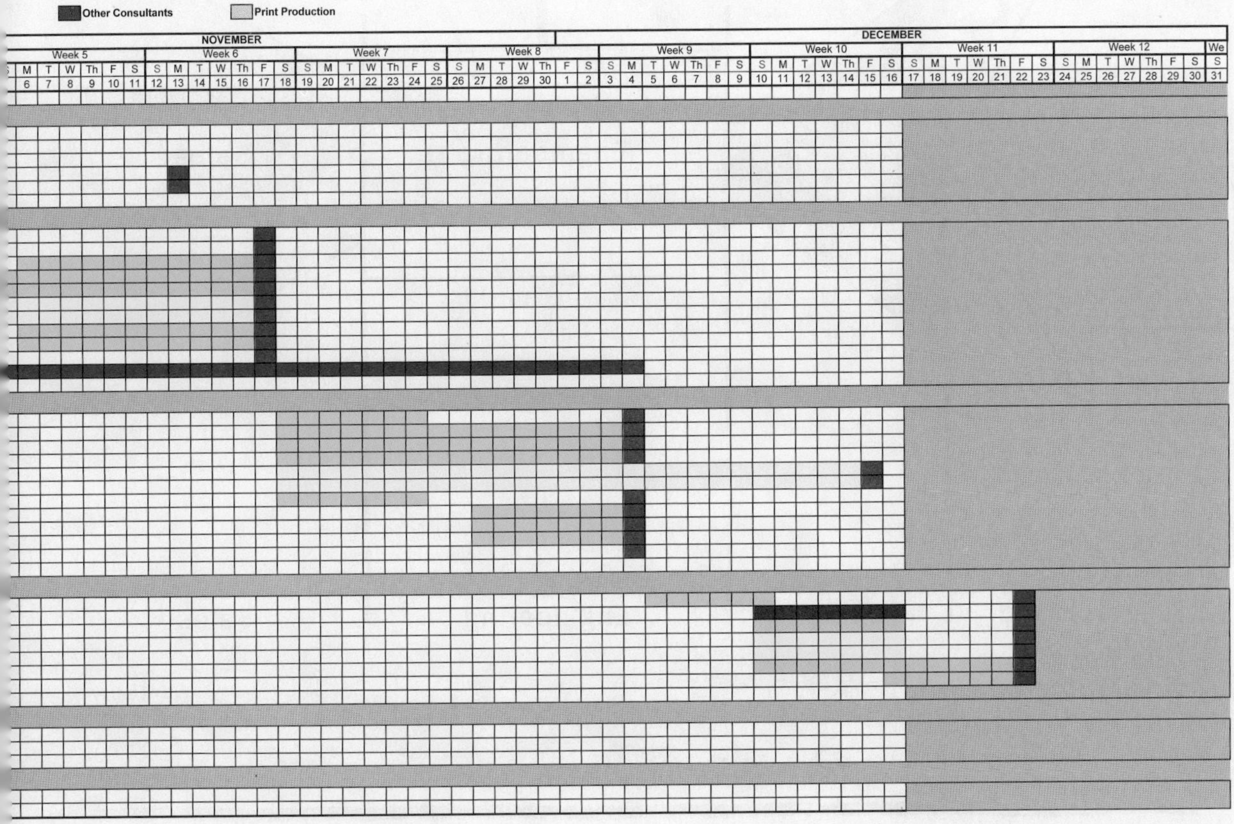

Other Consultants

Print Production

Month	Week	Day	Date
NOVEMBER	Week 5	S	
		M	6
		T	7
		W	8
		Th	9
		F	10
		S	11
	Week 6	S	12
		M	13
		T	14
		W	15
		Th	16
		F	17
		S	18
	Week 7	S	19
		M	20
		T	21
		W	22
		Th	23
		F	24
		S	25
	Week 8	S	26
		M	27
		T	28
		W	29
		Th	30
DECEMBER		F	1
		S	2
	Week 9	S	3
		M	4
		T	5
		W	6
		Th	7
		F	8
		S	9
	Week 10	S	10
		M	11
		T	12
		W	13
		Th	14
		F	15
		S	16
	Week 11	S	17
		M	18
		T	19
		W	20
		Th	21
		F	22
		S	23
	Week 12	S	24
		M	25
		T	26
		W	27
		Th	28
		F	29
		S	30
	We	S	31

wanted the book in print, right up to when we were speaking. It laid out the schedule for virtually every step of the book's production. The key point was, before even thinking about the details of design, he had to understand exactly how much time he had to do it and still meet the deadline. Stated another way, this spreadsheet let him know exactly how much time he could allocated to exploring different concepts without putting the project at risk. It also told him exactly when he had to make decisions, and meet specific milestones. It is precisely this kind of discipline that provides the freedom for the designer to explore the creative potential of the project.

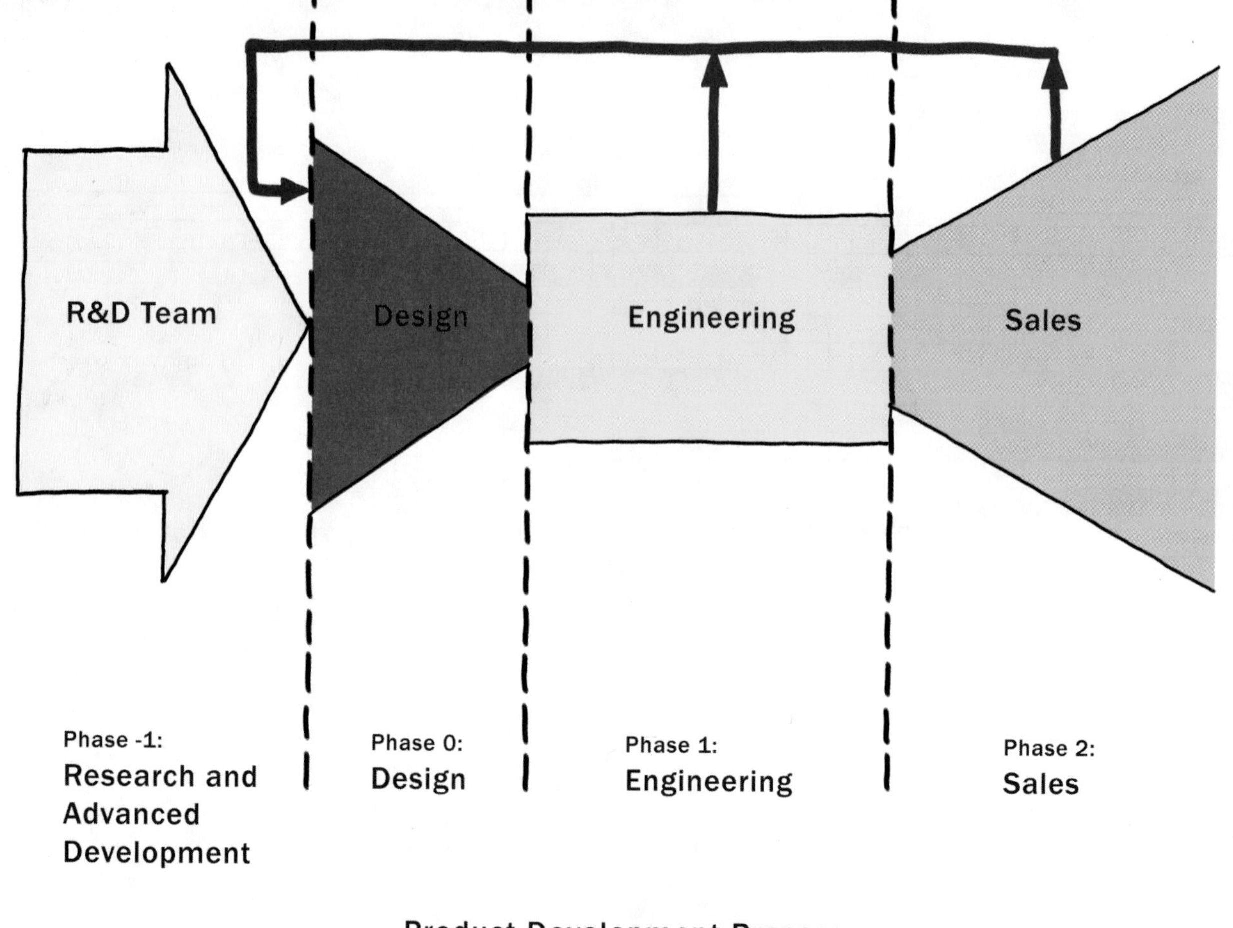

Figure 25: Research and Advanced Development Feeding Product Design

The Cycle of Innovation

Life is like riding a bicycle. To keep your balance, you must keep moving.
—Albert Einstein

I want to work through an example in order to illustrate the complementary roles of design and engineering. Since they are far enough away from computers to let us check our biases at the door yet still remain relevant, I have chosen to look at mountain bikes.

However, before going down this path I need to clarify something. Despite what is shown in Figure 23, engineering can precede design, and still not contradict anything that I have said thus far. To understand this, we need to make an important distinction. *Product engineering,* which is what is shown in the figure, is what is done to build the product once it receives a green light. *Research and advanced development* generally feed into design, as shown in Figure 25.

One of the worst product disasters that I have ever seen was due to a company that chose to ignore this. They committed to a product, despite knowing that there were still several fundamental unsolved research questions with the technology on which the product depended.

> **Hint: You can't schedule scientific breakthroughs, so it is a bad idea to have your product schedule depend on them. Save your gambling for the casino, and manage your research group separately from your development team.**

On a happier note, some of my favourite success stories involve creative engineers who solve research problems that then enable great product design. The story of Trek's Y-Bike is one of them.

There is a lot of debate as to who made the first mountain bike, and when and where the sport began. Some would argue that it goes back to 1887 in Scotland, or to 1896 with the Buffalo Soldiers in Montana. There were bikes with pneumatic tires and rear suspension as early as 1907. In the 1920s, the *Tour de France*—ostensibly a road race—started to include routes through the Alps that consisted of little more than dirt and stone paths. And in the 1950s there was a Velo Cross club in Paris. In the mid 1970s, one group of *tilters* in Butte, Montana were taking their bikes over 12,700-foot Pearl Pass and into Aspen, while a bunch of enthusiasts were getting their thrills flying down Mount Tamalpais in Marin County.

Regardless of who started what, when, or where, the intent was pure fun, and the modifications they made to their bikes in the process were intended to enhance that fun, not to start a new industry.

At the same time, in the mid 1970s, the mainstream bicycle industry was boring. It was a mature industry that was pretty much flat. In other words, it had peaked. There was no significant growth. There was no significant innovation. And the same traditional players were trying, with essentially the same types of products, to increase their own share of the same stagnant market.

Mountain biking changed everything.

It caught on and transformed the marketplace. Innovation thrived. New companies were born. Specialized shops opened everywhere. The prices that people were willing to pay for bikes skyrocketed, and most importantly, whole new populations started biking. Perhaps better stated, populations who had previously considered bikes as a children's toy started to adopt them as the ultimate toy for adults. This new enthusiasm for biking even spilled over into the previously stagnant sectors of the markets, such as road racers and commuter bikes!

The fact that all of this took place in a mature market makes this most interesting as a case study. It is also why, when trying to innovate in similarly mature industries, one of my most consistent mantras is to ask:

What is the mountain bike?

That is, what innovation could have the same impact on some other stagnant and/or saturated industry as the mountain bike had on its industry? Fundamental to my belief system, by the way, is that there is always a mountain bike. The problem is finding it. A large part of what we are talking about is how to do so. Looking at a case study of the actual thing seems like a reasonable place to start.

Among the small bicycle companies active at the time was one called Trek. Trek was founded in 1975 and made conventional handcrafted steel bikes. However, they were one of the first companies to fully grasp the significance of the mountain bike. Their first mountain bike, shown in Figure 27, was launched in 1983 and was one of the first commercially available. (The very first mountain bikes were homemade—that was part of the fun.) What is most interesting about the bike in Figure 27 is how normal it looks. This is not what we would imagine as the stuff that revolutions are made of. Yet it was, and there is a serious lesson here.

As the mountain bike phenomenon took off, sales grew. Therefore, so did the competition—as well as the resources and impetus for further innovation. One of the main areas ripe for innovation was around the question of suspension. Remember, the riders are barreling down ungroomed trails at break-neck speeds. It is a bumpy ride.

Adding shocks to the front forks was not a huge problem, since they could adapt the well-established technology used in motorcycles. For the rear suspension, however, that was not the case. The reason lies in the difference between motor versus pedal power.

Figure 26: In the mountains in the 1923 Tour de France.

Credit: Keystone Photo Archives.

Figure 27: Trek's First Mountain Bike

Released in 1983, the most interesting thing about this bike is how conventional it looks. From the perspective of today, it is hard to imagine that this bike was considered a radical innovation, and helped transform the industry.

Photo: Trek Bicycles

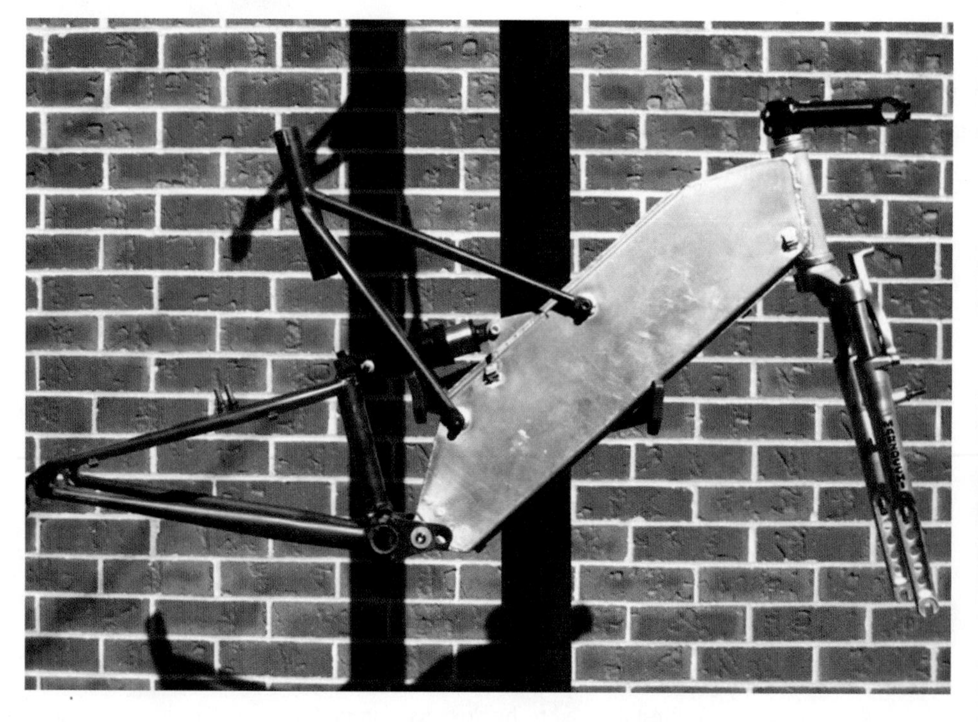

Figure 28: Engineering Prototype of Trek's Y-Bike

Photo: Trek Bicycles

Figure 29: Matt Rhodes' Design of Trek's Y-Bike
This is the result of applying industrial design to the prototype shown in Figure 26. The creativity of the designer complements and enhances that of the engineer, rather than competes with it.
Photo: Trek Bicycles

Figure 30: Specialized Epic Bicycle with Inertia Valve Shock
The rear shock on this bike has a patented inertia valve that is closed (rendering the shock inactive) when traveling over smooth terrain. When on rough terrain, the valve opens, thereby activating the shock.

When pedaling a bicycle, the basic forces applied to the pedals, especially when standing, are up-and-down. The whole goal of the system is to transform this up-and-down motion of pedaling into forward motion by means of intermediate rotary motion. The problem with motorcycle-type rear shocks and frame geometry is that, instead of being transformed into forward motion, a lot of the up-and-down pedaling action gets transformed into up-and-down bouncing on the rear suspension, and this constitutes a waste of precious energy. Furthermore, care must be taken since everything that is added to the bike, such as shock absorbers, adds weight. In case you need to be reminded, wasting energy while trying to pedal a heavy bike up a steep hill is not a good thing.

So, a lot of effort and innovation went into making a bike that was light, strong, and had frame geometry that minimized the rear-shock bouncing effect. This was a challenging problem, and one that took highly skilled engineers in advanced development to address. The result of Trek's efforts, led by Wes Wilcox, is shown in Figure 28. This is a photograph of the circa 1994 engineering prototype of what was to become Trek's Y-Bike.

There are a few observations worth making here:

The Y-Bike shown in Figure 28 is a radical departure from Trek's first mountain bike, shown in Figure 27. With today's eyes the latter looks very conventional, however it nevertheless launched the company on a radically different trajectory. But remember that in order to succeed, they did not need to leap directly to something as innovative as the Y-Bike.

The innovation reflected in the engineering prototype shown in Figure 28 clearly came from someone with a technical background, a mechanical engineer, rather than someone who studied art or design. Yet, it constitutes the central idea of the final design.

The engineering prototype shown in Figure 28 works. If you look at the photo carefully, you will see that if you added pedals, sprockets, wheels, a chain, brakes, and handle-bars to this prototype—all of which could be bought from any bicycle shop—it would be perfectly functional. You could ride it. You could sell it. And perhaps it would even perform extremely well. And no matter what the price or how well it performed, it is almost certain that it would be a commercial flop. Why? Anyone can see that the bike is not complete. Not because of the missing parts, but because the design is not complete.

What is obvious here with the mountain bikes is not obvious with software. My impression is that too often what we see in Figure 28 reflects the state in which software products ship. They kind of work, but are as far from complete as this version of the bike is.

Trek knew that they were in new territory with the Y-Bike. It was like nothing that the industry had previously seen. The prototype gave them additional confidence in the concept. But they also knew the same thing that you knew from looking at the picture.

Up to this time, since bicycles were a mature technology, design was largely a matter of convention. The inclusion of industrial designers in the process, for example, was rare. The Y-Bike, almost on its own, demanded that this change. Trek was smart enough to take the plunge.

Recognizing the need, they went to one of the top design schools in the world, *Art Center* in Pasadena, and recruited a fresh graduate, Matt Rhodes, to join the team. Now when I say fresh, I mean fresh. Matt had never done a product before. He came right out of school, and was given three months, a computer, and a brief to work with the team and bring the concept from prototype to product. The result is illustrated in Figure 29. Just looking at the figure, it is obvious that this was accomplished by the incorporation of a skill set that was very different from, but complementary to, that of the equally talented engineers who built the prototype shown in Figure 28. The result is more than just styling (design ≠ styling). It involved a refinement in materials, geometry, form, manufacturability, *and* styling.

Trek's Y-Bike was a huge success, and it made a significant contribution to the growth of the company. In fact, it was so successful that others wanted a piece of the action. Top among these was a company called *Specialized*.

However, in order to succeed in competing with Trek, Specialized would have to match or exceed the high standards of innovation that Trek had established. They set out to do just that.

Remember what I said earlier about strong up-and-down pedaling action causing the rear suspension to bounce, and therefore waste valuable energy? Well, Trek's design addressed the problem, but it did not eliminate it. This is the problem that Specialized set out to fix.

Before going further, however, it is important to understand something about the bike industry. As you would expect, there are key parts of the bike, especially the frame, that are built by the bike company itself. However, there are also a large number of other key components—such as derailleurs, cranksets, brakes, seats, wheels, and hubs, for example—that are sourced from third parties. In fact, it is very common to find the same components, such as derailleurs, on bikes from different bicycle manufacturers.

This is relevant to our story because shocks were one such component. Consequently, they did not fall within the core design, engineering, or manufacturing competence of typical bicycle companies. They were something that bicycle companies bought rather than designed.

Specialized's key innovation was to redefine how they viewed themselves. They chose to differentiate themselves from their competitors by involving themselves directly in the redesign of the rear shock absorber. They chose to change the rules rather than simply compete in the established game.

In 2003, what they did was revitalize and refine a concept first developed in the mid 1900s: the *inertia valve shock absorber* (Whisler 1943). As refined by Specialized, this is a shock absorber that is rigid when pedaling on smooth ground. That is, a valve prevents any of the fluid from flowing within the shock, which effectively makes your bike a "hard-tail." However, when bouncing downhill over rough ground (which is when you want a soft-tail), the valve opens, thereby letting the shock do its job of absorbing the bumps. Think of the result (shown in Figure 30) like an automatic transmission, except that what is automatic is the changing of the suspension characteristics to fit the terrain.

Having developed the concept and patented their innovations, Specialized then partnered with their main supplier, Fox Racing Shox (who added a few patents of their own), to complete the design and manufacture the new shock.

The result was that Specialized was able to innovate on a component that most of its competitors did not consider part of their direct business. By partnering with their supplier, they were able to innovate around a key component that is normally outside of their direct sphere of influence, while staying within their range of core competencies. And despite the shock being manufactured by a third party, they were able to retain exclusive use of the technology because of their patents.

Does that mean that Specialized won? Not at all. It just means that they are players who help keep the cycle of competition and innovation rolling and healthy. The real lesson is that to be a player, design and innovation have to permeate the company and its culture as a whole. A great design with a poor business model or unreliable engineering will not win. Excellence in design, engineering, sales, and the basic business model are all *essential,* but none is *sufficient* on their own. There are three interlinked lessons to be learned from what we have seen thus far.

First, the talents of research, design, and engineering are complementary, not competitive.

Second, there is room, in fact a need, for creativity throughout. The process illustrated in Figure 23 or Figure 25 is not at all intended to suggest that design (or research) does all the fun stuff, and then throws crumbs over the transom to engineering to be built by a bunch of uncreative drudge workers. Rather, implicit in all of this is that each group has its own problems to solve and opportunities for making a creative contribution to the final product.

Third, there can be no silos. Matt clearly had to collaborate with product engineering in order to get the manufacturable result shown in Figure 29. Doing so is consistent with the "no silos" concept shown in Figure 24.

The rest of this book is focused on the design component, but I have taken a good deal of time on this last point because it is so important. No matter how well one does design, no matter how brilliant and creative one person or sector of the organization, without appropriate integration among the various components, without a healthy entrepreneurial spirit, and without an organization-wide culture of innovation, all that the best design process and five dollars will buy you is a coffee at Starbucks.

And for those who have now been seduced by the whole question of bicycle design, you are highly encouraged to read Mike Burrow's, *Bicycle Design: Towards the Perfect Machine.* For a good survey of the history of the bicycle, see Bicycling Magazine's, *The Noblest Invention.* Noguchi-san (2000) is a history of the design of bicycles and their components. It does so by means of a collection of stunning drawings that will be of interest to anyone who has a passion for technical illustrations. To see an example from cycling of how business, performance, and design can be trumped by regulation, read Moll (2004).

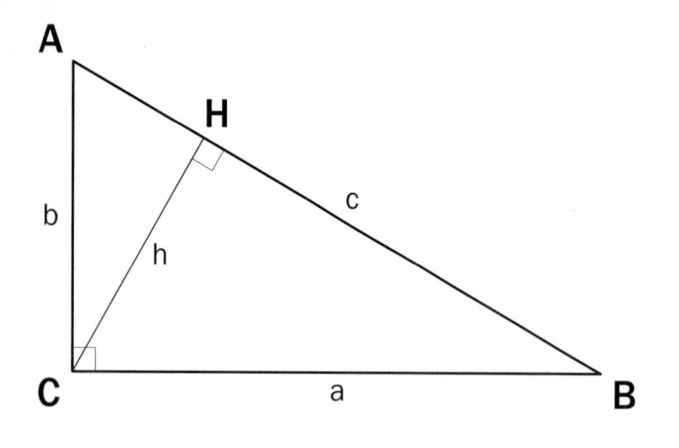

Proof of the Pythagorean theorem

Like many of the proofs of the Pythagorean theorem, this one is based on the proportionality of the sides of two similar triangles.

Let ABC represent a right triangle, with the right angle located at C, as shown on the figure. We draw the altitude from point C, and call H its intersection with the side AB. The new triangle ACH is similar to our triangle ABC, because they both have a right angle (by definition of the altitude), and they share the angle at A, meaning that the third angle will be the same in both triangles as well. By a similar reasoning, the triangle CBH is also similar to ABC. The similarities lead to the two ratios:

$$\frac{AC}{AB} = \frac{AH}{AC} \text{ and } \frac{CB}{AB} = \frac{HB}{CB}$$

These can be written as:

$$AC^2 = AB \times AH \text{ and } CB^2 = AB \times HB.$$

Summing these two equalities, we obtain:

$$\begin{aligned} AC^2 + CB^2 &= AB \times AH + AB \times HB \\ &= AB \times (AH + HB) \end{aligned}$$

In other words, the Pythagorean theorem:

$$AC^2 + BC^2 = AB^2$$

*** VALU-MART ON BLOOR ***
MANULIFE CENTRE

RH FOUR 4.19
0.240 kg Gross Weight
-0.010 kg TARE CREDIT =
0.230 NET kg @ $7.69/kg
MUSHROOMS WHITE 1.77
0.550 NET kg @ $5.49/kg
BRUSSELSPROU 3.02
REDDI SNACK ALMD 4.99
BUTTER 5.49
PINEAPPLE GO 4.99
4PK SPONGE 12 2.99
COCA-COLA ZERO 12 1.49
PC BM EGG OMG BR 3.79
NATREL MILK 4.69
4.48 1=GST 0.27
4.48 2=PST 0.36
TOTAL 38.34

----------TRANSACTION RECORD----------
NATIONAL GROCERS RETAIL 4909800
Bloor Valu-mart
55 Bloor St. W.
Toronto ON
STORE 00433 TERM L0043301
SLIP # 4585 REG 01
RETAIN THIS COPY FOR YOUR RECORDS

DATE TIME AMOUNT
02/11/2007 15:28:26 $ 38.34
APPROVED
Process amount to account shown

You could have earned 380
PC points with President's Choice
Financial MasterCard. Apply Today
Visit pcfinancial.ca

****
GST # 12404-8406 RT0001
THANK YOU, COME AGAIN
02/11/07 15:29 Felicia 1 04585

Figure 31: The difference between mathematics and arithmetic.
Being able to add up your grocery bill does not make you a mathematician. Likewise, decorating your home does not make you a designer.
Source: wikipedia.com

The Question of Design

Societies do not evolve because their members simply grow old, but rather because their mutual relations are transformed.
—Ilya Prigogine

If design is so important yet neglected, and if we should be taking steps to remedy that situation, then perhaps it makes sense to clarify what we mean by "design."

Here is where the trouble starts. Take any room of professionals and ask them if they know what design is, or what a designer is. Almost everyone will answer in the affirmative and yet practically everyone's definition will be different. That is to say, people's definitions are so broad that almost every act of creation, from writing code, building a deck, making a business plan, and so on can be considered design. If one goes to the literature instead of one's colleagues, the result will be pretty much the same.

The problem is, when a word means almost anything or everything, it actually means nothing. It is not precise enough to be useful. Take your typical company trying to develop a new product, for example. If those creating the business plan, planning the sales and marketing campaign, writing the software, performing usability studies, and such are all doing "design", then how can I be arguing that we need to incorporate design into the process? By that definition of design, it is already there at every level of the organization and every stage of the process.

Now I could be wrong about this. For example, the well-known writer and psychologist Don Norman has stated in an epilogue to his most recent book (Norman 2004):

> We are all designers.

I have the highest degree of respect for Don, but in my opinion, this is nonsense!

Yes, we all choose colours for our walls, or the layout of furniture in our living rooms. But this no more makes us all designers than our ability to count our change at the grocery store makes us all mathematicians. Of course there is a way that both are true, but only in the most banal sense. Reducing things to such a level trivializes the hard-won and highly developed skills of the professional designer (and mathematician).

If you are a nurse or paramedic, you can legitimately refer to yourself as a medical practitioner, but not a doctor. None of this is intended to discount the skills or professionalism of those who have medical skills but are not MDs. To the contrary. Their skills may well save a life. In fact, the more we understand and appreciate the nature of their specific skills, the more they help us understand and appreciate the specific skills that a doctor, or a specialist, bring to the table. And it is exactly this kind of awareness, in terms of the skills of the design professional, that I see as lacking in so many of those who profess to speak for the importance of design, or their own affinity or capacity in design.

I think that I do understand what people like Don Norman are trying to express when they say, "we are all designers." I accept that it is well intentioned. But statements like this tend to result in the talents, education, and insights of professional designers being discounted, or distinguished from everyday design decisions.

Perhaps the whole thing could be cleared up through a bit more precision in our use of language. Just as the term "medical practitioner" is more general than "doctor," we might distinguish between "design practitioner" and "designer." Or, perhaps we just need two distinct but related words, analogous to *arithmetic* compared to *mathematics.*

Regardless, in the sense that I use the term, everyone is distinctly not a designer, and a large part of this book is dedicated to explaining the importance of including a design specialist in the process of developing both things and processes, what their role is, and what skills they bring. But if now you are expecting me to give you a clear definition of design as I use the term, I am afraid that I am going to disappoint you. Smarter people than I have tried and failed. This is a slippery slope on which I do not want to get trapped.

What I mean by the term "design" is what someone who went to art college and studied industrial design would recognize as design. At least this vague characterization helps narrow our interpretation of the term somewhat. Some recent work in cognitive science (Goel 1995; Gedenryd 1998) helps distinguish it further. It suggests that a designer's approach to creative problem solving is very different from how computer scientists, for example, solve puzzles. That is, design can be distinguished by a particular cognitive style. Gedenryd, in particular, makes it clear that sketching is fundamental to the design process. Furthermore, related work by Suwa and Tversky (2002) and Tversky (2002) shows that besides the ability to make sketches, a designer's use of them is a distinct skill that develops with practice, and is fundamental to their cognitive style. This is something that we shall discuss in more detail in a later chapter.

I can also say what I do not mean by design, in particular, in the context of this book. I do not mean the highly stylized aesthetic pristine material that we see in glossy magazines advertising expensive things and environments. This is fashion or style that projects a lie, or more generously, a myth—a myth that can never be real. By "design" I don't mean the photographs of interiors of rooms where nobody could live, of clothes that nobody could wear, or of highly stylized computers or other appliances whose presentation suggests that they were "designed" as if they don't need cables and that they are to exist on perfectly clear desks without even a human around to mar their carefully styled aesthetics.

No, the type of design that I want to talk about in this book gets down and dirty. It is design for the real world—the world that we live in, which is messy and constantly changing, and where once a product is released, the designer, manufacturer, and vendor have virtually no control or influence over how or where it is used. Once sold, it leaves the perfect world of the glossy advertising photos. In short, I am talking about design for the wild. Carrying on our bicycle theme, contrast the renderings of the two mountain bikes illustrated in Figure 32 with that shown in Figure 33. Hopefully this helps make my point. The "design" that I want to talk about goes beyond the object and cannot be thought of independent of the larger physical, emotional, social, and experiential ecology within which it exists. (To further pursue other notions of "the wild," see, for example, Hutchins 1995 or Attfield 2000.)

I can offer another approach, one that makes an end-run around the whole dilemma. This option takes a lead from Fällman (2003a, 2003b). Rather than pursue the question "What is design?" (which probably none of us will agree on anyhow), let us ask a different (and perhaps better) question: "What is the archetypal activity of design?"

For Jones (1992), the answer would be drawing:

> ... the one common action of designers of all kinds (p.4)

Fällman's answer is similar, just a little more specific. It would be sketching. In agreeing with him, I am not alone. Others, such as Goel (1995), Gedenryd (1998), and Suwa & Tversky (1996) have come to the same conclusion.

In saying this, it is important to emphasize that I am not asserting that the activity of sketching is design. Rather, I am just making the point that any place that I have seen design, in the sense that I want to discuss it in this book, it has been accompanied by sketching. So, even if we can't (or won't) define design, we can perhaps nevertheless still gain some insights into its nature and practice by taking some time to delve into the nature of sketching.

Figure 32: Two Renderings of a Mountain Bike

The above view is expository. It shows the design in an objective way. In the one on the facing page, it was decided to render the bike in a stance that was less neutral—one that started to project some character (for me at least), a kind of embedded playfulness. Now contrast these representations to that in the following figure!

Images: Trek Bicycles

TREK
BONTRAGER
XTR

Figure 33: Down and Dirty (and Wet) in the Wild
The real test comes not where the rubber meets the road, but the mud, rocks, sticks, and yes, the water. Even though the images in Figure 32 have value, this is a rendering of what a mountain biker really buys. It is the aspiration (and hopefully the reality) of the experience. And despite being the best representation of what one gets with the product, unlike the preceding renderings, the bike is hardly visible. This is the wild!
Photo: Trek Bicycles

“We are NOT all designers.”

I can feel the hackles of some of my colleagues rising when I make such a dogmatic statement as, "we are not all designers." Especially some of those from Scandinavia. The reason is that there is an approach to design called "participatory design" (Greenbaum & Kyng 1991; Clement & Van den Besselaar 1993; Muller 2003) in which the lay-person is an active and essential participant in the design process. Rather than following the "Designer as God" model, where products come from "on high" like manna from heaven created by "The Designer," participatory design adheres to an ethic of "Designer as Facilitator." In this case, the role of the design professional is to work with the users/customers as a kind of combination coach/trainer, to help them come to an appropriate design solution themselves.

In the world of participatory design, therefore, we are all potential participants in the design process. However, a careful reading of my preceding words will show that there is no contradiction here. Yes, the lay-person can play a critical role in the design process. But if we are all designers, then why is a design professional required in participatory design? Why don't the lay-people just do it on their own?

My words are far less controversial if you grant me one small concession: that design as a profession is as rich as math or medicine. We have no problem accepting that although medicine is distinct from math, it is still rich enough to encompass disciplines as diverse as neurology, cardiology, podiatry, and so on. Likewise, mathematics embraces a diverse range of specialties. As we shall soon see, my dogma does not apply to some narrow definition of design. The view of design that I am discussing in this book is broad enough to encompass participatory design, among other approaches to design practice. I see the discipline as that rich. But by the same token, as with math and medicine, I do not see that as implying that "we are all designers" or that there is not a distinct profession called "design."

So, when I speak of design, I do mean something distinct from engineering, marketing, sales, or finance, for example. However, in so doing, by no means do I mean to take away from, or downplay, the value or importance of the other creative activities that are part and parcel of any of these other functions. I am just not referring to these activities when I use the term "design."

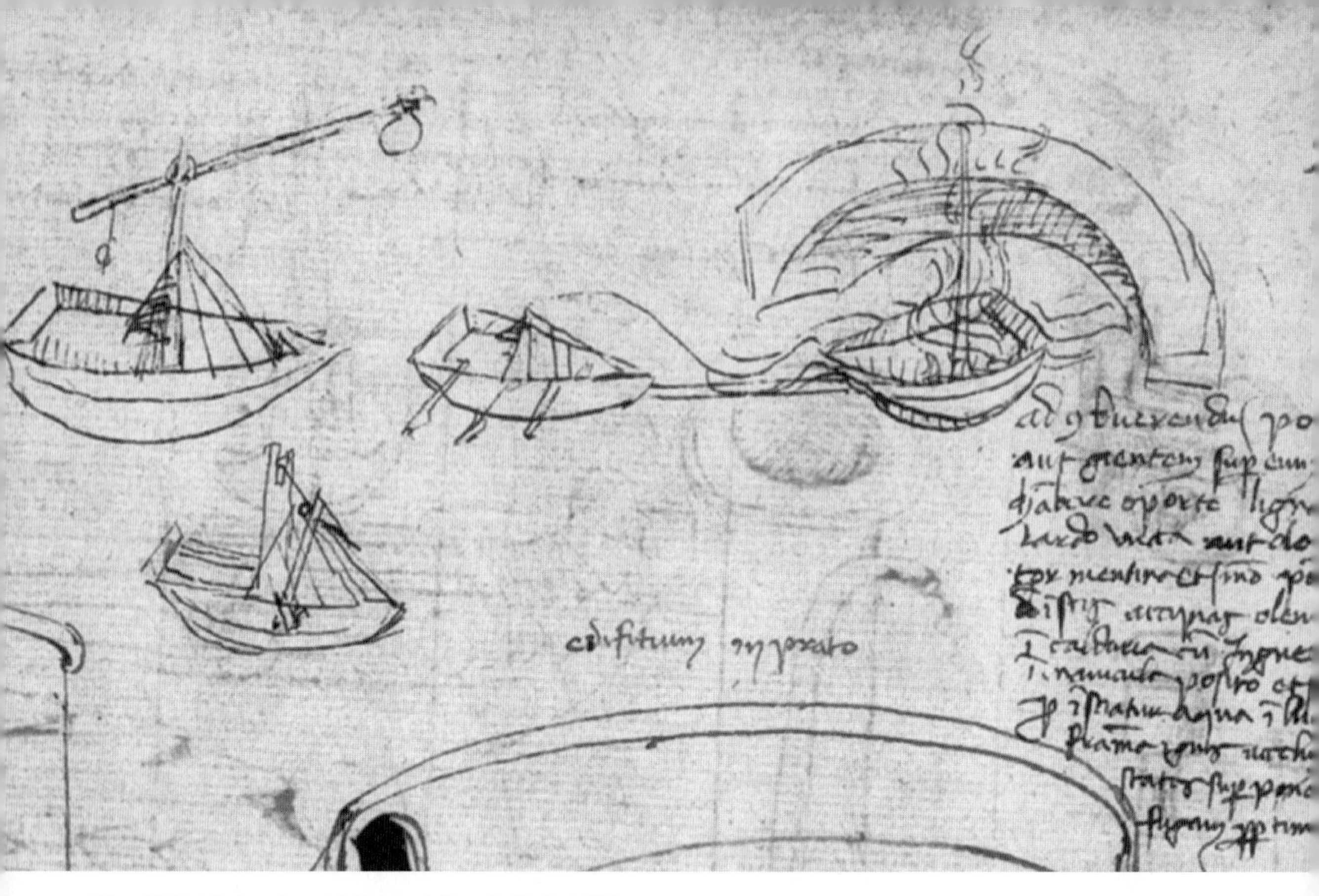

Figure 34: Details from Taccola's Notebook (from first half of C15)

Several sketches of ships are shown exhibiting different types of protective shields, and one with a "grappler." These are the first known examples of using sketching as a tool of thought.

Source: McGee (2004); Detail of Munich, Bayerische Staatsbibliothek. Codex Latinus Monacensis 197 Part 2, fol. 52'

The Anatomy of Sketching

The only true voyage of discovery is not to go to new places, but to have other eyes.
—Marcel Proust

Both sketching and design emerged in the late medieval period, and this was no accident. From this period on, the trend was toward a separation of design from the process of making (Heskett 1980). With that came the need to find the means whereby the designer could explore and communicate ideas. Sketching, as a distinct form of drawing, provided such a vehicle.

The first examples of sketching, as we think of it today, come from Siena, from Mariano di Jacobi detto Taccola (McGee 2004). In the first half of the fifteenth century, he embarked on a four-volume set of books on civil and military technology, called *De Ingenisis*. In a manner not unlike George Lucas and *Star Wars*, he completed volumes 3 and 4 first, and delivered them to the emperor in 1433. Volumes 1 and 2 were never completed. Rather, he went on to work on another project, *De Machinis*, which he completed in 1449.

This might seem like a little too much arcane detail, but you kind of need to know it in order to understand the following excerpt from a recent book about Taccola's work:

> What is significant for our purposes is that Taccola worked out many of the ideas he presented in *De Machinis* by filling the unfinished pages of Books 1 and 2 of *De Ingenisis* with hundreds of rough sketches, turning them into a sort of notebook. Examining these sketches and comparing them to the drawings in *De Machinis* we are able to follow a person actually working out technical ideas for the first time in history. (McGee 2004; p. 73.)

That is, Taccola's sketches, such as those seen in Figure 34, are the first examples of the use of sketching as a means of working through a design—sketching as an aid to thought.

For a discussion of the figure, we turn again to McGee:

> Here we see that Taccola has sketched three different kinds of protected attack boats: one with a stone dropper, one with a ram, and one with a large hook or "grappler" on the side. We immediately see that his technique has enabled him to quickly generate three alternatives. Using paper, he is able to store them. Stored, they can be compared. In short, Taccola's style provided him with a graphic means of technical exploration. (McGee 2004; p. 76)

Now let us move from the renaissance to the present. For the sake of argument, let us assume that design and sketching are related. Furthermore, let us assume that we can gain insights about design by way of cultivating a better understanding of sketching. Doing so is not too much of a stretch. For example, museums such as Boijmans Van Beuningen in Rotterdam exhibit sketches, models, and prototypes in their own right, as a means to inform us about the process of product design.

> In the past few years within the profession of industrial design there has been increasing attention on the story behind the object, in which sketches, design drawings, models and prototypes play a prominent role. They make possible a reconstruction of the interesting history of their origin. Above all they make visible the designer's contribution, which is often very different to what one might expect. (te Duits 2003; p.4)

In this spirit, I want to introduce a number of sketches that were generated in the course of realizing a product, in this case a time-trial racing bicycle designed for Lance Armstrong for the Tour de France. These appear as Figures 35 through 39. The first four images are in chronological order. The first three take us from sketch to engineering drawing. The visual vocabulary of all the figures is different, and it is important to keep in mind that these variations are not random. Rather, they are the consequence of matching the appropriate visual language to the intended purpose of the rendering. The conscious effort of the designer in doing so is perhaps most reflected in Figure 38, where the designer has gone to extra effort to "dumb down" the rendering in order to ensure that it did not convey a degree of completion that was not intended.

In looking at the drawings, keep in mind that they follow only one of the many concepts explored—the one that was eventually built. Early in the design process it would not be unusual for a designer to generate 30 or more sketches a day. Each might explore a different concept. The figures used are intended to show different styles of visual representation of just one of these, not to show the breadth of ideas considered.

Looking at them individually, we see that Figure 35 is clearly a sketch. Its visual vocabulary suggests that it was hand drawn, quickly and effortlessly, by a skilled artist. It says that it does not represent a refined proposal, but rather simply suggests a tentative concept. But what is it in the vocabulary that tells us all this? Largely, it is the freedom, energy, abandon, and looseness of the lines. It is the fact that the lines continue on past their natural endpoints. It tells us no rulers were used.

Even if the designer laboured for hours (or even days) over this rendering, and used all kinds of rulers and other drafting tools, it does not matter. The rendering style is intended to convey the opposite, because the designer made this sketch with the clear intention of inviting suggestions, criticisms, and changes. By conveying the message that it was knocked off in a matter of minutes, if not seconds, the sketch says, "I am disposable, so don't worry about telling me what you really think, especially since I am not sure about this myself."

Figure 36 is a refinement of the previous sketch. It has all the sketch-like properties of Figure 35, but includes rough shading in order to tell the viewer more about the detailed 3D form of the concept being pursued. As in the previous sketch, it would look at home on the wall of a drawing class. It says, "I'm thinking seriously about this form, but the ideas are still tentative. But as I am getting more serious, tell me now what you think."

Figure 37 is not a sketch. This is a "serious" piece of work. Because of the wireframe mesh on the surface, the precision of the lines, and the quality of the corporate graphics, this rendering says that it took a lot of care and work, and that it was done on a computer. It is a 2D rendering of an accurate 3D model of the entire frame. Compared to the previous two drawings, it says "expensive" and "refined" (although the retention of the wireframe mesh in the rendering also says "but not finished"). It says, "We have made some decisions and are seriously considering this path."

Let me put it this way: of the dozens of concepts worked up to the level of the first two sketches, very few would be taken to this stage. To any literate reader of drawings, this is implicit in the style of rendering itself. The funnel is converging.

Now we move to my favourite rendering, Figure 38.

This is a hybrid. What the designer has done is make a photorealistic three-quarter view rendering of the 3D model previously seen in Figure 37. He has then made a composite with it and the hand-drawn sketch seen in Figure 35. But why would he do this? He was working to a tight deadline. He had no time to spare, and this took extra work. He already had done the 3D model. He just could have used the photorealistic three-quarter view rendering on its own. The answer is in the figure itself. The extra effort was undertaken to imbue the resulting image with the quality of a sketch. To make it look all the more effortless. To say, "This isn't finished," and to invite suggestions and communicate that the design was still open to change.

Now look at Figure 39. By this stage it is clear that these are examples of sketches. These types of sketches are actually among the first ones done in a project.

Michael Sagan, the designer, describes his process and use of such thumbnail sketches as follows:

> Typically I do very loose thumbnails to capture a gesture or a theme to start out. Often I will jot down words or phrases that I use as a semantic guide. As a design review step I will have another designer evaluate my 3D work ... checking back against my thumbnails and semantic guide-words. If the designer hits any of the words I count that as a success. In the case of this sheet that I included here ... one designer picked out almost all of the words ... much to his surprise when I showed him these images.

Finally, note the following. First these thumbnail sketches were made in the course of designing what, at the time, was probably the most technologically advanced bicycle ever built. Second, stylistically speaking, they are completely in keeping with, and would be perfectly at home in, the sketchbooks of Taccola.

Figure 35: Early Three-Quarter View Sketch of Time Trial Bike

Although done on a computer, this is a freehand sketch. Notice that the representation is tentative. What tells you this? Contrast this to the representation in Figure 37.

Credit: Michael Sagan, Trek Bicycles

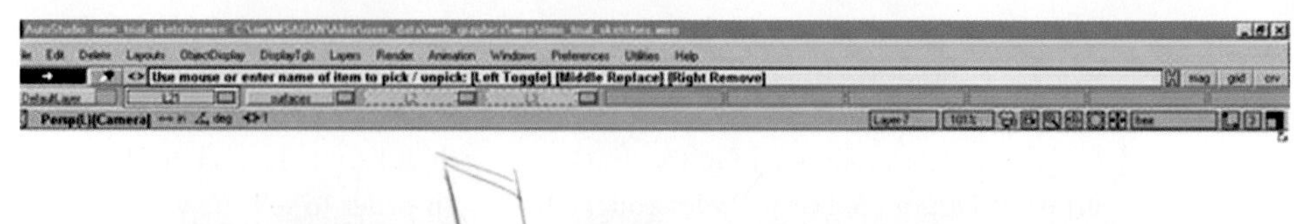

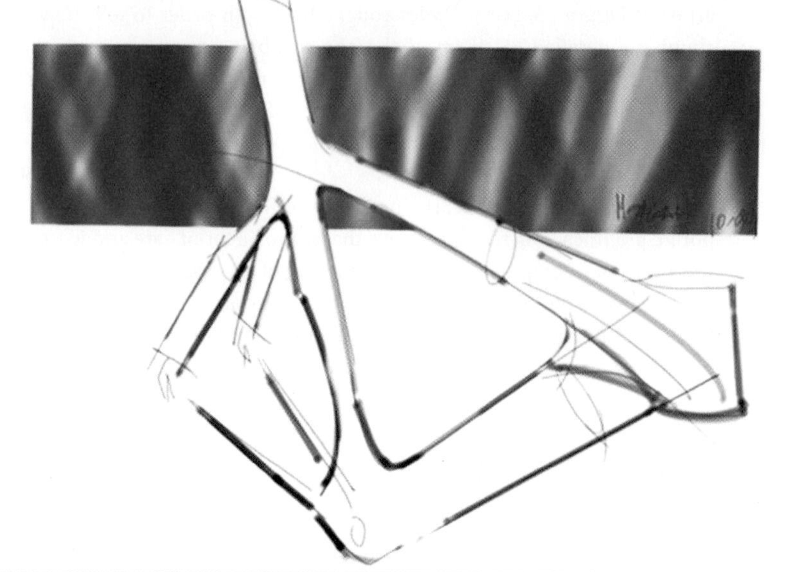

Figure 36: Shaded Three-Quarter View Sketch of Time Trial Bike

This is a refinement of the sketch seen in Figure 35. Through the use of shading, the sketch communicates more about the 3D form of the concept. Notice that despite this refinement lines still extend through the “hard points.”

Credit: Michael Sagan, Trek Bicycles

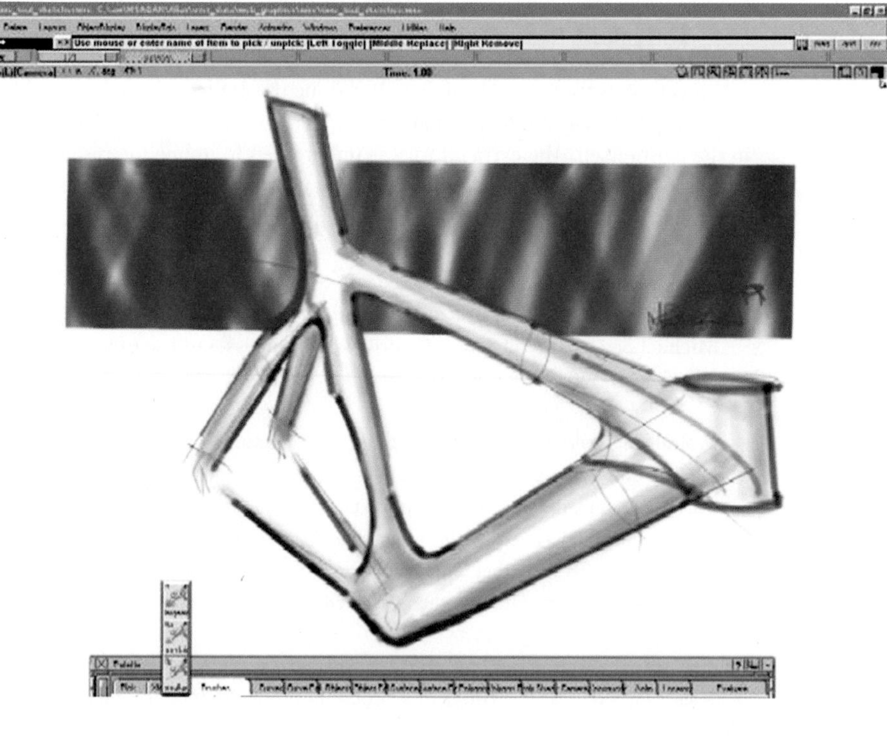

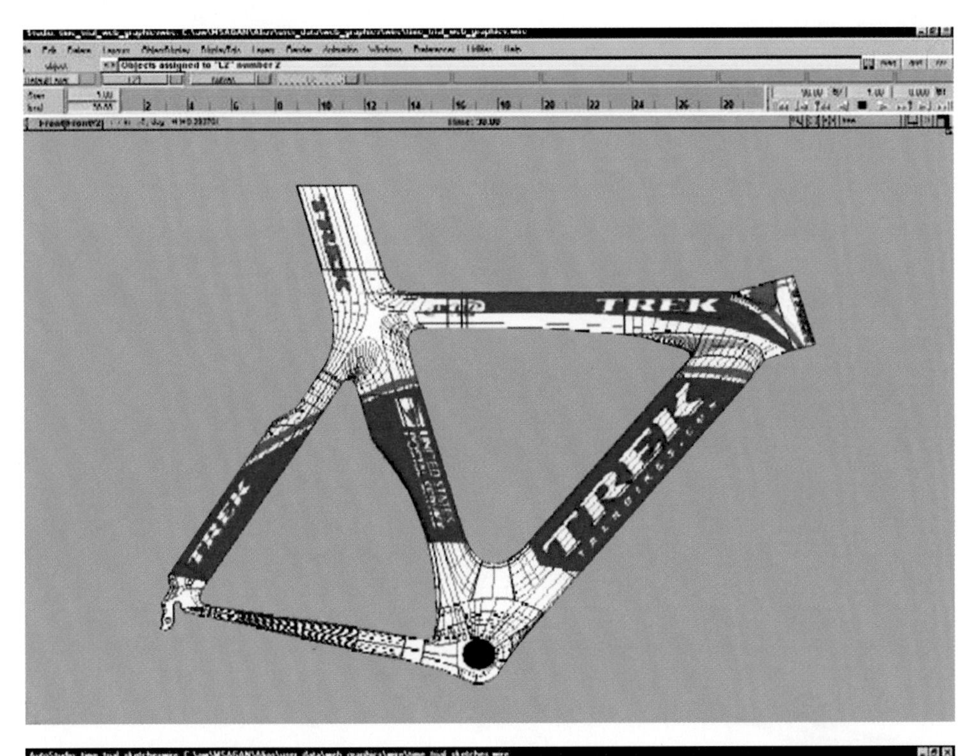

Figure 37: Side View of 3D Shaded Model of Time Trial Bike

This is a side view of the same bike seen in the previous two figures. Contrast this representation to that in Figure 36. Both are shaded to highlight the form. Ignoring the addition of the graphics for the moment, is it obvious, is it clear which of the two is more refined, closer to "final," and which took the most effort to create, and which will take the most effort to redo in the event of a change or suggestion. This image is clearly not a sketch.

Credit: Michael Sagan, Trek Bicycles

Figure 38: Accurate 3D Shaded Model Superimposed Over Three-Quarter View Sketch

This image is perhaps the most interesting. It is a composite of a three-quarter view of the 3D model seen in Figure 37 superimposed over the sketch seen in Figure 35. Given what we have seen thus far, ask yourself why the designer would do this.

Credit: Michael Sagan, Trek Bicycles

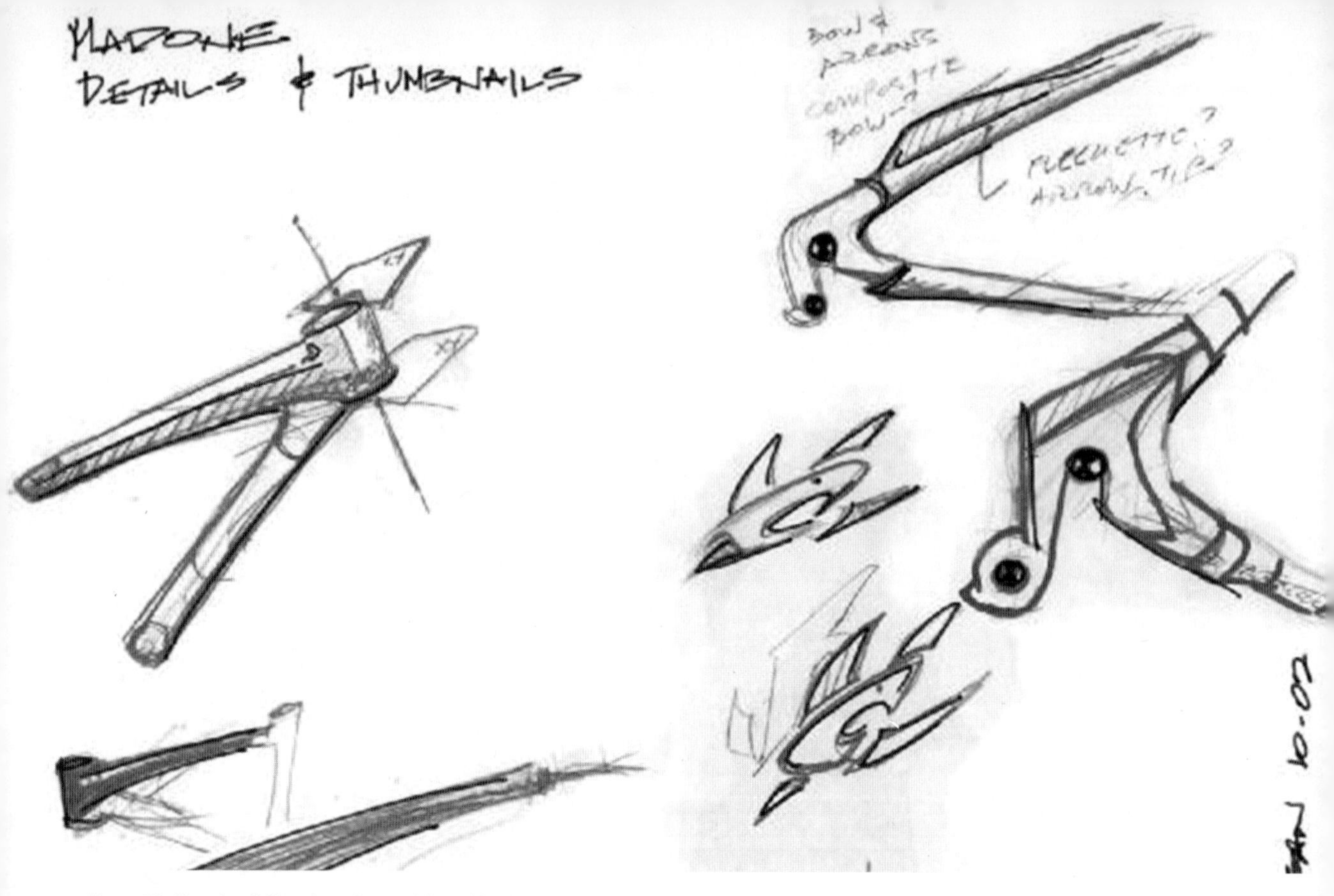

Figure 39: Thumbnail Sketches, Scanned from Sketchbook

In what century were these made? Yesterday? During the renaissance? You can't tell from the form, only from the content.

Credit: Michael Sagan, Trek Bicycles

Sketching is not only the archetypal activity of design, it has been thus for centuries.

Having come this far, what I would like to do now is step back and try to use what we have seen in these examples as a means to come to some characterization of sketches in general.What I am after here is an abstraction of sketches and sketching. What I want is to go meta and identify a set of characteristics whose presence or absence would let us determine if something is, or is not, a sketch —at least in the way that I would like to use the term.

Here is my best attempt at capturing the relevant attributes of what we have seen and discussed. Sketches are:

Quick: A sketch is quick to make, or at least gives that impression.

Timely: A sketch can be provided when needed.

Inexpensive: A sketch is cheap. Cost must not inhibit the ability to explore a concept, especially early in the design process.

Disposable: If you can't afford to throw it away when done, it is probably not a sketch. The investment with a sketch is in the concept, not the execution. By the way, this doesn't mean that they have no value, or that you always dispose of them. Rather, their value largely depends on their disposability.

Plentiful: Sketches tend not to exist in isolation. Their meaning or relevance is generally in the context of a collection or series, not as an isolated rendering.

Clear vocabulary: The style in which a sketch is rendered follows certain conventions that distinguish it from other types of renderings. The style, or form, signals that it is a sketch. The way that lines extend through endpoints is an example of such a convention, or style.

Distinct gesture: There is a fluidity to sketches that gives them a sense of openness and freedom. They are not tight and precise, in the sense that an engineering drawing would be, for example.

Minimal detail: Include only what is required to render the intended purpose or concept. Lawson (1997, p. 242) puts it this way, " ... it is usually helpful if the drawing does not show or suggest answers to questions which are not being asked at the time." Superfluous detail is almost always distracting, at best, no matter how attractive or well rendered. Going beyond "good enough" is a negative, not a positive.

Giro
CREDIT LYONNAIS
CL
Discovery
CHANNEL
AMD
TREK
10//2
7-21
Discovery Channel
TREK
TREK
SHIMANO
HED

Appropriate degree of refinement: By its resolution or style, a sketch should not suggest a level of refinement beyond that of the project being depicted. As Lawson expresses it, " ... it seems helpful if the drawing suggests only a level of precision which corresponds to the level of certainty in the designer's mind at the time."

Suggest and explore rather than confirm: More on this later, but sketches don't "tell," they "suggest." Their value lies not in the artifact of the sketch itself, but in its ability to provide a catalyst to the desired and appropriate behaviours, conversations, and interactions.

Ambiguity: Sketches are intentionally ambiguous, and much of their value derives from their being able to be interpreted in different ways, and new relationships seen within them, even by the person who drew them.

In the preceding, the notions of visual vocabulary, resolution, and refinement are really significant, and interdependent. Sketches need to be seen as distinct from other types of renderings, such as presentation drawings. Their form should define their purpose. Any ambiguity should be in the interpretation of their content, not in terms of the question, "Is this an early concept or the final design?"

> ... a sketch is incomplete, somewhat vague, a low-fidelity representation. The degree of fidelity needs to match its purpose, a sketch should have "just enough" fidelity for the current stage in argument building.... Too little fidelity and the argument is unclear. Too much fidelity and the argument appears to be over—done; decided; completely worked out.... (Hugh Dubberly of Dubberly Design Office; private communication)

> Some of the most serious problems occur if various parties—managers and/or customers and/or marketing—begin to view the early prototypes [read sketches] they see as the final product. (Hix and Hartson 1993; p. 260)

Finally, in its own way, our list is more than not like a sketch itself. It is tentative, rough, and has room for improvement and refinement. And also like a sketch, these same values may very well contribute to, rather than reduce, its usefulness.

Figure 40: Designing a Performance

The outcome of any design process is a desired effect. Sketches have to be understood as steps in this process. While the beauty or clarity of each individual drawing might be appealing to the designer, ultimately the goal is to attain the performance declare at the beginning of the design process. This awareness is what differentiate a dexterous designer from a proficient renderer.

Credit: Trek Bicycles

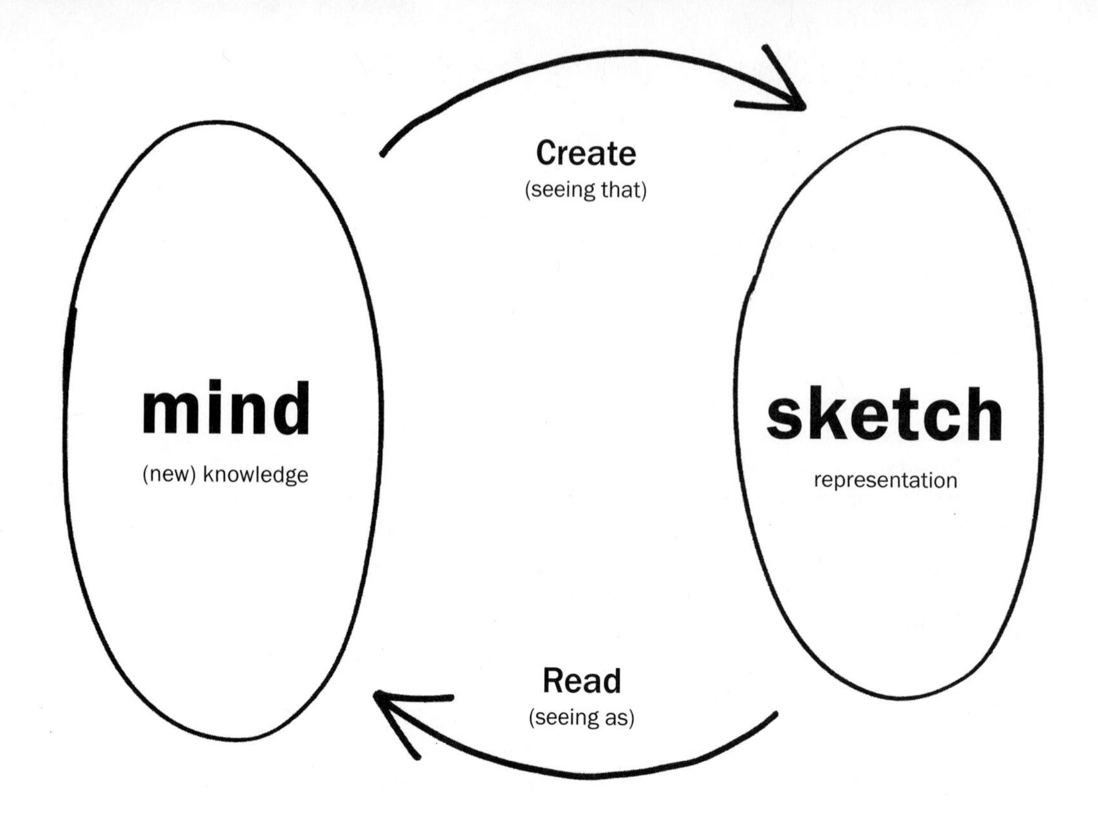

Figure 41: A Sketch of a Dialogue with a Sketch

The "conversation" between the sketch (right bubble) and the mind (left bubble). A sketch is created from current knowledge (top arrow). Reading, or interpreting the resulting representation (bottom arrow), creates new knowledge. The creation results from what Goldschmidt calls "seeing that" reasoning, and the extraction of new knowledge results from what she calls "seeing as."

Clarity is not Always the Path to Enlightment

My drawings have been described as pre-intentionalist, meaning that they were finished before the ideas for them had occurred to me. I shall not argue the point.
—James Thurber

In the preceding list, there is a reason that I left the attribute of ambiguity (Goel 1995) for the end. It is the least intuitive and perhaps the one needing the most explanation. The essence of the reason that this is so, is captured pretty well in the preceding quote by the cartoonist James Thurber, which I came across in Lawson (1997).

In an essay about ambiguity in design, Gaver, Beaver, and Benford, (2003) state:

> ... their use of ambiguity makes them evocative rather than didactic and mysterious rather than obvious.

For sure, designers want to get something concrete out of their sketches. As Gedenryd asks:

> Why would sketches be created from the very beginning of design, when just about everything is likely to change before you arrive at the final product? Hence, their purpose must be to serve the inquiry that creates them. And since they are created so early on, they have to serve this purpose right from the start. (Gedenryd 1998; p. 151)

The essence of this quote and that of Thurber's is that one can get more out of a sketch than was put into making it because of its ambiguity. The fact that the sketch is, well, sketchy—that is, leaves a lot out, or leaves a lot to the imagination—is fundamental to the process. My take on this is:

If you want to get the most out of a sketch, you need to leave big enough holes.

Ambiguity creates the holes. It is what enables a sketch to be interpreted in different ways, even by the person who created it. To answer Gedenryd's question, one of the key purposes of sketching in the ideation phase of design is to provide a catalyst to stimulate new and different interpretations. Hence, sketching is fundamental to the cognitive process of design, and it is manifest through a kind of conversation between the designer(s)

Figure 42: A Contrast in Skill: Two Drawings of a House

The top drawing was done by a 6-year old child and the one below by a professional designer. But I didn't have to tell you that. Drawing skill is obvious in the resulting artifact. However, skill in reading sketches is far less obvious. The artifact, such as it is, is in the mind and is not tangible. Yet, skill in reading is just as important as skill in rendering.

Figures: Keegan Reid & Michael Sagan

and their sketches (Schön & Wiggins 1992). This "conversation" is represented in Figure 41, and is captured—albeit in rather academic language—in one of the fundamental studies on sketching:

> ... sketching introduces a special kind of dialectic [conversation/dialogue] into design reasoning that is indeed rather unique. It hinges on interactive imagery, by a continuous production of displays [sketches] pregnant with the clues, for the purpose of visually reasoning not about something previously perceived, but about something to be composed, the yet nonexistent entity which is being designed. (Goldschmidt 1991)

This prompts us to view sketching as relating far more to an activity or process (the conversation), rather than a physical object or artifact (the sketch). Certainly the physical sketch is critical to the process, but it is the vehicle, not the destination, and ironically, it is the ambiguity in the drawing that is the key mechanism that helps us find our way.

> ... designers do not draw sketches to externally represent ideas that are already consolidated in their minds. Rather, they draw sketches to try out ideas, usually vague and uncertain ones. By examining the externalizations, designers can spot problems they may not have anticipated. More than that, they can see new features and relations among elements that they have drawn, ones not intended in the original sketch. These unintended discoveries promote new ideas and refine current ones. This process is iterative as design progresses. (Suwa & Tversky 2002).

These words that come from psychologists are reflected in the words of designers themselves:

> And so I did it; [learned to draw] ... I realized, however, that something else had happened along the way. Yes, I had learned to draw; but more importantly, I learned to think. My whole method of thinking experienced a complete switch. I began to see the world more clearly. As my hand sketched the lines, my mind revealed a whole new method of thinking that I had not known before. Being able to visualize things gave me a tool that I could use in all facets of life. What happened to my mind was much more important than the sketches I produced. (Hanks & Belliston 1990)

Likewise, speaking about artists rather than designers, we are told:

> ... the sketchpad is not just a convenience for the artist, nor simply a kind of external memory, or durable medium for the storage of particular ideas. Instead, the iterated process of externalizing and reperceiving is integral to the process of artistic cognition itself. (Clark 2001; p.149)

What is extremely important to understand is that it takes the same kind of learning to acquire the skills to converse fluently with a sketch as it takes to learn to speak in any other foreign language. Now you may have never thought of things in this way before, but having done so, this then becomes obvious. For example, the nondesigners among us are only too painfully aware of the contrast between our ability to draw and that of our friends who went to art school. But for those who need a not too subtle reminder, look at the two examples shown in Figure 42.

As the caption says, it is obvious which was done by a child and which was done by a professional. What is not obvious is that there is a similar difference on the other side of the equation. The figure highlights a difference in ability to draw a picture on paper. What it doesn't show is that there is a similar difference between laypeople and professionals in their ability to draw insights from that sketch on the paper. Both arrows in Figure 41 involve specialized skills. In short, the ability to participate in a conversation with a sketch involves competence in both reading and writing. Furthermore, the skill of extracting meaning from sketches continues to develop with experience. This was shown in a study conducted by Suwa and Tversky (1996). They showed that there are significant differences in competence even between student architects and those who have been in practice for a number of years. If the skill develops, that means that it is being used, and if it is being used, one can safely assume that it is important to the activity. Hence, it is worth paying attention to.

Here is the problem. Differences in our ability to create sketches are recognized and appreciated, largely because they are reinforced by a physical artifact—the sketch. However, in terms of our differing ability to read or extract new insights from sketches, there is no tangible reminder. Hence, this aspect of a designer's skill is seldom recognized, and almost never engenders the same respect and status as the ability to draw.

All of this brings us to a few key take-away messages:

Sketching in the broad sense, as an activity, is not just a byproduct of design. It is central to design thinking and learning.

Sketches are a byproduct of sketching. They are part of what both enables and results from the sketching process. But there is much more to the activity of sketching than making sketches.

Learning from sketches is based largely on the ambiguous nature of their representation. That is, they do not specify everything and lend themselves to, and encourage, various interpretations that were not consciously integrated into them by their creator.

Both the creation and reading of sketches are specialized skills that distinguish designers from nondesigners, and even experienced designers from student designers.

Designers have to be aware that what is "natural" to them, in terms of how they read sketches and what they see in them, is not obvious to others, and they must take that into account in how they educate others, and what representation they use to communicate ideas.

Those without design training, but who work with designers, need to be sensitive to this difference of skills as well, and beware of making uninformed judgments based on superficial readings of their materials.

In order to improve interactions, these same "foreigners" should do their best to gain some literacy in design representations, and designers should go out of their way to help them in this.

Conversely, designers should make a conscious effort to be able to read the "foreign" representations of these same colleagues (such as spreadsheets and business plans). Likewise, they should be helped by their "native speaking" colleagues in doing so.

At the executive level, it is important to establish a corporate culture that understands and respects the design plan and objectives, as much as the business plan and its objectives. They are two sides of the same coin, and the success of one depends on the success of the other.

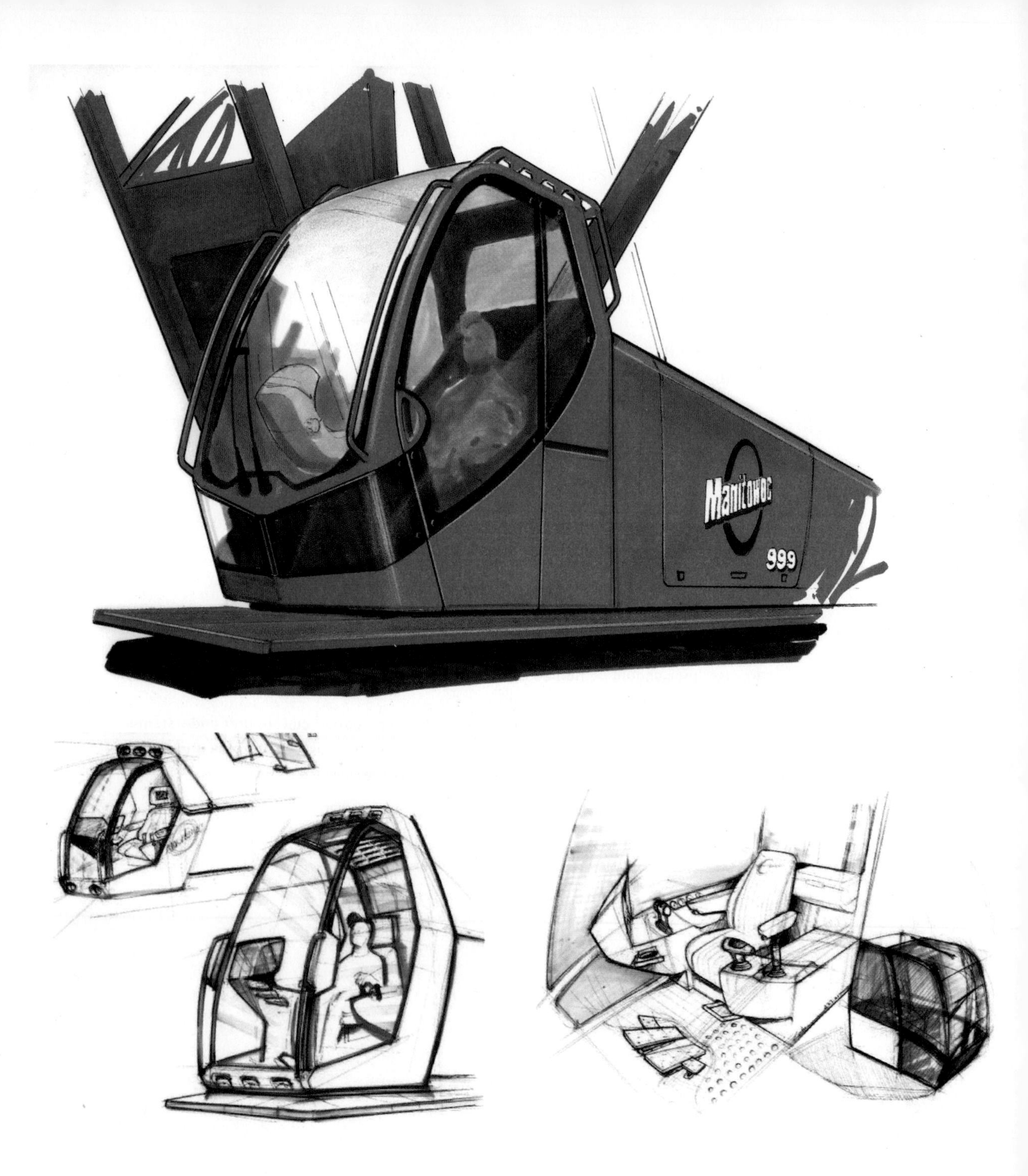

Figure 43: Variations in Rendering Style

Here are a variety of types of rendering of a particular product. In each, the style is chosen to suit the content and the intent. The point is that sketching is not the only rendering style relevant to design. Each style is best for something and worst for something else.

Credit: Brooks Stevens Design

The Larger Family of Renderings

The best way to a good idea is to have lots of ideas.
—Linus Pauling

With our focus on sketches and sketching, it is easy to lose sight of the fact that they represent only part of a much richer ecology of drawing and rendering types.

Trying to understand their place in this larger space is not always easy. This is complicated by the fact that, as with many things, there is more than one way that sketches can be categorized. For example, they can be grouped according to profession (architect, industrial design, fashion, graphic design, fine art, etc.), medium (pencil, markers, pastel, pen and ink, airbrush, etc.), or subject (life drawing, automobiles, buildings, nature, etc.).

I mention this because I want to emphasize how rich and diverse the space is, even with traditional media. This is valuable to keep in mind since what we are going to explore will expand this richness significantly.

Coming back to types of rendering, the way that we are going to categorize them is by intent. What is the purpose of the drawing? Why was it made?

Even here there are different analyses. In fact, there are even different analyses by the same person. Compare Lawson (1997) with Lawson (2004), for example. For our purposes, I am going to err on the side of simplicity, and stick to the types of general categories we might find in books on technique, such as Powell (2002). There is little penalty in so doing, since the result is not inconsistent with what others have written and it keeps things simple.

There are five types of rendering that I think are worth identifying:

Sketch
Memory Drawing
Presentation Drawing
Technical Drawing
Description Drawing

By breaking things down this way, I hope to achieve a few things. First, I hope to help further clarify what I am talking about when I discuss sketches, as much by pointing out what they are not as what they are. Second, I want to help expand our literacy. Explicitly pointing out these categories will hopefully help prompt you to always question if you are using the most appropriate rendering style for the purpose at hand. Finally, in terms of your thinking I want you to substitute the word rendering every place we have used the word drawing. That way all of this will generalize to the richer set of media that we are going to use in the future—and not just with sketches, but with the other four functions, as well.

Finally, a note for those interested in diving deeper into how different designers adopt different approaches to drawing, depending on project and personality. I highly recommend taking a look at *Sketch • Plan • Build* (Bahamón, 2005). It includes a collection of thirty projects from a broad range of architects, with photographs of the final project along with the renderings that led up to the final design.

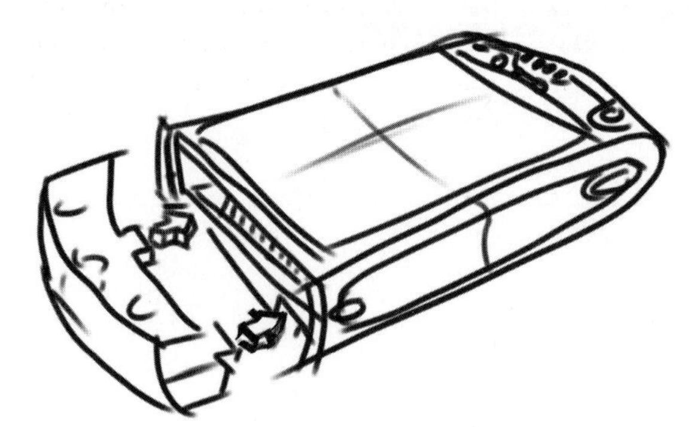

Sketch: This is mainly what we have been talking about. They are sometimes referred to as Thinking Drawings, and are described as Design Drawings by Lawson (1997). They are generally made by designers mainly for designers, and are central to the process of ideation.

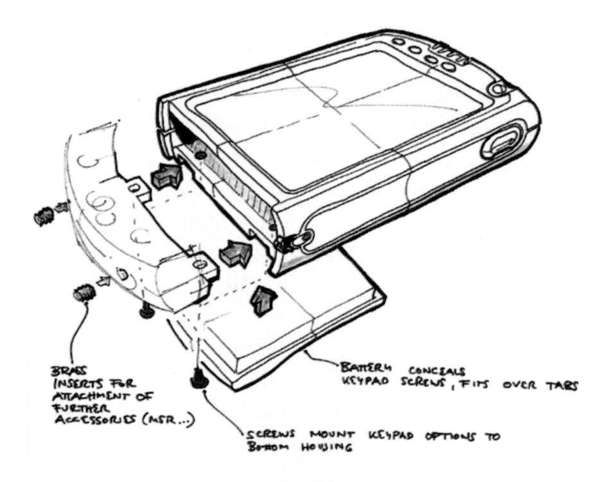

Memory Drawing: These are one of the oldest styles of drawing. These are renderings made to record or capture ideas. Think of them as extensions of one's memory—like a hand-rendered photograph recording a thought or something seen.

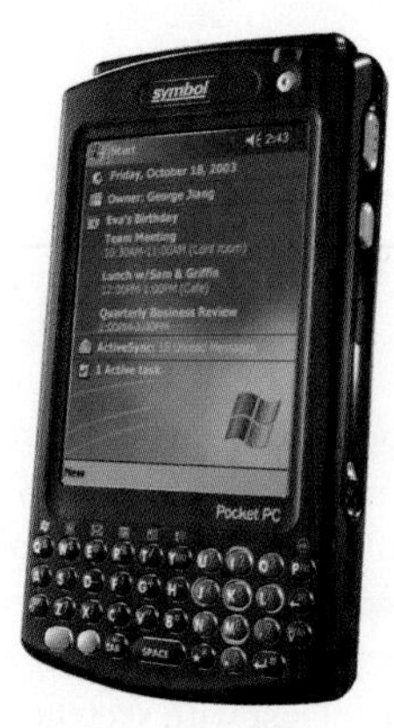

Presentation Drawing: These are drawings made for the customer, client, or patron. As stated by Powell (2002, p.6), these people "usually lacked the skill needed to read these drawings [sketches] and therefore understand what the product would be like before it was actually made." Hence, just as the value of a sketch is in its ambiguity, and the "holes" that it contains, the value of a presentation drawing is in its ability to communicate and represent what is being presented to the untrained eye.

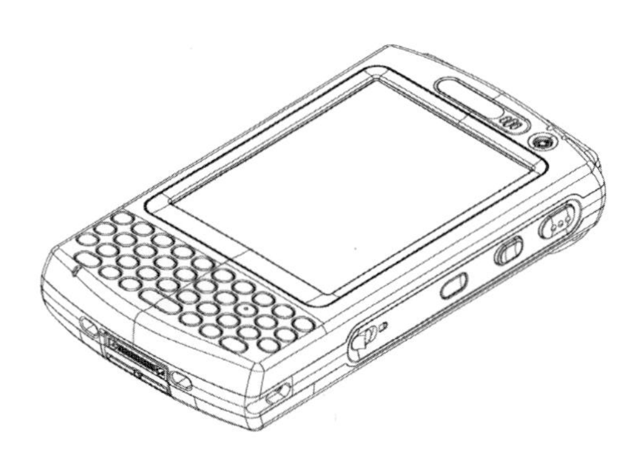

Technical Drawing: Technical drawings are a class of drawing that are primarily intended for those who actually are going to build what is drawn. They are typically accurate and are at the drafting and blueprint end of the scale, rather than that of sketching.

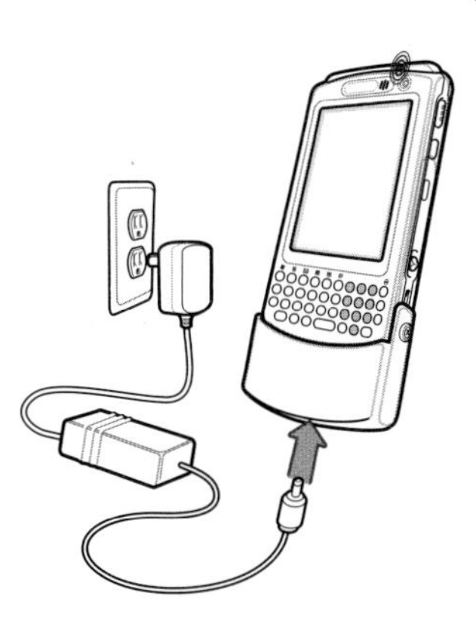

Description Drawing: This class of drawing is intended to explain something, such as a how something works, or is constructed. It would include things like cut-away or exploded-view drawings, or it could be broken up into frames, like a cartoon, as with the emergency cards that one finds in the seat-back pocket on airplanes.

Figure 44: Examples of Rendering Styles

Credit: Symbol Technologies

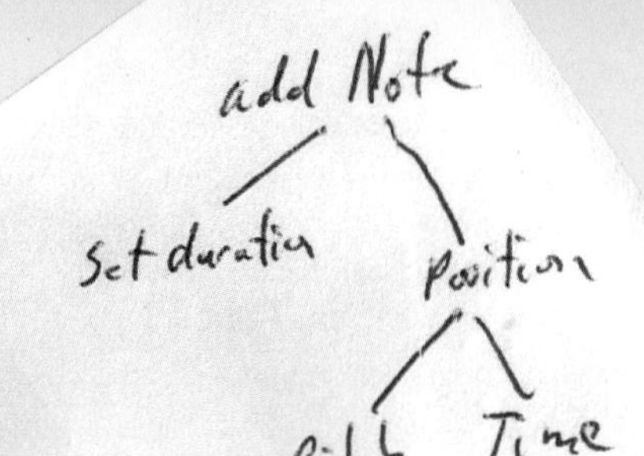

Music by Menu or Gesture

Is it more complex for the user to enter music using a novel single stroke shorthand or using a more conventional approach where the note duration is selected from a menu, and then placed on the staff at the appropriate pitch and point in time? See Buxton 1986 for an analysis

One Hand or Two

Let's say that we want diagonally scroll from location A to location B, which is off screen. What is the difference if we use the mouse to both select the points as well as operate the scroll bars, compared to having one hand select with the mouse and the other scroll using a touch pad? See Buxton 1990 for a more detailed discussion.

There is far more to design than just what we are talking about. In short, theory and science can underly what we sketch, thereby letting us start at a much higher level. This is encapsulated in this other form of sketching—the proverbial sketch on the back of an envelope. See Card, Moran and Newell 1983 for the classic book on the science behind interaction that can feed such sketches

Figure 45: The CitrusMate Electric Juicer

The noise that it made jangled my nerves in the morning.

Experience Design vs. Interface Design

...for they had gone out on a limb—but isn't that where the fruit is?
—Jeff Lowe

For years we have spoken about the human–computer interface and user interface design. This has always been about where the human and the computer meet. There are dozens, if not hundreds, of books on the subject.

Nevertheless, or perhaps because of this, the term interface has been falling into disfavour.

For a few years, it seemed that Interaction Design was going to be the new buzzword. Various schools started to have programs in the area. People had Interaction Designer as the job description printed on their business cards. But the most significant indicator of the currency of the term was the number of people who claim to have coined the term.

We can leave them to fight that one out. The issue is moot, since yet another term has entered the fray: Experience Design.

I like it. The main reason is that it is the most human-centric of the lot. Despite the technocratic and materialistic bias of our culture, it is ultimately experiences that we are designing, not things. Yes, physical objects are often the most tangible and visible outcomes of design, but their primary function is to engage us in an experience—an experience that is largely shaped by the affordances (Gibson 1979) and character embedded into the product itself (Norman 1988; Gaver 1991). Obviously, aesthetics and functionality play an important role in all of this.

Stepping out of the world of technology for a moment, this shift from a techno-centric or materialistic perspective to a human-centric one is mirrored in the following lines from the Italian climber Reinhold Messner, in speaking about the transformations in modern mountaineering:

> ... in the first 200 years of alpinism, it was the mountain that was the important thing. ... But for some years now, especially on my own tours, it is no longer the mountain that is important, but the man, the man with his weaknesses and strengths, the man and how he copes with the critical situations met on high mountains ... (Messner 1979; p.60)

Perhaps all this sounds a little too ethereal. Let me give a concrete example to illustrate the difference between interface and experience design. At the same time, let's just agree that from now on, I will use interaction design and experience design interchangeably. Why? Simply because the former is more in use, and after going through this example, I will have loaded it with the meaning that I intend, which is all that I really care about.

There are two things that you need to know for me to set up this example. First, I split my time between a house in Toronto and a cabin north of the city. Second, what gets me out of bed in the morning is fresh squeezed orange juice.

Figure 46: My Mighty OJ Manual Juicer

My first manual juicer, loaded and ready to go.

Figure 47: The OrangeX Manual Juicer

My second manual juicer. Its feel was a revelation.

For years, the thing that enabled this pleasure was the *CitrusMate*, shown in Figure 45. The only problem was, it was in the city, so it was of little use to me when we were staying at our place in the country. To put a stop to my whining in the morning, my wife took the initiative and bought me a juicer for the cabin. (This was not a totally selfless act, since I make juice for her, too.).

She bought the *Mighty OJ* juicer, shown in Figure 46. And I loved it. That is the good news. The bad news is, my experience with it made me hate the juicer in the city—that same juicer that only weeks before had been one of my most precious possessions. Why?

It is pretty clear from looking at the photos that these two products are very different. Nobody would confuse one with the other. They use different technology (electric vs manual). The user interface for each is also clearly different. But that was not the problem. Despite these differences, the usability of each was comparable, and the juice from each tasted the same.

For me, the difference that shaped my change in opinion was in the contrast in the overall experience of using the two products. Simply stated, having experienced the silent simplicity of the Mighty OJ, I grew to hate the noise that the electric CitrusMate made, especially first thing in the morning.

Thus did I emotionally separate from what had been a cherished possession.

So, after listening to my continual complaints about the noise of the electric CitrusMate, my wife got the hint. For my 55th birthday she bought me another manual juicer to replace it. The new one, shown in Figure 47, was from yet another company. It was the *OrangeX Manual Citrus Juicer*, designed by Smart Design in New York (Industrial Designers Association of America, 2001).

At this point, it is important that you compare the photos of the two manual juicers. Notice how similar they look. They both work by the same basic principle. They have the same user interface. You pull back the lever and place half of an orange face down in the "jaws." You then pull the lever in order to squeeze the juice into the container.

If you can use one, you can use the other. The juice tastes the same from each, and takes the same amount of time to prepare. And, from a distance, you might even mistake one for the other.

However, despite these similarities, from the perspective of experience, there is no comparison between the two.

The two manual juicers, the OrangeX and the Mighty OJ, have the same user interface. Yet, despite their similarities, the quality of experience using the OrangeX is as much better than that of the Mighty OJ, as the Mighty OJ's is to that using the electronic CitrusMate.

It is not that the Mighty OJ is bad. In absolute terms, it hasn't gotten any worse. It is just that, relative to it, the experience in using the OrangeX is so much better. And, with that improvement comes a whole new standard of expectation or desire, which, perhaps predictably, also led to a whole new bout of whining on my part with the equally predictable result that my wife eventually (sooner than later) replaced the MightyOJ with yet another OrangeX.

So what is it in the OrangeX that brought about such a difference?

Although the OrangeX is significantly heavier, the meaningful difference is not due to weight. Rather, my pleasure is due to the feel of the action when pulling the lever down. There is a cadence in the action that is almost musical. This is something that no drawing or photograph can capture, since it has to do with feel, and it takes place over time. The point is, I just can't use it without a smile.

At this point I want to reiterate that both of these manual juicers have the same user interface. From this I want to emphasize that *usability has nothing to do with their differences.* It is the quality of experience that marks their difference. This brings us to the next point that I want to emphasize: *this difference did not come about by accident.* It was the result of conscious design.

Figure 48: Gear Mechanism of the Mighty OJ
Rotary motion of the arm raises and lowers the jaws of the juicer by means of a rack and pinion gear. The gear ratio is constant.

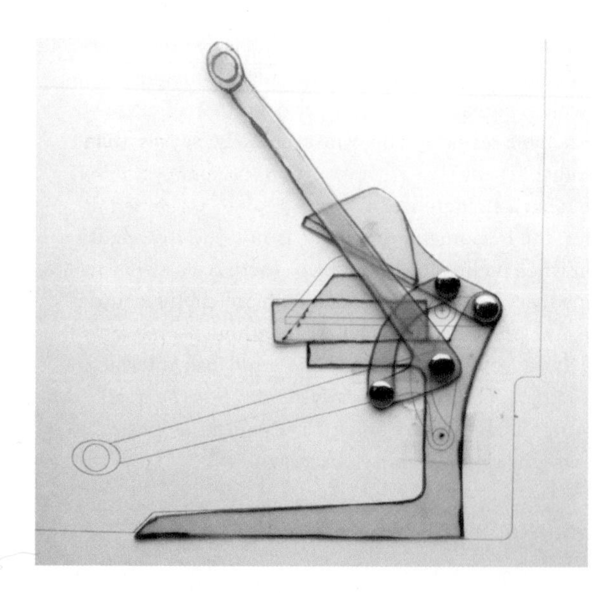

Figure 49: Two-Dimensional Study of OrangeX Mechanical Linkage
Cutting the parts out of Perspex and pinning them onto a board enabled quick testing of the linkage, as well as marking time lapse ghost images on the background. A chronological sequence of renderings (including this one) is shown in Figure 44.
Photo: Smart Design

If we look more closely (see Figure 48), we see that the Mighty OJ has a direct linkage between the lever and the jaws. This is by way of the simple rack-and-pinion gear mechanism seen in the figure. This gives the unit what is best described as a constant gear ratio, where maximum force must be applied at the end, or bottom, of the stroke.

In contrast, the quality of the OrangeX action is due to the subtle difference of its leverage mechanism. By the nature of the linkage between the arm and the jaws, there is a kind of camming effect. This is what delivers the cadence that I so love. The effect of the linkage design is to vary the gear ratio, so to speak, so that at the end of the squeeze—where with the Mighty OJ you have to push the hardest—the pressure required is reduced, and you come to a gentle conclusion of the squeeze. You can actually take your hand off the lever at the bottom of the stroke and watch the final drops of juice drip into the container.

The workings of this mechanism can be seen by looking closely at Figure 47, or even better, at the mechanical study shown in Figure 49.

There are a couple of things that really interest me about what is shown in Figure 49. First, it has a lot in common with the engineering prototype of the Trek Y-Bike seen previously in Figure 28. Both illustrate a mechanical innovation that lies at the heart of the experience of final product.

The second thing to notice in the study shown in Figure 49 is how economical it is. It is just some Perspex cut out and pinned together at the points of articulation. Furthermore, as is seen in the figure, these were mounted on boards, which enabled the designers to trace key positions of the mechanism onto the background, thereby achieving something like the effect of a time-lapse photograph. It is, in fact, a 2D dynamic exploratory sketch of the mechanism.

But this is only one of a number of studies that led to the final design. A selected sample illustrating the process can be seen in Figure 50.

By means of working through such a series of renderings and studies, the team was able to achieve the dramatic experience that I have just described.

At this point step back and remember that here we are just speaking of orange juicers. Yet even with this type of mechanical product, we see the attention to detail, and the special techniques that are involved in achieving that level of experience.

The lesson to take away from this can be gained by contrasting these orange juicers with the behavioural complexities of the types of electronic appliances that many of us are involved in bringing to market today.

Consider this:

If it takes this much effort and detail to achieve this standard of quality with such relatively simple things as juicers, why would we expect to get a similar quality experience from our new-world information appliances without, likewise, adopting very explicit and deliberate processes directed at doing so?

So, if we do decide that we want to strive for a comparable standard of experience in the products that we are designing, and therefore adopt an appropriate process for doing so, what might that process be? Making a contribution toward answering this question is at the heart of what follows.

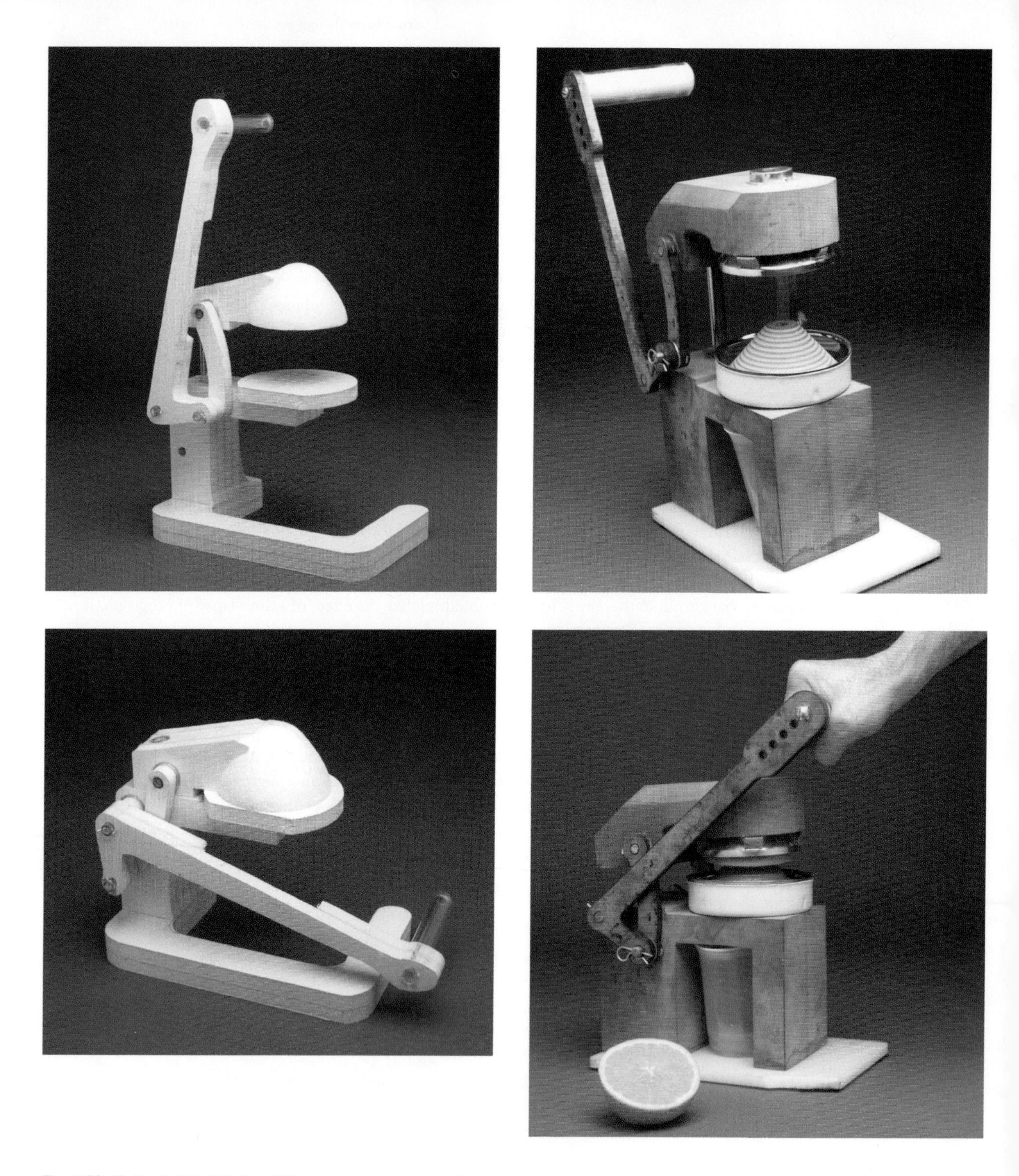

Figure 50: 3D Renderings Exploring Different Aspects of the Design

The two renderings in the left column explore the overall mechanism. The next two illustrate testing concepts for the squeezing mechanisms in the "jaws". The two with the orange background are sketches exploring variations of the overall form. Finally the two in the right column are renderings of computer models of what was to become the final form. These two are not sketches.

Photos: Smart Design

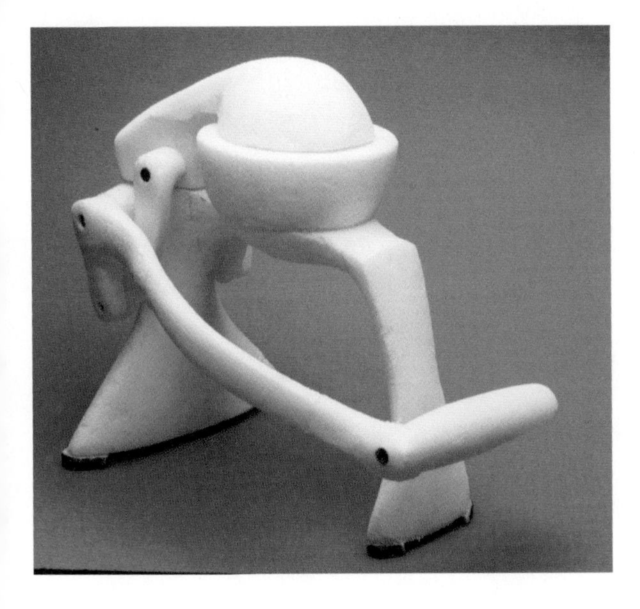

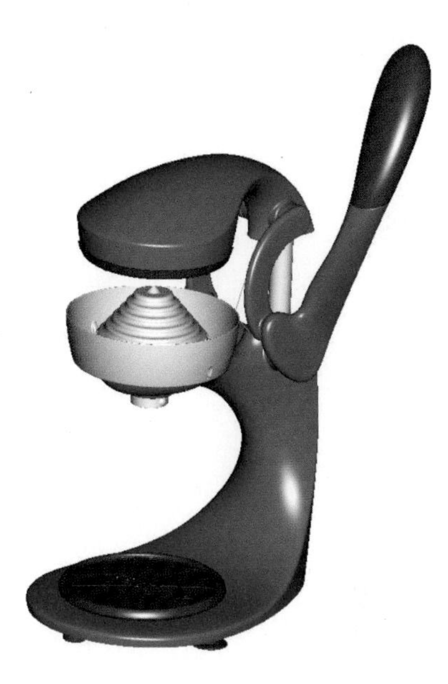

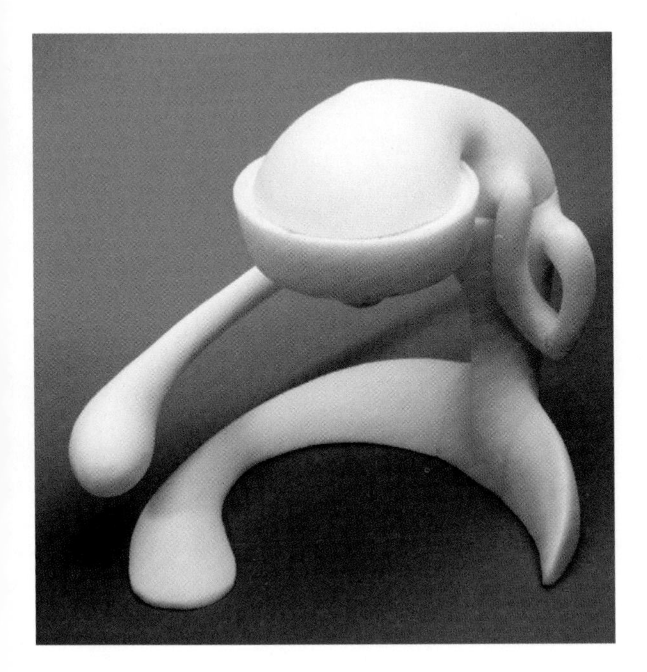

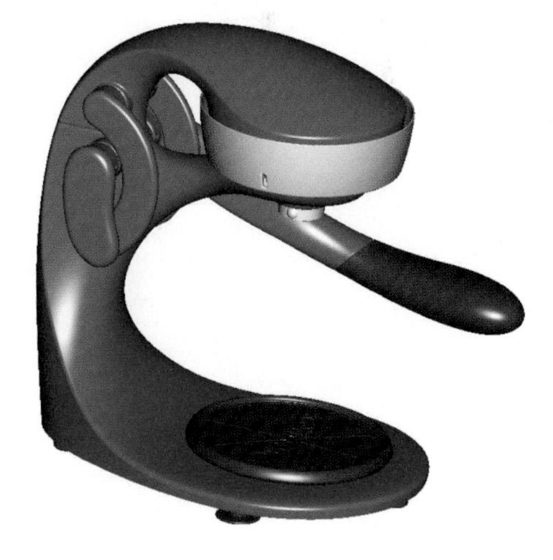

Sketching Interaction

When you come to a fork in the road, take it!
— Yogi Berra

So how might we find a process that enables us to design computer-based products with the same attention to user experience that we saw in the design of the OrangeX? I think that the answer lies in the OrangeX example, itself. Hence its importance.

It illustrated that the activity of sketching could be extended to other forms than just pencil on paper. The key here is to understand that sketching as I mean it has more to do with exercising the imagination and understanding (mental and experiential) than about the materials used. Hence, one might use pencil on paper, but one might also use a jar lid, a stick, and a piece of plasticine. It may even involve a computer. With the OrangeX example, the underlying process and objectives were the same, but the sketches themselves took on a more physical form than we have seen thus far. As I shall say more than once, the importance of sketching is in the activity, not the resulting artifact (the sketch). If sketches can take on physical form, be they 3D or sculptural, perhaps they can take on even more extended forms that will help us in our quest.

But how do we go deeper than this? If there are new forms of sketching, how can we pursue them?

One thing that we know is that sketches for experience and interaction design will likely differ from conventional sketching since they have to deal with time, phrasing, and feel—all attributes of the overall user experience. How rich is that?

> Experience is a very dynamic, complex and subjective phenomenon. It depends upon the perception of multiple sensory qualities of a design, interpreted through filters relating to contextual factors. For example, what is the experience of a run down a mountain on a snowboard? It depends upon the weight and material qualities of the board, the bindings and your boots, the snow conditions, the weather, the terrain, the temperature of air in your hair, your skill level, your current state of mind, the mood and expression of your companions. The experience of even simple artifacts does not exist in a vacuum but, rather, in dynamic relationship with other people, places and objects. Additionally, the quality of people's experience changes over time as it is influenced by variations in these contextual factors. (Buchenau & Suri 2000; p 424)

In light of this, let us ask again:

What is the nature of sketching in interaction design?
How do you sketch interaction?
What is to an interactive system what the early sketch in Figure 35 is to Lance Armstrong's time trial bike?
What are the fundamental skills required for sketching interactive systems?
What is the underlying process that one should follow to do this effectively and consistently?
What should be included in Sketching 101 in an Interaction Design curriculum?

The tack that we are going to pursue is that sketching in interaction design can be thought of as analogous to traditional sketching. Since they need to be able to capture the essence of design concepts around transitions, dynamics, feel, phrasing, and all the other unique attributes of interactive systems, sketches of interaction must necessarily be distinct from the types of sketches that we have looked at thus far. Nevertheless, to be considered sketches, they must be consistent with the attributes that we discussed earlier, namely:

Quick
Timely
Inexpensive
Disposable
Plentiful
Clear vocabulary
Distinct gesture
Minimal detail
Appropriate degree of refinement
Suggest and explore rather than confirm
Ambiguity

From our analysis of sketching in traditional design, we are able to find a compass that can help guide us in our exploration of sketching in this new domain. Although the surface of the renderings will be different, the underlying properties should be the same. Therefore, not only do we have a compass, we have a litmus test that helps us categorize examples that we encounter.

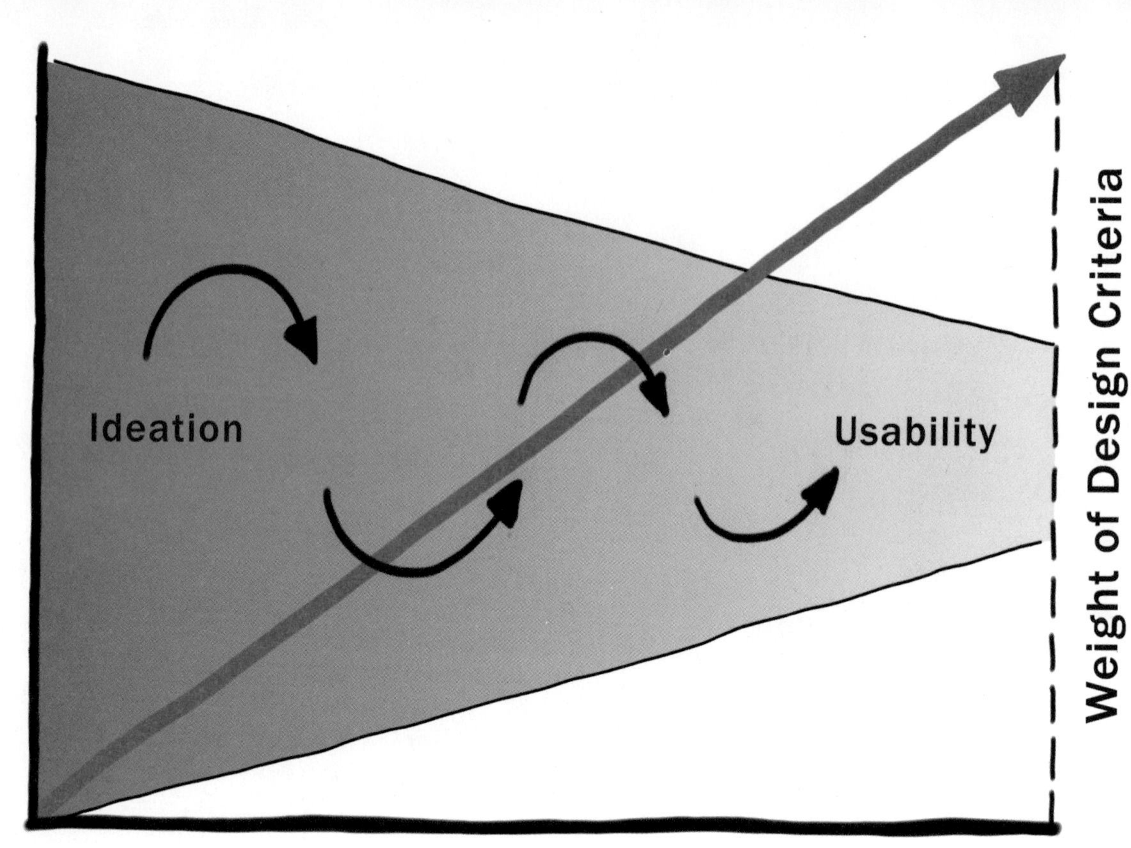

Figure 51: The Dynamics of the Design Funnel

The design funnel begins with ideation, and ends with usability testing. The former is largely dominated by sketching, which enables ideas to be explored quickly and cheaply. More refined (and expensive) prototypes provide the basis for the testing at the later stages of design. Where testing is a key concern, the most dominant artifacts are more refined (and expensive) prototypes. The transition from one to the other is represented by the transition from orange to yellow in the figure. As we progress, our overall investment in the process grows. This is indicated by the rising arrow and the y-axis label on the left. The y-axis label on the right side of the figure emphasizes that as our investment increases, so should the weight of the criteria that we use to evaluate our design decisions. In other words, you don't manage ideation the same way, or with the same rigor, as usability. Finally, the circular arrows are a reminder that we include users throughout the iterative process, not just during usability testing.

Sketches are not Prototypes

Practice is the best of all instructors
— Publius Syrus

Now that what I mean by sketching interaction is becoming a bit more clear, it is inevitable that someone is going to ask something like, "Isn't what you are calling a sketch just another word for prototype or low-fidelity prototype?" The answer is emphatically, "No!" The distinction between a sketch and a prototype is—for me at least—one of the most interesting things to emerge as I went down this path.

Sketches and prototypes are both instantiations of the design concept. However they serve different purposes, and therefore are concentrated at different stages of the design process. Sketches dominate the early ideation stages, whereas prototypes are more concentrated at the later stages where things are converging within the design funnel. Much of this has to do with the related attributes of cost, timeliness, quantity, and disposability. Essentially, the investment in a prototype is larger than that in a sketch, hence there are fewer of them, they are less disposable, and they take longer to build. At the front end of the funnel, when there are lots of different concepts to explore and things are still quite uncertain, sketching dominates the process.

These notions are captured graphically in Figure 51. The circular arrows reinforce that the whole design phase is an iterative, user-centred process. The coloured change reflects a transition from a concentration on sketching at the front to one concentrating on prototyping at the back. Related to this, and signified in the colour coding, is the accompanying transition from ideation to usability testing.

From the management perspective, perhaps the most important component of Figure 51 is the ascending red arrow. What this says is that the weight of the criteria by which ideas or concepts are injected or rejected varies with the investment made in them. Stated simply, at the beginning, ideas are cheap, so "easy come, easy go" and "the more the merrier." As we proceed, we have more and more invested in the concepts in play, hence we need to adopt increasingly formal or explicit criteria for evaluating what goes, what stays, and where we invest our resources.

Because the investment in the product is low, the front end is the one time in the product pipeline when one can actually afford to play, explore, learn, and really try and gain a deep understanding of the undertaking. In fact, too much concern for quality too early may well have a negative effect. I found a wonderful example illustrating what I mean by this referred to in a blog from someone called Bill Brandon:

SKETCH		PROTOTYPE
EVOCATIVE	→	DIDACTIC
SUGGEST	→	DESCRIBE
EXPLORE	→	REFINE
QUESTION	→	ANSWER
PROPOSE	→	TEST
PROVOKE	→	RESOLVE
TENTATIVE	→	SPECIFIC
NONCOMMITTAL	→	DEPICTION

Figure 52: The Sketch to Prototype Continuum

The difference between the two is as much a contrast of purpose, or intent, as it is a contrast in form. The arrows emphasize that this is a continuum, not an either/or proposition.

> The ceramics teacher announced on opening day that he was dividing the class into two groups. All those on the left side of the studio, he said, would be graded solely on the quantity of work they produced, all those on the right solely on its quality. His procedure was simple: on the final day of class he would bring in his bathroom scales and weigh the work of the "quantity" group: fifty pounds of pots rated an "A", forty pounds a "B", and so on. Those being graded on "quality," however, needed to produce only one pot—albeit a perfect one—to get an "A." Well, came grading time and a curious fact emerged: the works of highest quality were all produced by the group being graded for quantity. It seems that while the "quantity" group was busily churning out piles of work—and learning from their mistakes—the "quality" group had sat theorizing about perfection, and in the end had little more to show for their efforts than grandiose theories and a pile of dead clay. (Bayles & Orland 2001; p. 29)

Baxter (1995) argues that because the investment is so low and the opportunity to explore options is so high at the start, that this is also the stage in the product development lifecycle when you have the potential to realize the highest return on investment. Of course, this is a double-edged sword. It is also the point in the process where the consequences of an undetected bad decision, or an opportunity missed, can cost you the most (in real dollars or missed revenue). So, as the saying goes:

> Fail early and fail often.

And learn.

But adequate investment at this stage happens too infrequently, especially with software companies. The paradox is that those same firms that can't afford a relatively small planned investment in design at the front end, seem quite able to afford the far higher unexpected and unbudgeted (but predictable) high back-end costs that result from a bad product being late and underdelivering on its potential.

Jumping in and immediately starting to build the product, even if it does get completed and ship, is almost guaranteed to produce a mediocre product in which there is little innovation or market differentiation. When you have only one kick at the can, the behaviour of the entire team and process is as predictable as it will be pedestrian:

> You cling ever more tightly to what you already know you can do—away from risk and exploration, and possibly further from the work of your heart. (Bayles & Orland 2001; p.30)

Robert Cooper (1993; 2001) compares managing product development costs in terms of the type of risk analysis that one would use at the poker table, or in managing an investment portfolio. Mike Baxter summarizes this in terms of the following Gambling Rule:

> When uncertainties are high, keep the stakes low. As the uncertainties reduce, increase the stakes. (Baxter 1995; p.10)

In summary, what all this says is that we must manage the front-end of the process differently than the back-end, regardless of whether we are looking at things in the large (the overall product pipeline—design, engineering, sales, etc.) or in the small (within the design funnel itself, where we must manage the sketching and ideation phase differently than we manage the back-end prototyping stage).

Where is the User in All of This?

It is only the future if it can't be made.
— Ross Lovegrove

At this point we risk being so bogged down in discussions about processes, roles, responsibilities, diagrams, and so forth, that we lose sight of why we are doing this in the first place: to design products that people want, need, like, and can use. Yet, it was only around the discussion of the Design Funnel shown in Figure 51 that I really said anything about user-centred or usage-centred design, usability testing, or any of the other associated buzz words. At least for anyone who knows me, this may seem especially curious. Therefore, let me inject a few comments.

First, this is largely explained by the fact that I simply take it for granted that the user is both considered and involved throughout the process. Arguing for the need for user involvement in a modern book on product design is as pointless as a discussion about the need to know the rules of arithmetic in an advanced mathematics textbook.

Second, there are various traditions, such as user-centred design and participatory design, that have a well-established literature on their approach to user involvement in the design process. Rather than put forward an alternative to these traditions, I would like to view what I write in this volume as complementary to, and compatible with, these traditions. Instead of trying to write a comprehensive handbook on design, my hope is to contribute something that augments the approaches already in place.

Third, it is precisely a concern for users that underlies the value of the approach to design that I discuss in this volume. Techniques fundamental to the design phase, such as sketching and prototyping, mean that iterative user involvement, participation, testing, and validation can occur much earlier than is often the case. Furthermore, this can happen in a form that captures the interactive nature of the system. The consequence is that user input can begin early enough to influence the design of the product. (Participatory design is one obvious approach to this, but not the only one. But it certainly lends support to what I am saying.) User involvement then obviously continues through the engineering phase, in the form of usability testing.

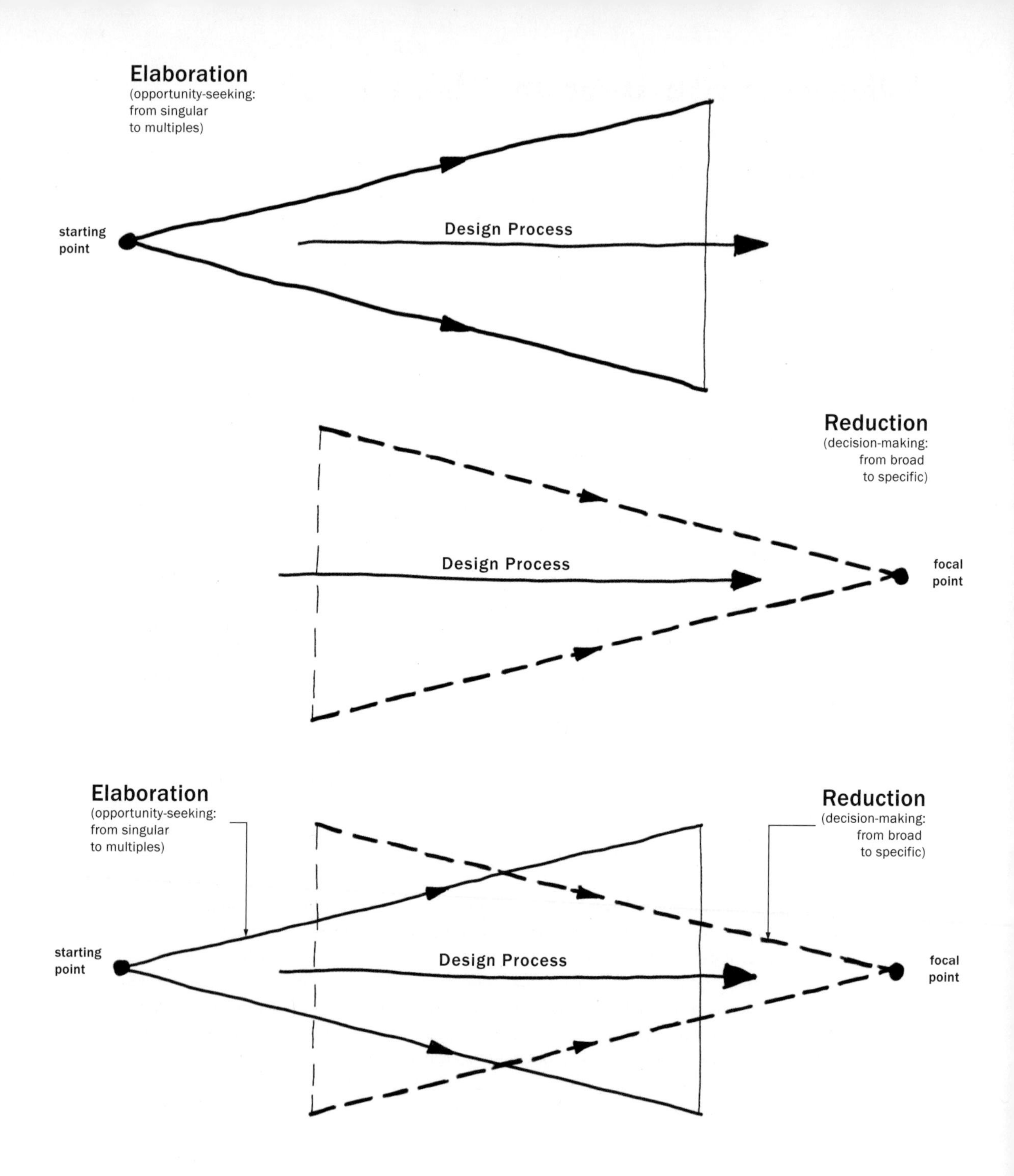

Figure 53: Overlapping Funnels

The reduction that results from decision making is balanced by the constant generation of new ideas and creativity that open up new opportunities to improve the design.

Source: Laseau 1980; p. 91

You Make That Sound Like a Negative Thing

There is no limit to what you can accomplish if you don't care who gets the credit.
— Ronald Reagan, Sign on his desk

The attentive reader will notice something about how I have represented the design process in Figure 51. It is a funnel that is narrower at the end than at the beginning. What that implies is that no matter how many great ideas get tossed into the hopper, in one sense, there is less at the end than at the beginning.

Of course, from another perspective, that is not true. At the beginning all we have are perhaps some ideas, some hope, and some ambition. At the end, ideally, we have something concrete—a design for a well-formed product.

Laseau (1980) has a nice way to capture this. He has an alternative representation of the process that is made up of two superimposed and opposing funnels, as shown in Figure 53. This representation prompts me to do something that earlier I said I would not do, namely give you one of my definitions of design. So here goes:

> **Design is choice, and there are two places where there is room for creativity:**
> 1. **the creativity that you bring to enumerating meaningfully distinct options from which to choose**
> 2. **the creativity that you bring to defining the criteria, or heuristics, according to which you make your choices.**

This formulation of the process makes explicit the fundamental importance of both the generative and reductive aspects of the design process. It also has a fairly close correspondence to a quote from the French poet Paul Valéry:

> Invention depends on two processes. The first generates a collection of alternatives, the other chooses, recognizing what is desirable and appears important among that produced by the first. What one calls "genius" is much less the contribution of the first, the one that collects the alternatives, than the facility of the second in recognizing the value in what has been presented, and seizing upon it. (Translation: Bill Buxton)

Each of the two processes or agents corresponds to one of Laseau's funnels. The expanding funnel represents the generation of the possible opportunities, or options

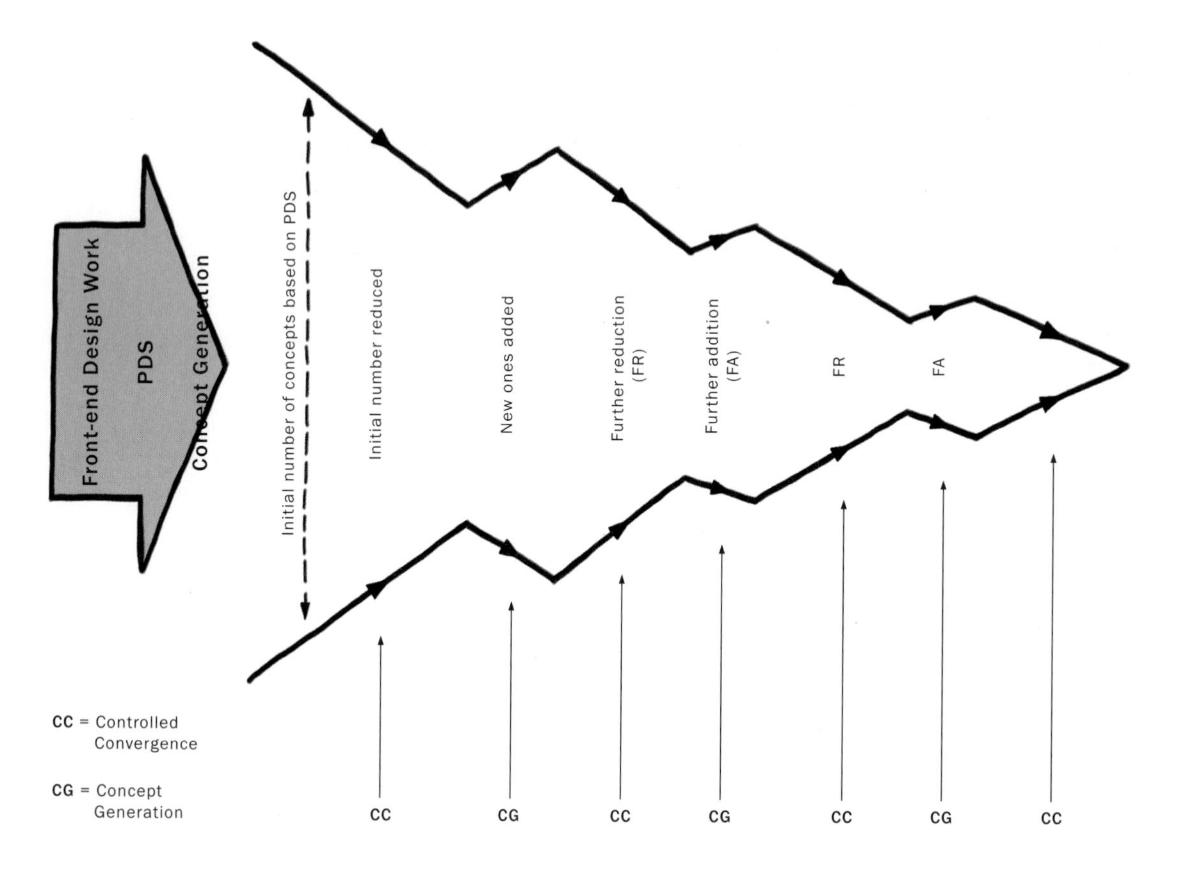

Figure 54: Flexible Approach to Concept Generation and Selection

This is yet another variation on representing the design funnel. After the front-end design work and the Product Design Specification (PDS), we see the process alternating between concept generation (CG) and concept convergence (CC), with the overall process gradually converging to the final concept.
Source: Pugh 1990; p. 75

from which one can select. The converging one represents the making of choices from among those options, and the gradual convergence onto the final design.

I like this figure for what it adds. And yet, it has its own weaknesses. Although it graphically highlights that the generation of new ideas is ongoing throughout the process, it doesn't reflect how the range of things that we can consider narrows as we converge on the final design.

The other thing that it doesn't show, but that is important to keep in mind, is that for the process to work, we must generate and discard much more than we keep. This is what Pugh (1990) has called controlled convergence. This leads us to yet another representation, that seen in Figure 54. The key take-away from this figure is how it illustrates the design funnel as alternating between adding and eliminating concepts—Concept generation (CG) which expands the scope of the funnel, and concept convergence (CC), which narrows the funnel.

Let me quote what Pugh says about this, since I don't think that I can improve upon his words:

> ... it allows alternate convergent (analytic) and divergent (synthetic) thinking to occur, since as the reasoning proceeds and a reduction in the number of concepts comes about for rational reasons, new concepts are generated. It is alternatively a generative (creative) and a selection process. An essential feature of this approach is the comparison of each alternative concept, in turn, with a peer concept in such a manner to render a fixed viewpoint on any one concept impossible. (Pugh 1990; p 74)

This is all very much in keeping with the notion of design as choice that I articulated earlier. It also reinforces the concept of design rationale (MacLean, Young & Moran 1989); that is, constantly being able to articulate the *reason* for your decisions. And, the notion of the peer concept reinforces the notion that design is not a straight path. We always need things to which we can compare any option. That is, the question is not, "Do I want this?", but rather, "Do I want this rather than that, and why?"

Ultimately, the common thread in any of these funnel-shaped representations, regardless of their variations, is that they converge. That is, not all ideas survive. More get tossed out (hopefully for good reason) than kept.

And guess what? Some of the ideas that get thrown out will be yours. Furthermore, these may include some that you believe to be among your best. To succeed, you not only have to learn to live with this, you have to learn to live for it. Here is how I see it:

> **People on a design team must be as happy to be wrong as right. If their ideas hold up under strong (but fair) criticism, then great, they can proceed with confidence. If their ideas are rejected with good rationale, then they have learned something. A healthy team is made up of people who have the attitude that it is better to learn something new than to be right.**

It has to be this way. Hence, one of the mantras of a healthy group is:

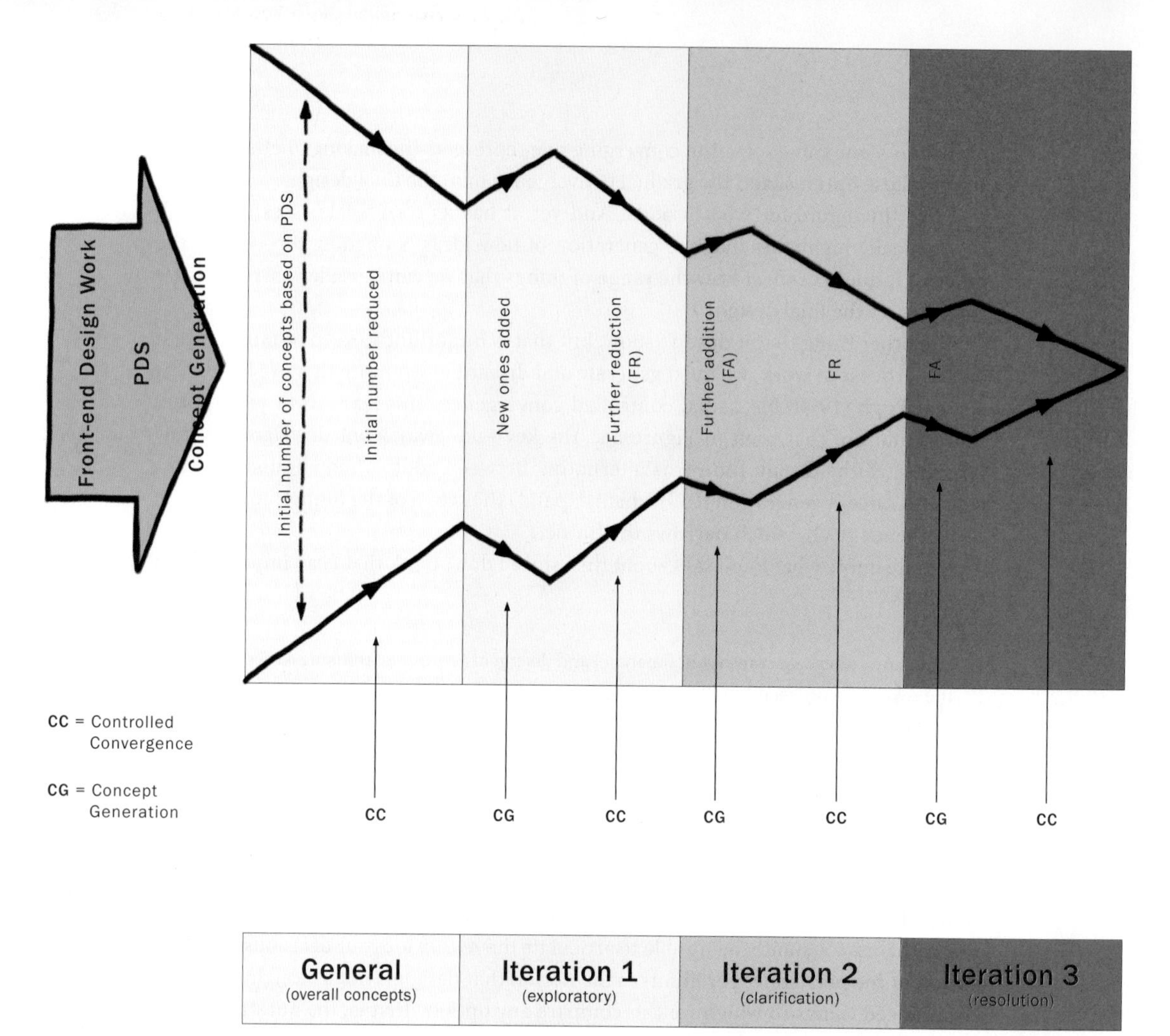

Figure 55: Another View of the Funnel

This variation on Pugh's illustration emphasizes the iterative nature of the process

Source: Pugh 1990; p. 75

There are no dumb questions. There are no ideas too crazy to consider. Get it on the table, even if you are playing around. It may lead to something.

This freedom to consider anything needs to be balanced by an understanding that choices have to be made, and that not everything considered will find its way into the final design. But remember, what does find its way into the design may well have gotten there precisely because of some other idea that did not. As with a novel, the whole plot may turn on a character (in this case, an idea) that got killed off before the *grande finale.*

There is one other point that I want to make regarding otherwise great ideas not making it into the final product. To do so, I need to give you another one of my definitions of design:

Design is compromise.

I know that this flies in the face of the image of the designer genius, whose ideas should not be tampered with. But as we have already discussed, the design team represents a number of stakeholders, such as design, engineering, product management, marketing, and so on. Each has its own legitimate priorities, and these priorities will often come in conflict with each other. An example might be making the appropriate trade-off between the "right" interface from the designer, and the "possible in my lifetime" reality confronting the engineer.

Scott Jenson (2002) calls this Yin/Yang Design, and makes two crucial points. First, the sooner that these issues are recognized and addressed, the less impact there will be on the final product. Second, the only way for this to happen is if all the stakeholders are part of the design team from the start. It is their engagement in the type of interactions that we are describing here that can lead to the most effective balance between the various yin/yang forces.

For example, as Figure 52 indicated, sketches serve to suggest, propose, and question. Part and parcel of this is to provoke scrutiny and criticism of the ideas that they represent. They need to be challenged and tested from all angles. However, if this is not managed well, egos can get bruised, tempers can flare, and serious damage can be done.

What keeps things healthy and stimulating, as opposed to the source of resentments and bruised egos, is the process by which these choices are made. Central to this is the need to be as clear about the rationale for various decisions as we are about the decisions themselves.

Being explicit about the *design rationale* accomplishes at least two things. It helps guide the process away from decision by bullying, browbeating, or seniority to one where the reason for the decision is understood, and can be articulated by anyone on the team.

Understanding the rationale for a decision is also a wonderful remedy to being a prisoner of your own decisions. That is, after a decision is made, you might learn something new. If you know why you made a previous decision, then it becomes much easier

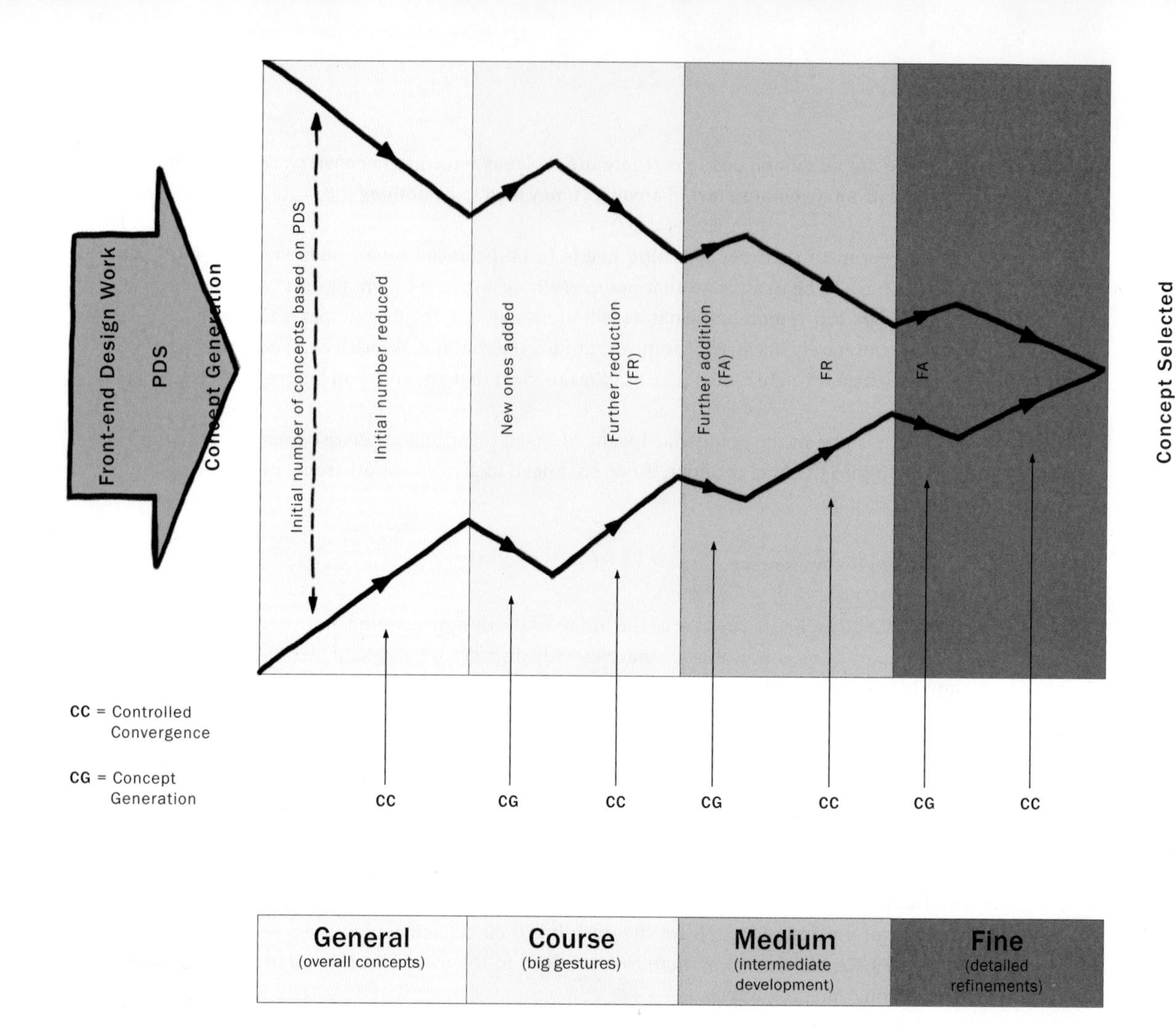

Figure 56: And yet Another View of the Funnel

This variation on Pugh's illustration emphasizes the fact that the process is the same as one converges to ever finer levels of granularity of design issue.

Source: Pugh 1990; p. 75

(and safer) to determine if it should be changed. As well, when the project goes to engineering, it is often of great help if those doing the implementation understand the why, as well as the what, of a design decision.

Successful execution of a design depends on communication, and capturing the design rationale is an important component in this.

Your partners on the design team must be your strongest critics. One of the most important reasons for having a team with diverse skills and experience—from design through technology, business, manufacturing, and marketing, for example—is the richness and breadth of perspective that they bring to evaluating the ideas on the table. Here is my spin on this:

From "the glass is half full" perspective, the thing to remember here is that one of the most positive forms of criticism is a better idea, and frequently, that better idea would never have come about were it not for the idea that it replaces.

Scott Jenson talks about "criticizing your way to a solution." I like this way of putting it.

Anyone who has taken a drawing class knows that the group critique at the end of each session is as important, if not more, than the time spent drawing. Learning how to give and take criticism is as much a part of the pedagogical intent as is the development of drawing technique.

This practice is fundamental in art and design education; however, one cannot assume that it is part of the normal training of other disciplines, such as computer science, that make up the interaction design team. This is why I am making such a big deal about it here, and will come back to it later.

This difference in background is one of the things that needs to be taken into account in how the design group is managed. As will be discussed in the next section, one way to facilitate this is through the design of the design environment itself—the physical and social ecology within which the team works. With care, this is one of the most powerful tools that can be deployed to encourage and facilitate collective discussion, debate, criticism and the exploration of ideas.

Figure 57: Design Echoed in Elementary School

Source: Queensferry Primary School

If Someone Made a Sketch in the Forest and Nobody Saw It...

Thoughts exchanged by one and another are not the same in one room as in another.
—Louis I. Kahn

I have to say it: sketches are *social things*. They are lonely outside the company of other sketches and related reference material. They are lonely if they are discarded as soon as they are done. And they definitely are happiest when everyone in the studio working on the project has spent time with them.

Sure, the act of creating a sketch can help an individual designer work through concepts and refine ideas. And sometimes, that is all that is required. The sketch can be discarded as soon as it is finished. But more often than not, a significant—if not the greater—part of the value comes in encouraging its social life. And for such encounters, the sketch's favourite meeting place is the wall-mounted corkboard.

I will go further: a design studio without ample space to pin up sketches, reference photos, clippings, and the like, such as those illustrated in the photo spread of Figure 58, is as likely to be successful as an empty dance club.

> Common inquiry must be rooted in a history of shared experience at many levels ... (Ivan Illich 1971)

This aspect of the physical and social ecology of the design studio serves a range of important functions. You don't make a decision on whom to marry based on first impressions. So why would you want to do so when selecting the design concept that you want to pursue, and therefore live with? Yet that is too often what happens in design reviews, where management is seeing the concept for the first time. (I'm not saying that first impressions are not important, or not sometimes right. I'm just saying that you can't rely on them alone.)

> Humans create their cognitive powers by creating the environments in which they exercise those powers. At present, so few of us have taken the time to study these environments seriously as organizers of cognitive activity that we have little sense of their role in the construction of thought. (Hutchins 1995; p. 169)

Now I suspect that my kindergarten teacher would take exception with this, given that she clearly spent a significant amount of her time organizing the environment precisely for the purpose of stimulating learning. Likewise, as the illustrations accompanying this chapter clearly show, designers are extremely conscientious about doing the same thing.

> It is common for designers to pin up part-finished drafts around the area in which they are working so that they are open to their own reflections (even at times when attention is not specifically directed to them) and to responses from colleagues. (Black 1990; p. 286)

These examples don't negate Hutchins' comment. Rather, they illustrate what is at risk if we do not pay adequate attention to the design of the environments in which we work.

Hanging work in the environment lets it "bake in." It is there in the background, and becomes part of the ecology of the studio. You live with it for a while, and with familiarity grows either insight or perhaps contempt. Displaying the work in juxtaposition with other material helps in the discovery and exploration of new relationships. Not only that, it provides the opportunity for your fellow designers to get to know, and comment on, this "partner" with whom you all might be spending a considerable amount of time.

However, the corkboard is no more what it appears to be on the surface than is a sketch. Its importance lies less in the object itself, than in the social and cognitive behaviours to which—through its affordances—it provides the catalyst (Gaver 1991). It isn't just something into which you can easily stick map pins. It is an awareness server: a technology that affords a sense of shared awareness of common references among the design team. Thereby, it provides an important and efficient means for communication and collaboration. It is as important to the design process as sketching itself.

That is not quite right, since that statement still treats sketching and corkboards as separate entities, as objects in the material sense. They are not. The notion of sketching should be, and henceforth will be, considered to embrace the larger notion of the term. It encompasses the social and physical ecological aspects that we have discussed, such as shared awareness, baking in, collaboration, communication, juxtaposition, and critique. And it does so within and through explicitly designed spaces and locations in the design studio.

All of this is the norm, and hardly even needs to be said in a traditional design studio. But what we are talking about in this book is not the traditional design studio, but a studio populated by people who come from other traditions than, for example, graphic or industrial design. This needs to be taken into account, as is highlighted by the following comments that Scott Jenson made to me on reading an early draft of this chapter:

> You make a really good point here but there is a deep social phenomenon to take into account. I did create a pin up board space in a central location.... It was dismal failure. People never felt their sketches were 'good enough' to put up on the board. There was something daunting about the physical act of leaving their cubical space and going into the global shared space to post something. Some went up of course but I received feedback that it seemed like a 'bold move' to do it. It's like the refrigerator door in the family kitchen. It's only cool for Mom to put things up. Now this could be British versus American values but I was a bit stumped by this reluctance.
>
> However, I did notice that people were more willing to put sketches up within their own cubicle. This made me try to reconfigure the cubes so that we had four cubes around a communal space with low 4-foot corkboards in the center (to keep the informal social interaction in place). It also was not a 'wall' but a small posting area so it didn't cut into the space as much. I was only able to get this to work for a single group (crazy British office politics) and I can't really claim too much here but this group certainly used this space actively. I should note they also just talked more in general as they were easier to engage on many levels.
>
> In another group, they couldn't get a corkboard but they had a communal white board and it was used in much the same way. There were impromptu meetings held around the board and once two people started talking, it was a very obvious, impromptu gathering and inevitably pulled in one or two more. Those moments made my heart glad, I just wish I could have figured out ways to do it more.

The reason that I shared these comments is that they are so effective in making the point that—in contrast to some of the articles that I have read in the popular press recently—simply plunking a bunch of corkboards or foamboards around your work space does not magically turn it into a design studio. These are artifacts with certain affordances, but their effective use requires as much attention to the cultivation of the culture of the studio as to the detailing of the architectural space.

Figure 58 (Following Spreads):

A Sampling of Design Spaces and Ecologies

Credit: Bruce Mau Design

MASSIVE CHANGE PROJECT MODEL
Research
Body of Knowledge
Professional Network
Outcomes
Book
Exhibition
Product
Website
Film
Radio
Marketing
Consumer Network
Massive Change
The Project
Massive Change
The Movement?
Face Book

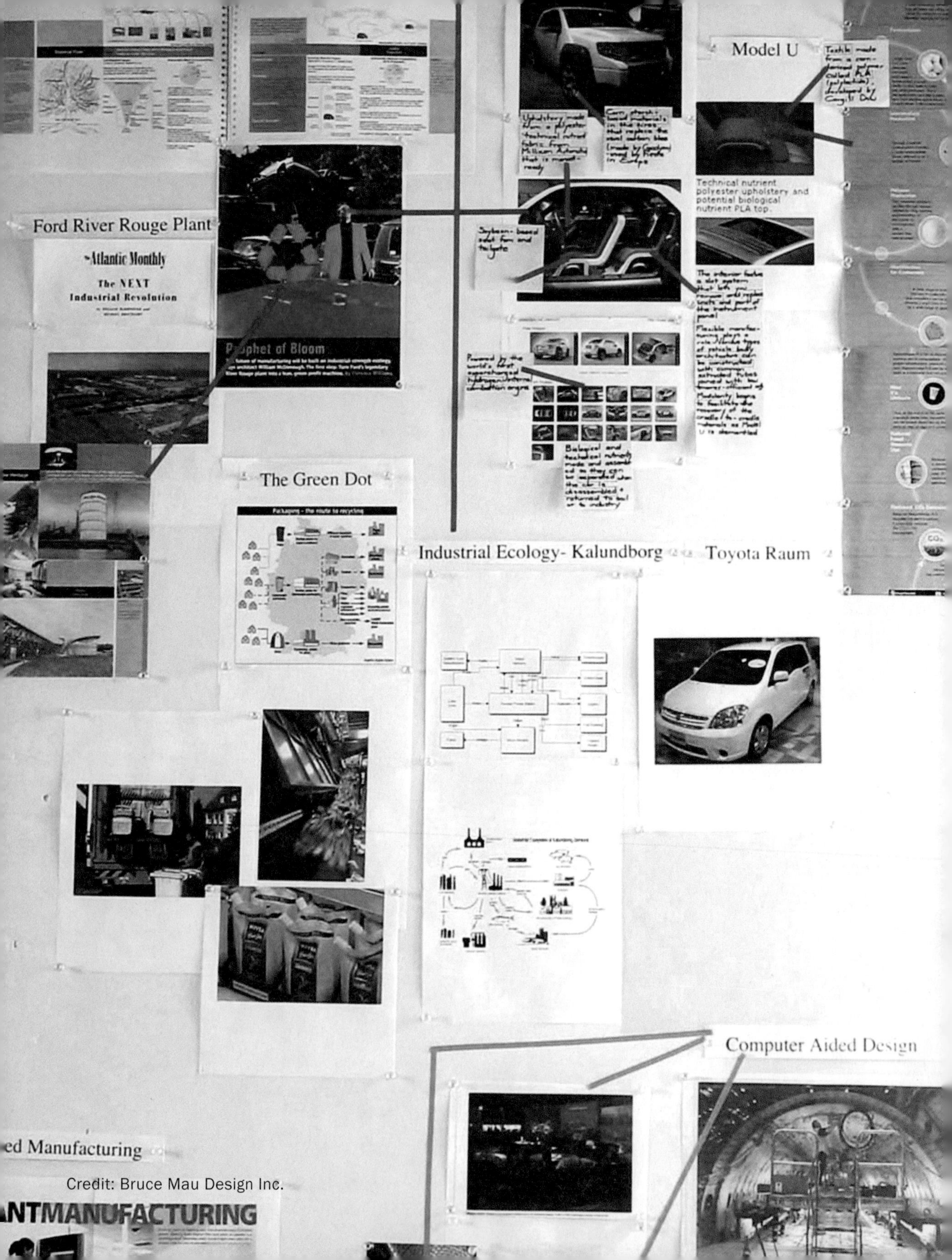

Credit: Bruce Mau Design Inc.

Credit: Microsoft Corp

Credit: Research In Motion

Credit: Microsoft Corp

Portfolio Wall

Even if we have gotten the cultural and architectural details right, we still may have a problem in terms of adequately supporting interaction design. The difficulty lies in the nature of media that make up our sketches and much of our reference material. Almost by definition (and as the examples that we will examine later will reaffirm), much of interaction sketching has to do with temporal phenomena. Hence, many of the sketches will be interactive, or cinematic in form. How does one mount this on a corkboard?

This is where we employ a variation of the old "It takes a thief to catch a thief" rule. We have a problem. It is technology that has partially created the problem. It is also technology that we are designing. Good design solves problems. So, we can apply what we know to solve this one, too. One attempt to do so is what I call The *Portfolio Wall,* shown in Figure 59 (Buxton, Fitzmaurice, Balakrishnan & Kurtenbach 2000).

The approach follows one of my fundamental views on design:

> **One of the metrics for the design of any $generation_n$ technology is to solve the problems introduced by $generation_{n-1}$, while retaining the benefits.**

This is not unrelated to Kransberg's second law:

> Invention is the mother of necessity. (Kransberg 1986)

The Portfolio Wall is essentially a first-generation interactive digital corkboard. It takes advantage of the emergence of inexpensive digital data projectors and large flat panel displays to address the problem of, "Within the studio, or from studio to studio, how do we share electronic imagery, whether it be still, cinematic, or interactive?"

Its genesis is in automotive styling rather than interactive design. On a visit to General Motors, the VP of Design, Wayne Cherry, expressed to me his frustration in trying to get a sense of what was going on in the studios. Although computer-aided industrial design had become an important part of their process, it was also the source of Wayne's frustrations. His practice was to walk through the studios each morning at about 6:30 A.M. In the "old days" of working on paper and drawing boards, this practice gave him a good sense of what was going on. Based on what he saw, he was able to make suggestions, provide encouragement, and in general, do his job. This is all because the work was visible.

With the adoption of computers, this was no longer the case. The work ceased to be visible, with the exception of that which the designer chose to print out and mount on the physical cork-boards. Work in progress, which is the stage where it is most receptive to comment, remained invisible to him. To see what someone was working on would require interrupting the designer (assuming that they were even there—a rare thing for designers at 6:30 A.M.).

The *Portfolio Wall* enabled designers to "throw" things up on the wall electronically, including turntable animations and such, thereby making them visible. Consequently, Wayne (or anyone else in the studio) could both see the work and actively explore it without necessarily interrupting the designer, due to the interactive nature of this communal technology.

Now some might argue that the same purpose could be achieved simply by having a project web site, where images could be shared among the community. Understanding why this is not the case is central to the point that I am trying to make. Though cheaper and well understood, a conventional web-based approach is far inferior, since it does not have the property of visible persistence or provide the same sense of background awareness. With it, one must take explicit action to look at the work. It does not come for free by simply being in the studio.

Having said that, the *Portfolio Wall* does have some serious weaknesses that can be picked up by a simple comparison of Figure 59 with those seen in Figures 58. The problem with the Portfolio Wall is that all of the images appear at a uniform size and in a rectilinear layout. It forces a way too tidy layout and presentation. This is in marked contrast with its traditional counterparts, where the size and layout of images are key cues to guide the eye and to help establish relationships among what is being displayed. This is something that absolutely needs to be fixed, and will be, inevitably, either in future versions or by a competitive technology.

Finally, the significance of the *Portfolio Wall* to our current discussion has less to do with the specifics of its design, and more to do with its being an example of how interactive and cinematic sketches, prototypes, and other electronic reference material can be shared, on a public wall-mounted display, thereby preserving the best of the traditional studio, while benefiting from the best of the new technologies.

Figure 59: The Portfolio Wall

This is an interactive digital corkboard. It can be implemented using a large format plasma panel, as illustrated, or with a projector. As with a traditional corkboard, it can display sketches or other reference images that a designer wants to share. Unlike the traditional corkboard, the images can be dynamic, such as displayed on turntables, or as animations or videos, and they can be interacted with, such as enlarged, as illustrated in the figure.

Photos: Azam Khan

Procedure

The Object of Sharing

The strength of the pack is in the wolf.
— Rudyard Kipling

In reality, sketches are just one of the things that are shared in order to create a common set of references to support the design activity. You will already have noticed this if you looked closely at the figures in the previous section. The foamboards illustrated contain a mix of both sketches and found material (magazine clippings, photos, etc.).

However, what isn't included are 3D physical sketches, such as we saw with the OrangeX juicer in Figure 50. Nor do we see 3D objects such as toys or other products that are the equivalent to the found reference material that we saw on the foamboards. Most 3D objects simply don't lend themselves to being pinned on the wall. We could photograph or videotape them, and then display them on something like the Portfolio Wall. But then one loses the ability to get a sense of their tangible properties: their weight, feel in the hand, balance, and so on. You also lose the ability to really explore them by examining them from all sides.

In contrast, look at the work spaces illustrated in the photographs on pages 160 and 161. What you see instead are all kinds of curious objects sitting on top of monitors, tables, shelves—anyplace that there is a flat surface on which to place them. Some of these objects are sketches. Others are reference material that may relate directly or indirectly to a project. They may be inspirational decoration, or simply something to play with as a diversion.

Figure 60: It's a Messy Business

Sketching might be the closest that you come to going back to kindergarten other than having a five year old child, or becoming an elementary school teacher. Think about your favourite shirt. It has a great feel. It is about the texture of the material, the cut, and how you feel in that particular colour or pattern. I don't care how good your computer graphics are, or the quality of the photography in a catalogue, you cannot get a sense of these things without actually trying the shirt on. If this is true for the consumer, it is even more true for the designer. It has to be hands on. And that is a messy process, in the most wonderful way.

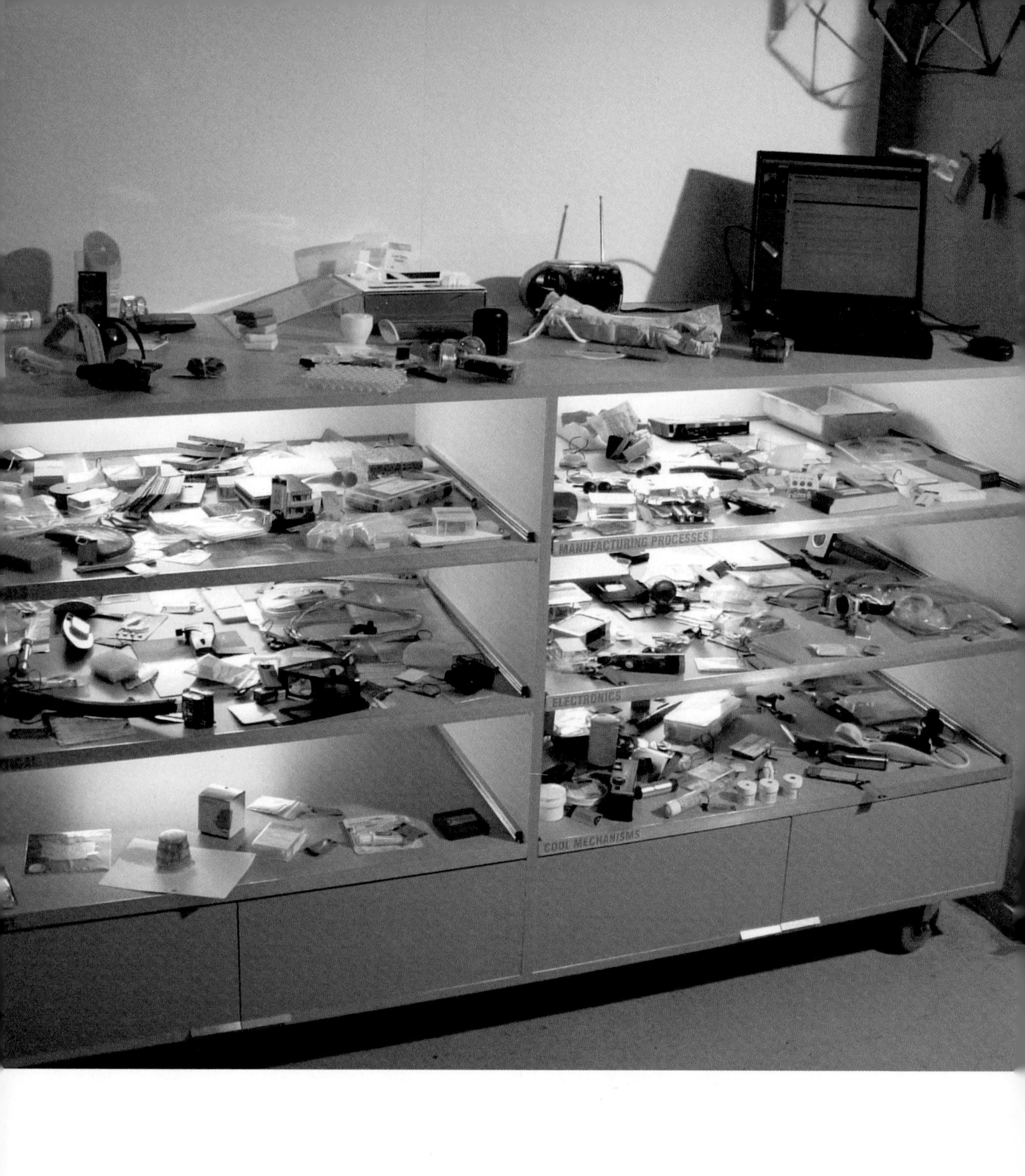
MANUFACTURING PROCESSES
ELECTRONICS
COOL MECHANISMS

Tech Box

Some studios take this practice further, technologically and organizationally. For example, the design firm IDEO has something that they call the *Tech Box* (Kelley & Littman 2001). The *Tech Box* at IDEO's London studio is shown in Figure 61.

It consists of hundreds of gadgets. Most are laid out on open shelf-like drawers. Some are toys, and are just there because they are clever, fun, or embody some other characteristic that may inspire, amuse, or inform (or perhaps all three). Others might be samples of materials that could be useful or relevant to future designs. This might include flexible cloth-like fabric that can also be used as a touch pad, or rubber that does not bounce.

Since the *Tech Box* is a kind of mini library or museum, it has someone from the studio who functions as its "curator" or "librarian." And like conventional libraries, all of the objects in the collection are tagged and catalogued so that supplementary information can be found on the studio's internal web site.

As an indication of how much store the company puts in having its employees have a shared set of references, there is a *Tech Box* in every one of their studios worldwide. Furthermore, even though anyone can add things to the collection, if you do, you must get one for the *Tech Box* in every one of the studios. These are circulated to the other studios by the local curator, who also makes sure that the appropriate entry is made to the associated web database.

The value of the *Tech Box* has been such that IDEO has expanded their deployment outside the company, as one of the services that they offer their clients.

Of course, it is important to keep in mind that the *Tech Box* is just one way that physical objects are shared as reference material within the studio. They are no different than any other studio, in that the space occupied by both projects and individual designers are cluttered with all forms of paraphernalia.

Also, the idea of something like the *Tech Box* is not unique (although I am not aware of anyone else formalizing its use to that degree). Nor is the basic practice hard to adopt and adapt to your own situation. One of my favourite manifestations was shown to me by Jonas Löwgren, a friend who teaches at the School of Arts and Communication in Malmö, Sweden.

Figure 61: Tech Box

A wonderful example where innovation in the design ecology can improve support of design.

Credit: IDEO

CoWall

Sweden's School of Arts and Communication's equivalent of the *Tech Box* is something they call *CoWall,* shown in Figure 62 (Löwgren 2005). They describe it as a "tangible project archive" (Ehn & Linde 2004).

With *CoWall,* the shelf-like drawers are replaced by Plexiglas cubes, within which the objects are displayed. Like the *Tech Box,* information about each object displayed in the cubes is stored in the computer. What is neat about the design is that each object also is tagged with either a barcode (just like your groceries at the supermarket), or with something called an RFID tag (sort of an electronic alternative to a barcode). When an object is picked up and placed on a particular shelf called the organizing zone, the tag enables the computer to recognize what object it is, and then to display its associated information on the large projection screen seen in the figure. This information can take the form of text, video, audio, and/or still images. Furthermore, an integrated printer enables you to take away information about the object of interest on paper.

What I like about the *CoWall* is that it provides such a simple, elegant, and seamless mechanism for bridging between the physical and the digital. The object itself is the user interface to its associated information. We are going to encounter RFID tags again in the context of our discussion of toolkits that let designers mock up physical devices simply and quickly. My opinion is that an excellent first project to undertake using such tools is to build your own version of the *CoWall.*

It will be a great learning experience that is not too difficult. But it will also result in something that could become a really central part of the larger ecology of your workplace. I just love working with things that embody the very essence of what I talk about and do. Design is all about "walking the walk," and this is a great way to do it. And, by the way, just like IDEO spreads the *Tech Box* among their studios, having a *CoWall* demonstrating the innovations of your company outside the executive boardroom or the corporate visitor's centre is a pretty good way to both demonstrate and take advantage of your company's commitment to design and innovation.

Figure 62: CoWall

A mixed-reality (physical and virtual) approach to sharing physical artifacts.

Credit: Per Linde, Jonas Löwgren

IDStudioLab
IDStudioLab

Cabinet

Finally, due to its simplicity, one of the most elegant solutions that I have seen for sharing both objects and images is something called *Cabinet,* developed by Ianus Keller (2005) at the Technical University of Delft. In many ways, *Cabinet* is modeled on a conventional overhead projector. A storyboard summarizing the concept is shown in Figure 63b.

It is a horizontal surface on which one can place objects or images and have them captured by an overhead camera. Once captured, the material is then projected back onto the same surface. In this sense, it is a type of scanner. However, the surface is actually a large graphics tablet, which means that once captured, material can be moved, categorized, and annotated. In general, it can be shared, especially if a *Cabinet* in one location is linked to one in another.

In many ways, the *Cabinet* embodies many of the sharing features of the *Portfolio Wall,* but also provides the designer full freedom to determine the size, orientation, and layout of images, as well as the ability to capture. It goes well beyond being just an interactive display.

I especially like it for what it doesn't do. The approach is simplicity. It knows what it is, but more importantly, it knows what it isn't. No, it is not perfect, and it is not a product. But concepts like the *Portfolio Wall, Tech Box, CoWall,* and *Cabinet* are as important to the experience designer as foamboard (which they complement, rather than replace) has been in the past. However, I have to reiterate that more important than supporting technology is cultivating the social practice of sharing that defines the culture of design.

Figure 63a (Left): Cabinet
A form of telepresence that supports collaboration around shared objects.
Credit: Ianus Keller

Figure 63b (Right): Schematic Drawings of Cabinet
Credit: Ianus Keller

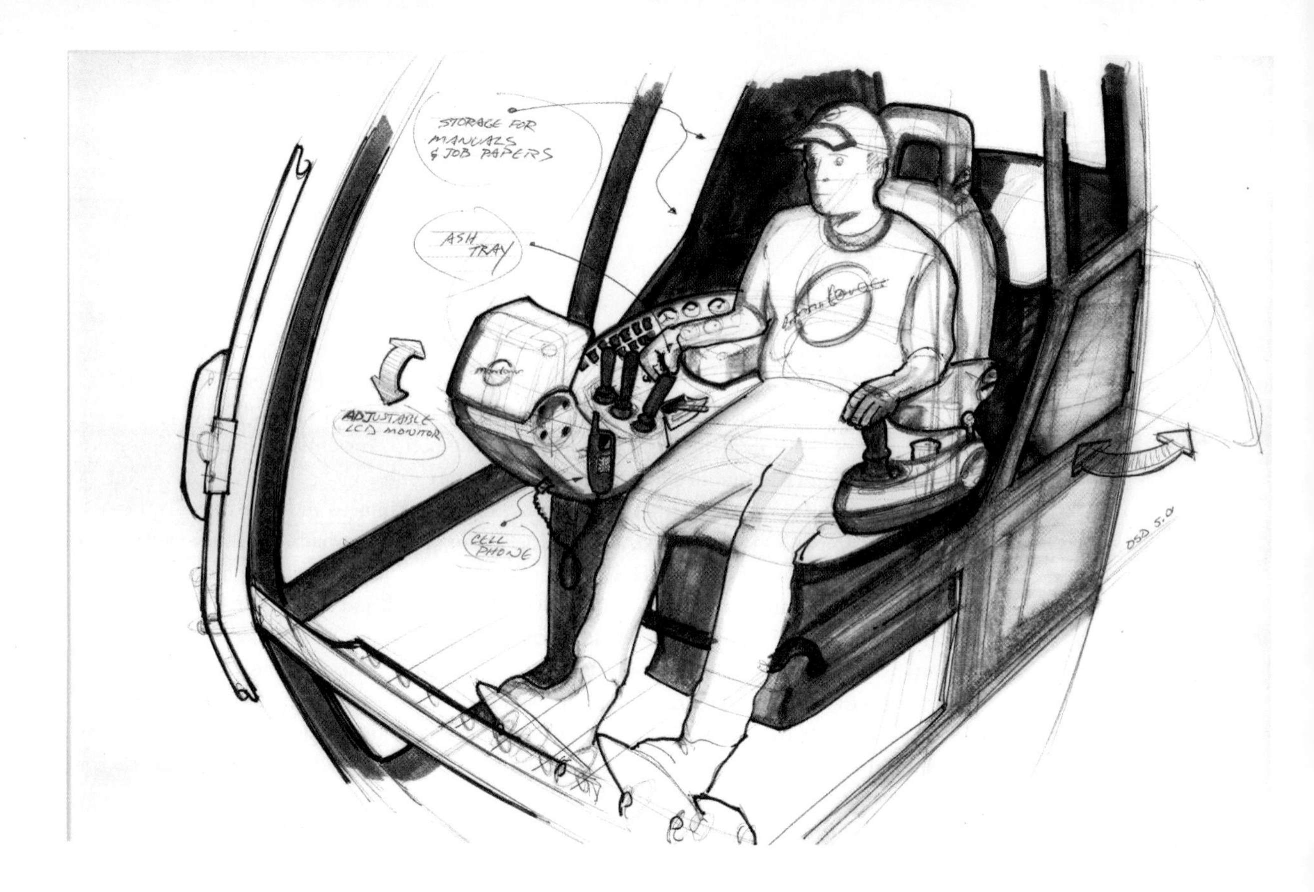

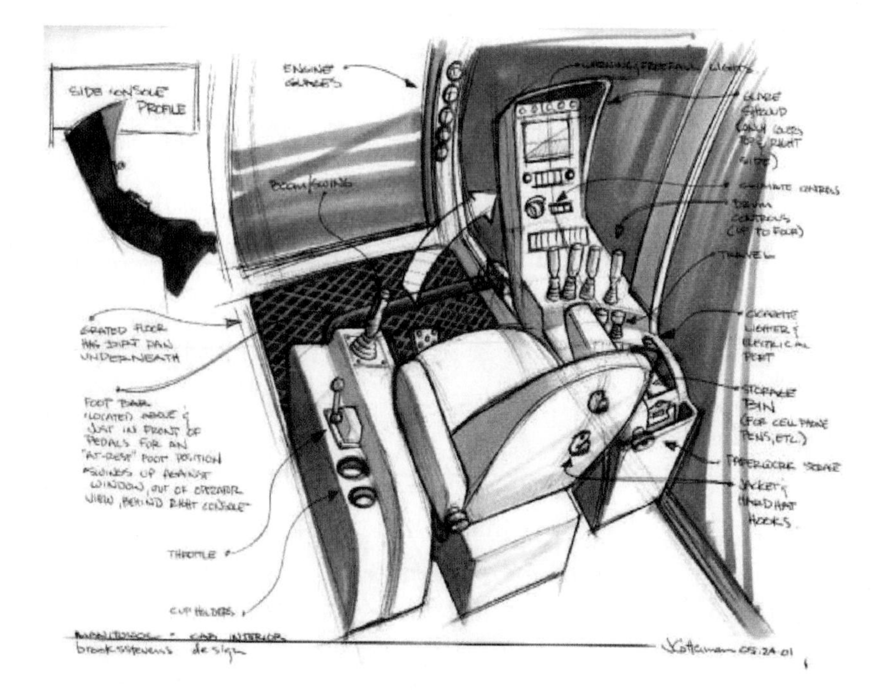

Figure 64: Annotating Sketches

From the designer's point of view, Annotation helps the viewer understand the specific ideas that are not readily apparent in the illustration itself. From the critic's perspective, annotation allows an outsider to comment on the particulars of the design.

Credit: Brooks Stevens

Annotation: Sketching On Sketches

Making a plan and sticking to it guarantees a sub-optimal solution.
—Andrew Fitzgibbon

Imagine that you are a traditional designer working on your drawing board. The concepts that you are developing catch the eye of a passing designer and this evolves into an animated conversation. At some stage, your colleague wants to express an idea that can be effectively communicated only by drawing on your work.

Alternatively, imagine that you want to augment a finished sketch with some notes and/or graphic material that explain things like the colour scheme or some other detail, yet you don't want to alter the original.

These are just two examples of situations where you may want to draw on or otherwise mark up and annotate a rendering, but you don't want to change the original. That way your work remains intact and can be viewed with or without the annotations.

There are lots of other examples that crop up in the design process, but these two should be sufficient to make it clear why this capability is worth discussing. With traditional media, the standard way that this is done is by placing a layer of velum over the original, and making the additional marks on it. Since velum is much like the tracing paper we used as children, the original drawing can still be seen, and what is drawn or written on the velum appears as a layer on top of the underlying rendering. An example of this is illustrated in Figure 64.

This is fine for traditional media, but what do you do if what you want to mark up or annotate is not a static image on a piece of paper, but a video or animation? This is a very relevant question given that the materials we use in interaction design are often cinematic and involve time, for example. If annotation and layered mark-up are important in traditional media, then the assumption should be that it will also be so when sketching with other media as well.

One answer to the question of annotating dynamic media comes from the film industry itself. It lies in how—in the predigital era—film editors used to indicate when and where transitions such as dissolves, fades, and wipes were to occur in a movie. To do so they would draw directly on the relevant portion of the film using a grease pencil, commonly called a China Marker (Chandler 2004). This is illustrated in Figure 65.

Although the editor's grease pencil marks were made directly on the film rather than on a separate layer, the basic idea can be adapted to the digital domain in audio and video editing, where one would draw over some form of timeline.

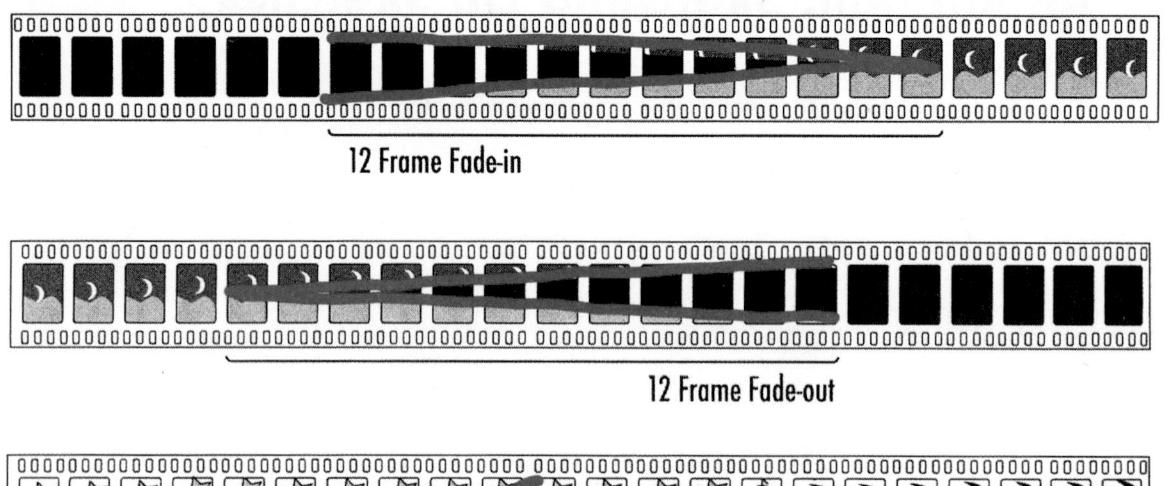

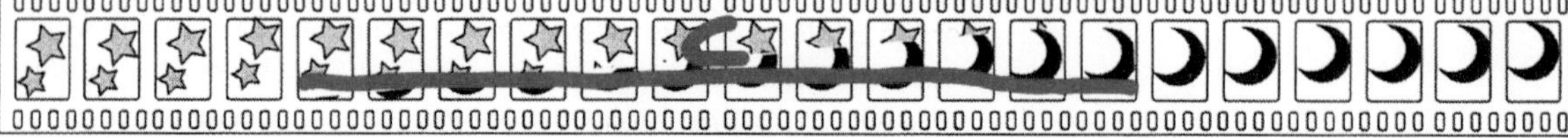

Figure 65: Grease Pencil Annotation on 35 mm Film
With conventional film, one drew directly on the film to indicate when and where effects such as dissolves, cross fades, etc. were to occur. In the first two examples the red marks indicate a twelve frame fade-in from black and fade-out to black, respectively. The third example indicates a twelve frame dissolve, and the "C" indicates the centre point between the two shots where each is equally visible.
Images: Based on Chandler (2004) Excerpt from Cut by Cut: Editing Your Film or Video provided courtesy of Michael Wiese Productions, www.mwp.com.

Figure 66: Winky Dink and You
Drawing with grease pencils on acetate over the TV screen in order to help the characters in the episode solve problems.
Photo: CBS

There are other ways to think about drawing on top of video, such as drawing directly on frames that are presented sequentially, like during video playback, rather than spread out along a timeline. To clarify what I mean by this, compare the type of annotation illustrated in Figure 65 with that seen in Figure 66.

The latter is drawn from a children's TV show called, *Winky Dink and You*. Produced by CBS, this 1953 program is probably the first example of free-hand drawing on video. In this case, children drew with grease pencils on acetate overlays on the TV screen in order to help the characters in the episode solve problems.

Of course what the children drew had no actual impact on what happened to Winky Dink, except in their minds. But that is not the point. What matters from our perspective is that the basic approach is valid and can be applied when implemented using modern digital tools rather than a classic black-and-white television from 50 years ago.

Examples of this taken from both animation and videogame production can be seen in Figure 67. There are two things in particular that I think are worth noticing in these images. First, notice the ease with which you can distinguish between figure and ground; that is, the foreground annotation layer and the background image that is being annotated. Second, despite the images being still frames (albeit from a dynamic animation), notice the way in which the use of line type and arrows convey movement, energy, and the nature of the intended experience.

So far we have seen how annotation can be used on static and dynamic media. However, in the examples discussed so far, the annotations themselves have been static. They may indicate motion, but they themselves did not change over time. There are examples from both practice and research that illustrate that the annotation itself can be cinematic or dynamic.

We don't have to go very far to find an example. Simply listen to the director's commentary while watching one of your favourite DVDs. That secondary voice track is absolutely an example of annotation, and like the original use of velum over a sketch, it is a separate layer (in this case audio) over the top of the original—and a layer that can be turned on and off.

Likewise, most of us have seen a combination of voice and graphic annotation in action while watching sportscasts on television. In this case, we have a commentator drawing over the video while giving a verbal analysis of some football or hockey play.

The drawings may be crude, but that is not the point. Like sketching, it is the timeliness of the commentary and its quality and relevance that is important. As it is in sports and film making, so should it be in interaction design.

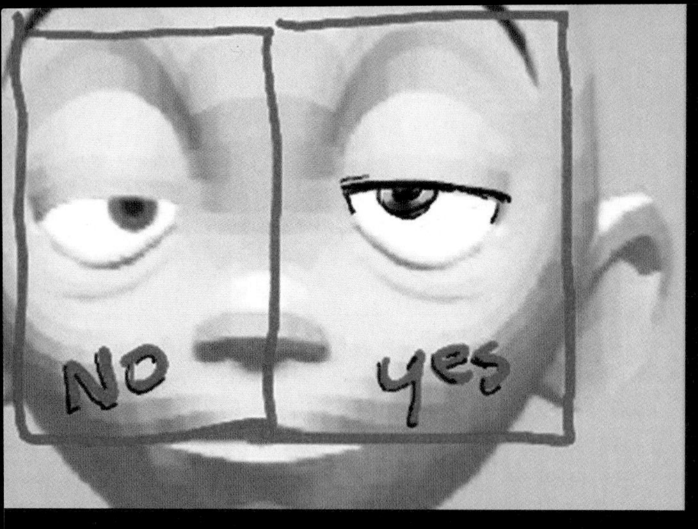

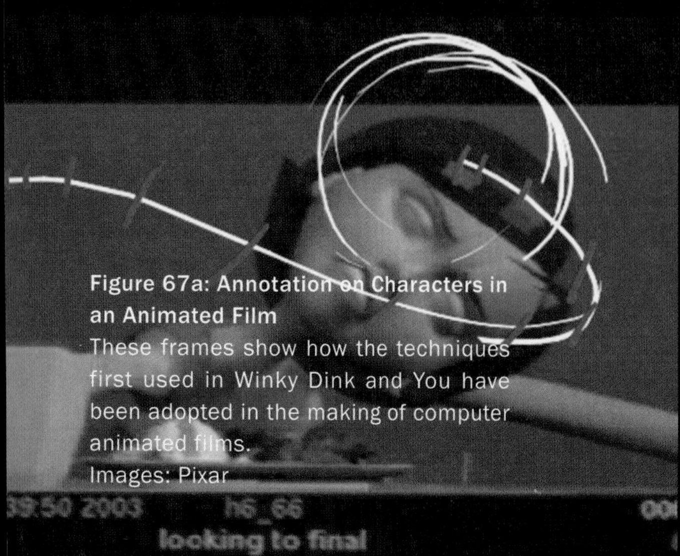

Figure 67a: Annotation on Characters in an Animated Film

These frames show how the techniques first used in Winky Dink and You have been adopted in the making of computer animated films.

Images: Pixar

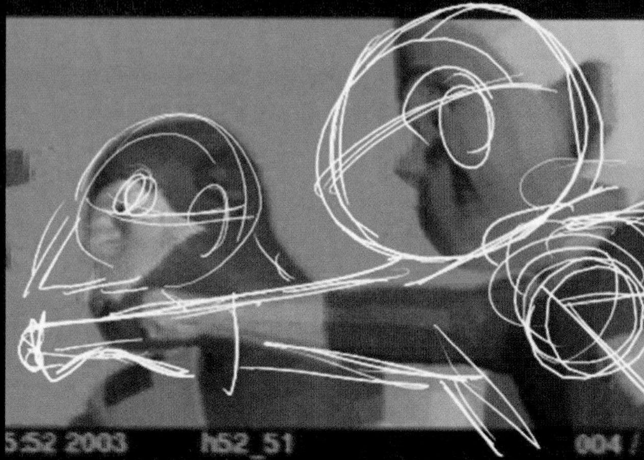

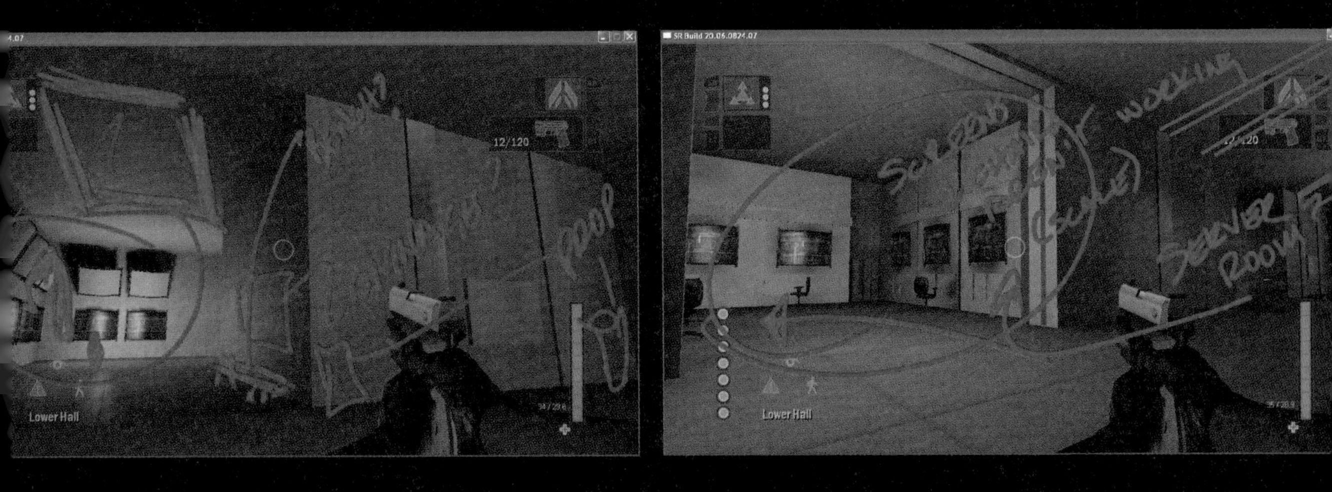

Figure 67b: Annotation in Videogame Production
In these images we see examples of how annotation is used to give notes and commentary on the virtual environments to be used in a videogame.
Images: Microsoft Corp.

Figure 68: Video Annotation

It is now common to see sports broadcasts where the commentator draws on top of the video to explain a play, or some matter of technique – as illustrated here. One of the things that we need are simple and readily available tools to do the same thing with the videos that we use in design. Source: Pat Morrow

My firm belief is that this kind of capability should be part of the standard repertoire of technology and techniques in any studio involved in interaction design, for production, education, or research. Furthermore, the barrier to entry in terms of using such tools must be very low. They must enable fluent and immediate commentary by intermittent and regular users alike.

Yet, despite such systems being commercially available, this is not the case. One of my primary motivations here is to change this.

I want to be clear that in making this point, I am not suggesting that one can go out and just purchase the perfect tool for the job. Yes, some tools exist today, and they are better than nothing, so there is no excuse not to have something workable in place. But that does not mean that there is not room for substantial improvement. Hence, my second agenda is to point out the importance of this type of tool to the design process, and the value of investing in the development of better and more appropriate solutions.

I conclude this section with a brief description of one such effort to do so. This is to illustrate the kind of things that I think are needed and can be done. The underlying point is that we need to be as conscientious in the design of our own tools as we are in the design of the products that we make with them.

Around 2000, the group that I was involved with became interested in how designers could add voice and marking annotations to 3D digital models. How could they do so with the same type of fluency that they had previously exercised with traditional media (Tsang, Fitzmaurice, Kurtenbach, Khan & Buxton 2002).

For example, how could the studio head walk around a digital model of a car and visually indicate things while verbally commenting on them. When appropriate, how could this be accompanied by the ability to manually sketch on the surface of the model? Could one do so with the same fluency and using the same skills as traditionally used on velum? Could this be done in the studio without having to wear cumbersome gear like head-mounted displays, special gloves, and such?

The simple answer is Yes to all these questions. The prototype system that we built to demonstrate this was a kind of virtual video camera called *Boom Chameleon.* It is illustrated in Figure 69.

The console of the device consisted of a microphone-equipped 17" touch screen that was mounted on a counter-balanced articulated boom. All the articulated joints were instrumented with shaft encoders that could be read by the computer in real time. Hence, as you held the screen and walked around the central pedestal supporting the boom, the position and orientation of the screen could be sensed by the computer controlling what appears on the screen. Hence, the view of the model

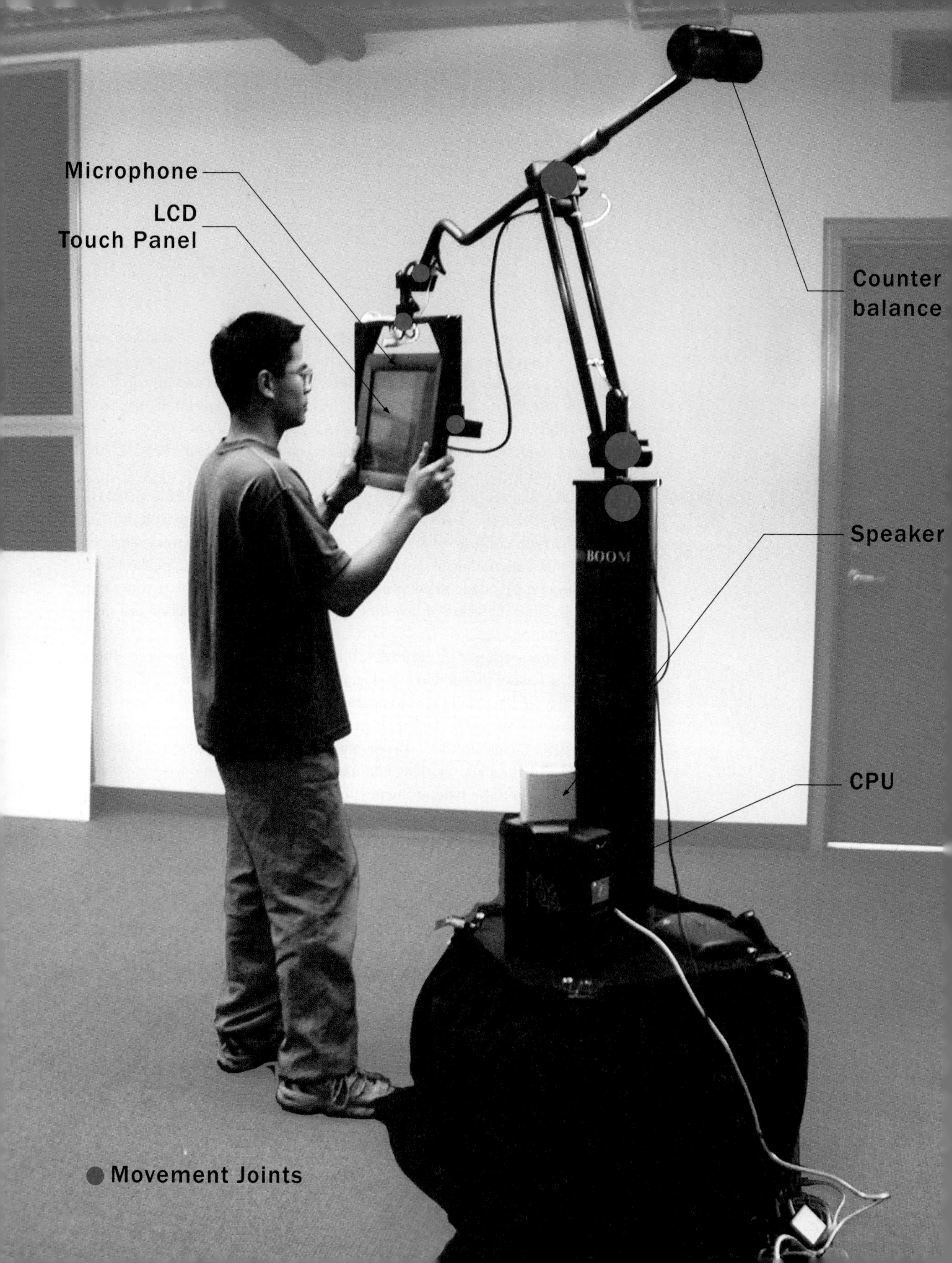
Microphone
LCD
Touch Panel
Counter
balance
BOOM
Speaker
CPU
Movement Joints

Figure 69: Boom Chameleon

A virtual window that permits the user to view and annotate a 3D digital model graphically using a finger on the touch screen, and simultaneously by voice. One can walk around the object to see it from all sides, much as if the screen were a camcorder.

Source: Tsang et al. (2002)

could dynamically change in a manner analogous to the way it would if you were walking around it holding a video camera. Just like with the camera, you could move up and down, or toward or away from the pedestal, which would result in being presented an upper or lower view of the model, or a close up or distant view, respectively. Furthermore, at any time, one could use the touch screen to highlight a particular aspect of the model using a virtual flashlight. Likewise, one could draw on the model's surface. This is illustrated in Figure 63, and can be seen in more detail in the associated video.

As well as capturing what you pointed at and the marks that you made, the system was capable of recording the movements of the display as well as your verbal comments while doing so. In this sense, it was like a virtual video camera with pointing and marking capabilities.

Making and recording annotations is only half of the equation, however. Others need to be able to know that they exist, where to find them, and how to view them. The key to this is illustrated in Figure 70.

Annotations were indicated as virtual snapshots suspended above the 3D model. Selecting any one of these initiated full-screen play-back of the annotation. So that things didn't get too cluttered, the annotation indicators could be collapsed into just a simple horizontal bar, or made invisible for unobstructed viewing of the model.

Is this the perfect annotation tool? Of course not. There is no such thing. This is just an example of one approach to developing new tools to support the design process.

Ultimately, there are two major take-away messages from this section.

First, in attacking new classes of design problems and adopting new technologies, we need to be vigilant that we don't inadvertently drop important and proven aspects of the design process along the way, just because they are not well supported by the tools that exist today.

Second, as I have already stated, we need to apply the same skills to the design of our tools, environment, and process as we do to the products that all these are intended to serve.

Fluent, simple, and spontaneous annotation is not the only example where this applied. It is simply the one that I have chosen to make the point.

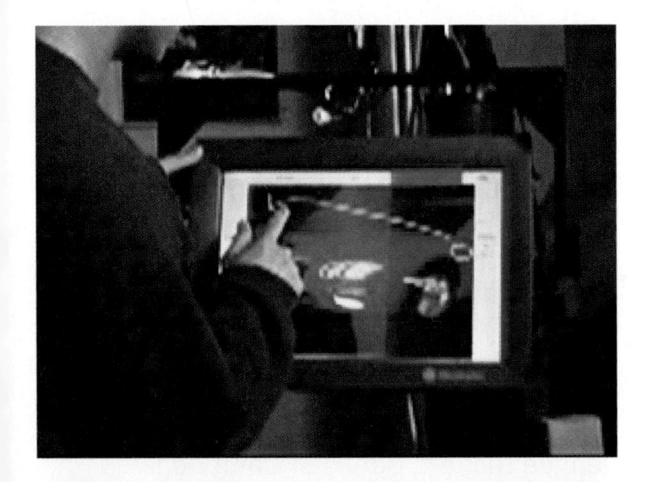
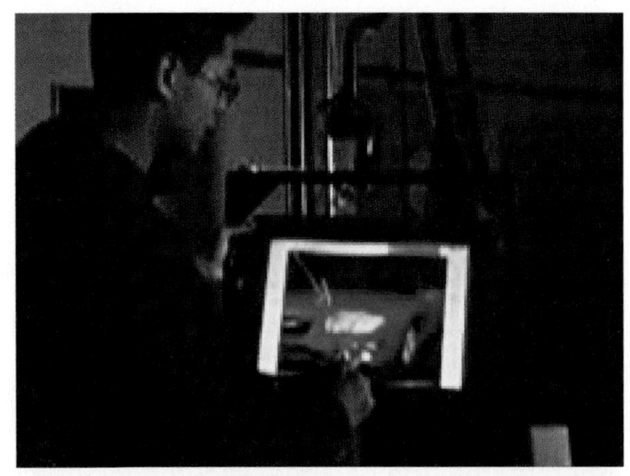
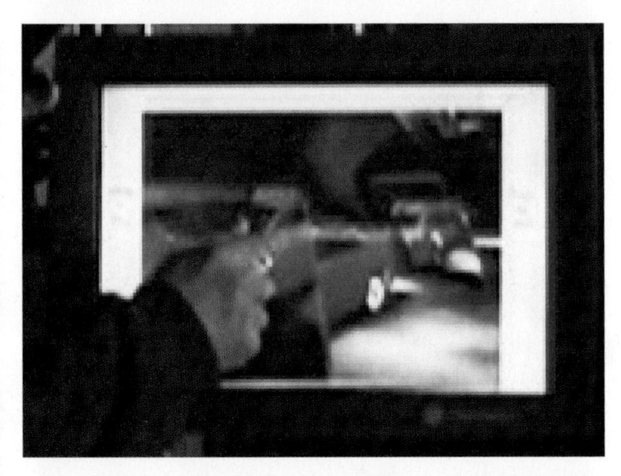
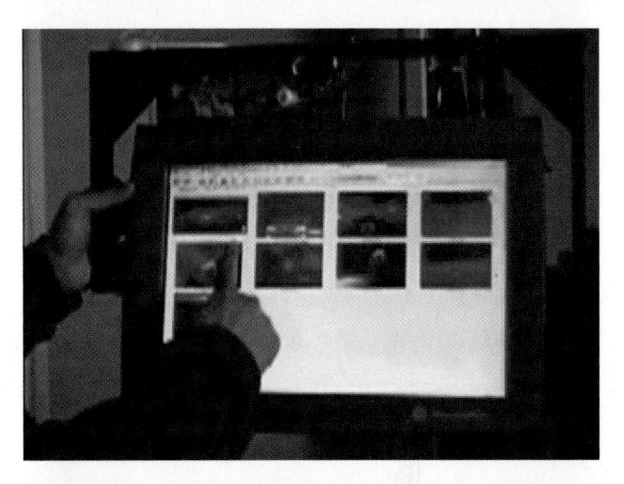

Figure 70: Annotate, and Record

During a design session, the users can move around the virtual model, highlighting areas of interests, annotate using their fingers. The entire session can be recorded, and key frames can be saved for reference. Source: Tsang et al. (2002)

On Hunters, Gatherers, and Doodlers

The preceding discussion has made it pretty clear that designers are both doodlers, by virtue of their sketches, as well as hunters and gatherers. The latter is made evident by the types of paraphernalia that they collect for inspiration and reference material.

It is important to point out that not all of what is sketched or collected is for public consumption. Some is for one's self, and is often private, like a journal, diary or scrapbook.

To give you a feel for the richness behind that simple term, I want to show you some excerpts from the personal sketchbook of my colleague Richard Banks of Cambridge. In so doing, I am showing you a heritage that predates Richard, by a long shot. His notebook itself is from a company named Moleskine, "the legendary notebook of Van Gogh, Chatwin, Hemingway, Matisse and Céline."

When my book designer Hong-Yiu Cheung was laying out this section, he insisted that we reproduce the pages full size. His comment was:

> ... the hardcover sketchbook is one piece of original technology that is designed "for the wild." If you think about it, the sketchbook is the artist/writer version of the mobile studio.... the hardcover sketchbook was designed precisely to be portable, casual, and treated roughly, for field work. ... a place to deposit notions, ideas, intents, and half-formed thoughts.
>
> We are often very self conscious about proposing ideas because they are not thought through and hence embarrassing ourselves in front of others. The sketchbook is an environment where it can be safe for us to articulate these thoughts to ourselves.
>
> The value ... does not come from the degree of finish, but the authenticity of the tracking of a person's ideas. A sketchbook used only in the studio environment will probably not be as rich as a sketchbook that followed a person that has travelled out in the field, in a variety of contexts. By carrying it around with you all the time, it also allows you to reference back in time in a moment's notice, to connect previous thoughts with the present situation. This I think is also the power of the sketchbook. A kind of long-term development of criss-crossing streams of consciousness.

Such is the designer's sketchbook. And such is the reason that we give this one such prominence.

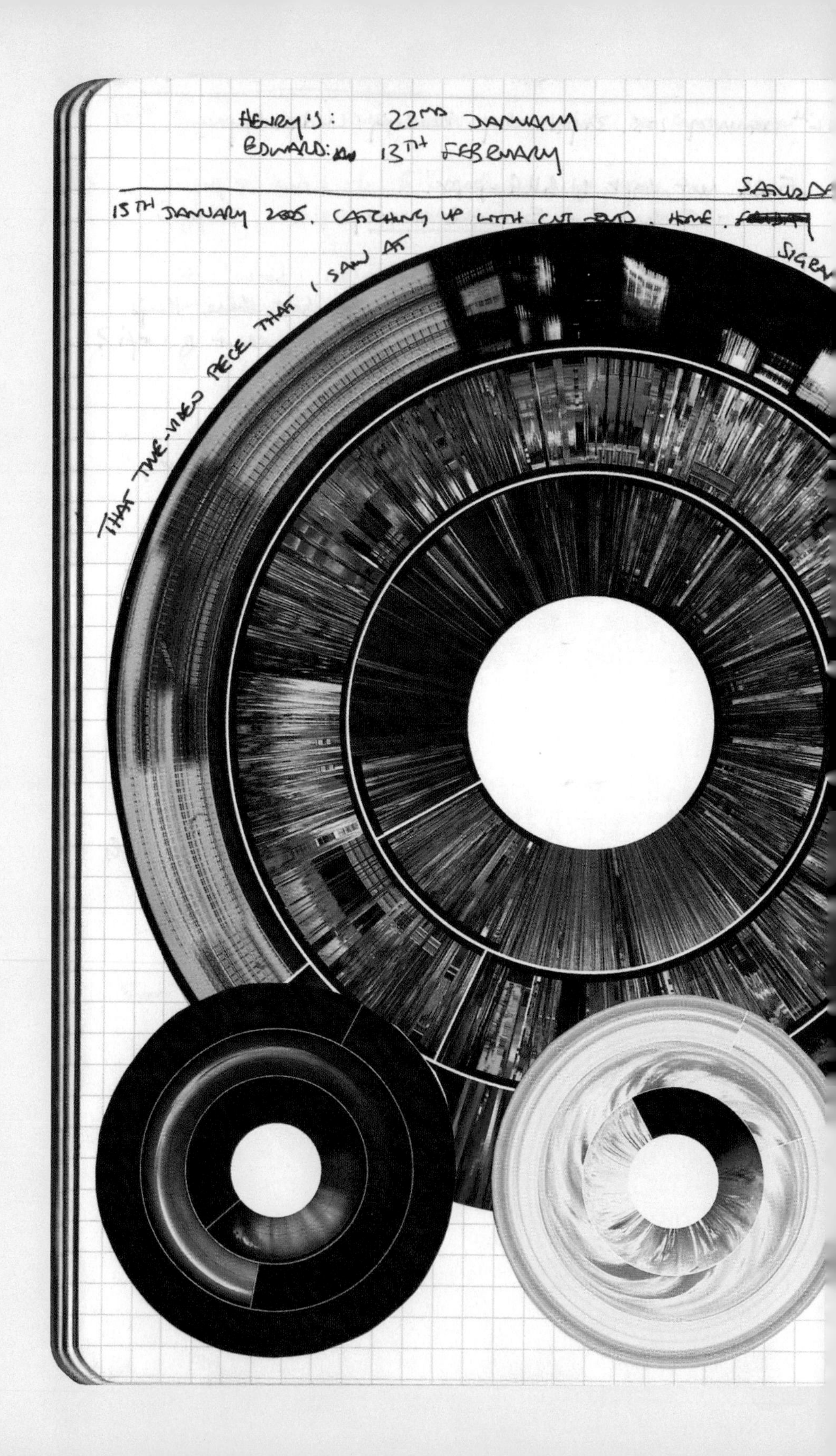

HENRY'S: 22ND JANUARY
EDWARD: 13TH FEBRUARY
15TH JANUARY 2005. CATCHING UP WITH
HOME.
THAT TIME-VIDEO PIECE THAT I SAW AT

L // THE OLYMPIC STADIUM > BEIJING

HERZOG +
DE MEURON
STRIKE AGAIN!

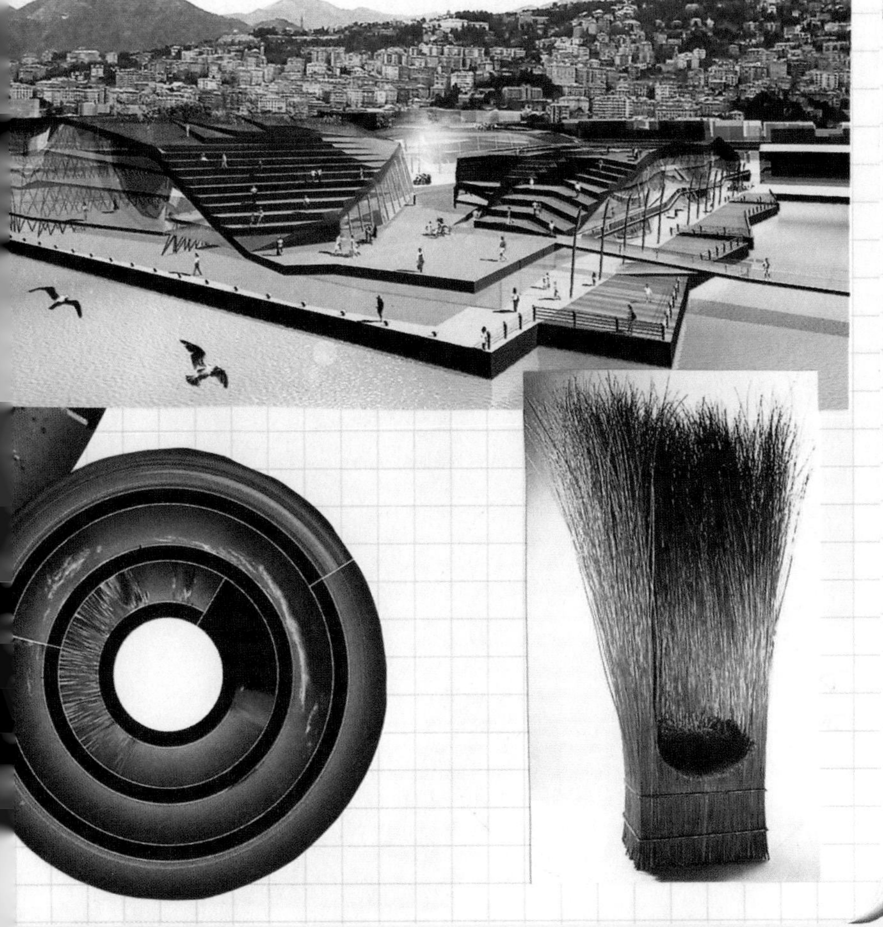

MSX DESIGN PRESENTATION : WIKI'S — WHAT → WIKIPED[IA] → LONDON OF

WORK : VISUALIZATIONS

: STORAGE .

PATENT MEETING :

Big meeting : me: TRENDLINES : Making best guesses

TRENDS BLOG : | 600+ articles | cos referenced

BC themes | alias | watching for

bleeding edge

DESIGN : Primarily focused on rapid prototype

generation. Slow pace towards scena[rio]

presentation

THEMES : Structuring the work.

How these were generated | what th[e]

big groups are

work : rapid sketches to try and draw

conclusions | package of work | d[e]

VISUALIZATIONS

STORAGE

Hand over to Dan : emotional UI + f[ollow]

up.

SMART :

RJ04 KDN

uly 13th 2004. Need ideas for wireless. TUESDAY. READING.

What is wireless?

- mobility
- freedom from cables / constraints
- distance
- work/play location choice.
- access from anywhere.
- information at any time.

wireless is to the internet what the mobile phone is to the landline.

Location. Location. Location.

Freedom. Liberation. Control.
Fun. Connection. Play.

In the park.
In the car.
In the home.

virtual display.

Role

Location 1

Location 2

WORK

SHOPS

HOME

NOVICES — INTERMEDIATE — ADVANCED — EXPERTS

GO THROUGH STEPS

May stop anywhere on this line, which is *fine*!

Physical object interactions

Mouse, keyboard, screen

Physical Software interactions

What things are on screen.
Where things are.
States.

LEARNING THE BASICS

Navigation

Right/left click
Backwards, forwards,
opening, closing,
saving, undoing.

REGIONS

Titlebar, toolbar,
Taskbar

THIS IS A TASKBAR
I'm not a novice
START

WAYS TO TEACH THEM STUFF.
LEARN AS YOU GO
LEARN BY EXAMPLE
HOW DO USERS GET CONFIDENT

Confidence meter.

How do you ask someo
"Is this your first time
using a pc?"
without getting annoying

what about OEMs
overriding everything...?

If you need to know *one* thing its this...

PSST...
Wanna know sm cool?

(shades of the office assistant)

THINGS USERS ARE WORRIED ABOUT.

SHOW ME

s there any way of establishing a users experience?

Ask them → Annoying

Try and guess → unpredictable

Do you need help with a concept?
Do you need help from a friend? → Network of friends.
New User support group

Not knowing the basics
↓
Not knowing how to set something up. → Not online ∴ problem.
↓
Ignoring warnings

Problem 1: figuring out the expertise of someone.
Problem 2: knowing what they need help with.
Problem 3: Building a UI that grows as they go.

This is a taskbar

Taskbar bounces on screen as first element. Introduce each element.

Friend sharing screen.

START | 0000

Simple UIs.

START |

Even simpler UI

13TH APRIL 2005. IN LONDON TO SEE KEVIN SPACEY. WEDNESDAY. LOND

TO DO:
TODAY

TREND PRESENTATION
- TALKING POINTS
  - LIST of PAGES + NOTES
- GIRL FROM IPENEMA
  - SPEAKERS + BEAMS
- ~~DATES FOR MS CATCH UP~~
- ~~LINKS TO LAUNCH ITEMS~~
  - ~~ROCKETS~~
- LOOK UP BETTER EXAMPLES FOR THE FUTURE
- ~~REFERENCE PAGE~~
  - ~~SITE~~
  - ~~THEME~~
  - ~~ATLAS~~
- ~~FACES~~
  - ~~# of Articles DATE OF CONCEPTION~~
  - ~~REMOVE "ARCHITECTURAL analogy"~~
  - ~~FIX NON-CIRCULAR CIRCLES~~
  - ~~22: Add Blackcomb information Atlas~~

TODAY

PAGES + NOTES

- 20: launch site
- 26: WORD DOC of market/lifestyle trends
- 27: Photos of stacking process
- 28: link to flash movies.
- 29: Excel spreadsheet for Cams that matters
- 30: WIRELESS EXEC SCENARIO

3: WELCOME
4: INTRODUCTION + OVERVIEW — PREMISE / PRACTICALITIES / METHODS / QUESTIONS
5:
6: Common tools. Common methods. Common understanding. Personas. Data sharing etc.
7: Need to marry knowledge of customers with knowledge of what's coming next. No process. Ad hoc. Mostly powerded. Technical, Architectural ~~analysis~~. analogy. Good at e
8: Interaction with Trend blog (and others like it): Provide a reason for new stuff. Start building new methods for taking advantage of it. New tools. New tech matters, but so does how people are using it.
9: Missing the boat on new ideas.
10: How many of these will we be late to?
11: TBR. If you need some inspiration this is where you can lo
12 And you can build it into your product development process.

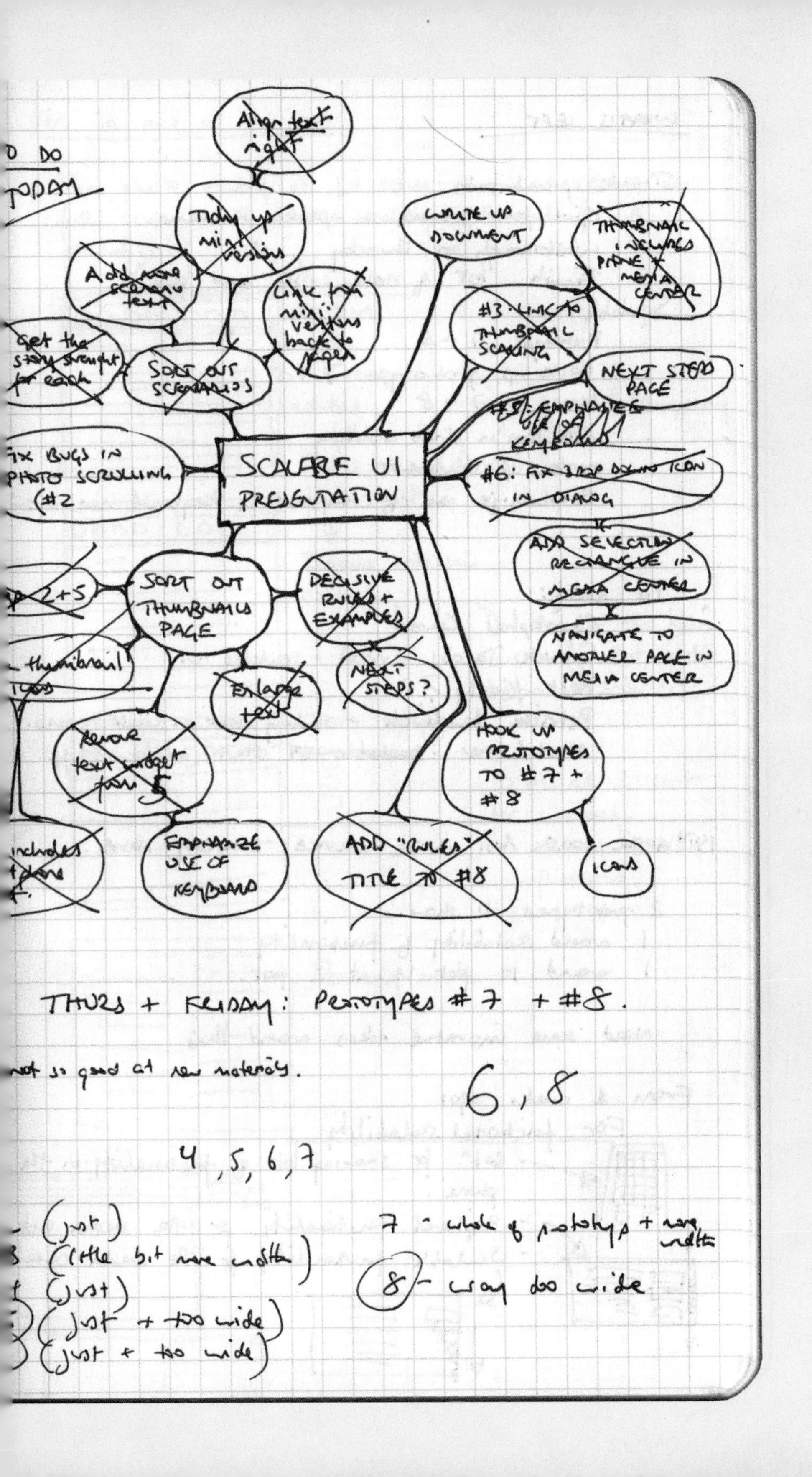
TODAY
SCALABLE UI PRESENTATION
Align text right
Tidy up mini version
Add more scrolling text
Sort out scenarios
Get the story straight for each
WRITE UP DOCUMENT
#3: LINK TO THUMBNAIL SCALING
NEXT STEP PAGE
#6: FIX DROP DOWN ICON IN DIALOG
ADD SELECTION RECTANGLE IN MEDIA CENTER
NAVIGATE TO ANOTHER PAGE IN MEDIA CENTER
FIX BUGS IN PHOTO SCROLLING (#2
SORT OUT THUMBNAILS PAGE
DECISIVE RULES + EXAMPLES
NEXT STEPS?
Enlarge text
HOOK UP PROTOTYPES TO #7 + #8
ICONS
EMPHASIZE USE OF KEYBOARD
ADD "RULES" TITLE TO #8
THURS + FRIDAY: PROTOTYPES #7 + #8.
not so good at new material.
6, 8
4, 5, 6, 7
(just)
(little bit more width)
(just)
(just + too wide)
(just + too wide)
7 - whole of prototype + more width
8 - way too wide

LOCATION	HOME	CAR	COFFEE SHOP	PARK	CAR	OFFICE	MEETING	CORRIDOR	CAR
NETWORK	NEIGHBORHOOD (shared)	HIGHWAY	MUNICIPAL	MUNICIPAL	HIGHWAY	BUSINESS	BUSINESS	BUSINESS	HIGHWAY
PEOPLE	GUY, WOMAN, GIRL, BOY, CAT	GUY	GUY, CLIENT 1, BARISTA, CUST 1	GUY, CLIENT	GUY	GUY	GUY, COL 1, COL 2, COL 3	GUY, COL 4	GUY, COLLEAGUES
DEVICES	clock, fridge, screens,	Phone, screens, music, mike, GPS	Phone, screen, pen	Phone, screen, pen, personal area network	phone, screen, speakers, mike	Screen, phone	Whiteboard, conferencing	Phone	Phone, screens, speakers, mike, GPS
ACTIVITY	Morning ritual	Checking for next job	MEETING A CLIENT	MEETING A CLIENT	Going to work	Catching up	meeting w/ team	Ad-hoc conversation	End of day
	TIME CLOTHES CLEANING Coffee machine that knows your mug. Wireless health check	Finding the stop Getting contact info.	Community board: who's in, what everyone is doing, chess roles etc. Advertise for a cat sitter cool: flows of people display. Indication of free	Creating personal space for the two of them. Sharing through personal area network. "Notes" left in the park	Sharing music with woman at home. Sending a photo home.	Checking up on home w/ camera Some smart card for home, work + club (in place).	Whiteboard defaults to digital photo wall.	Some kind of "lots of small" sensors "thing"?	Traffic jams: reporting back to traffic authorities Cool: map of car flow ahead + weather 20°C 19°C

LOCATION	CLUB	WORKOUT	CAR	HOME	KITCHEN	DINNER	ENTERTAINMENT	BED	SLEEP
NETWORK	MUNICIPAL	MUNICIPAL	HIGHWAY	NEIGHBORHOOD	NEIGHBORHOOD	NEIGHBORHOOD	NEIGHBORHOOD	NEIGHBORHOOD	NEIGHBORHOOD
PEOPLE	GUY, COLS, FRIEND	GUY, TRAINER	GUY	GUY, WOMAN	GUY, WOMAN	GUY, WOMAN, BOY, CAT	GUY, WOMAN, BOY, GIRL	GUY, WOMAN	GUY.
DEVICES		Equipment, headphones, display..		VOIP					
ACTIVITY	Checking in and charging	Exercising	HEADING HOME	UNPACKING	COOKING	EATING	TV/GAMES	RITUAL + READING	SLEEP
	Checking in, personal locker, use phone to open. Touch a poster to "pick up" event details.	Remembers routine; wireless headphones w/ audio streamed from home. Equipment is a server. Networked up to other machines. Shoes talk to equipment, offering real time advice	Streaming audio, phoning ahead, To cars around. Car calls mechanic	Knows it's the guy, some key for car + house. Phone switches network to IP.	Picking recipes based on whats in the fridge (tags). Fridge tells you which shelf stuff is on and where the frozen stuff is.	Lighting. Girl doesn't want to be "found" now (can see she is fine, though) - ambient object.	Streaming shows. 2nd wireless display for navigation. House knows which people are in which room.	Moving TV upstairs.	Heat sensing + regulating. Checking water temperature. Monitoring environment where there are people. List of light bulbs about to go out.

Figure 71: The Design Critique - Group Think or Firing Squad?

There are two extremes of design critique. One is the design jury, generally preceded by a sleepless night and followed by jangled nerves on the part of the person whose work is being reviewed. It probably has its roots in the Spanish inquisition! The other is a respectful and constructive commentary and review of the material under consideration. It can be with peers, subordinates, supervisors or all three. The objective is to find the best idea or understanding, not to score points or show how clever you are. Individuals don't win, the design does. And not being honest about problems or concerns is not an act of friendship. The critique of this sort is the lifeblood of design, and shows design to be one of the best types of team sport.

Credit: Harry Mahler

Design Thinking and Ecology

My experience is not authoritative because it is infallible. It is the basis of authority because it can always be checked in new primary ways. In this way its frequent error or fallibility is always open to correction.
— Carl Rogers, On Becoming a Person

At this point I want to step back and try to tie some things together. I do so at the risk of incorporating some redundancy.

I have been talking about a few things:

The nature of design
The importance of sketching
The plurality and variation of approaches that sketching affords
The capacity for fluid, simple, rich, and spontaneous annotation
The social nature of design, in terms of both people and sketches
The importance of having persistent displays that enable work to "bake in" to the shared consciousness
The fundamental role of critique in design education and process

To me, a designer is someone who integrates these points into their thinking and working life.

Due to its speed and low cost, sketching affords us the luxury of exploring a range of different solutions to any problem that we are confronting. In this process, the sketches do far more than just capture ideas. Through the process of their creation, and the conversations that they facilitate, they are also a key vehicle for their generation and communication.

The ability to spread these alternatives out and view them holistically in various juxtapositions is also fundamental. This is critical to facilitating the ability to step back and jump up a level—to think about, discuss, and hopefully understand the big picture: the meta-view.

Doing so using displays that are both shared and persistent (i.e., where things can be left in place), easily viewable in the background by us and others, enables us to live with our ideas. It gives us time to have second thoughts and conversations. As in our personal lives, with sketches we want time to get to know each other—to become familiar and feel each other out before making a commitment.

Taken together, all this serves the purpose of affording, encouraging, and enriching the conversations that we have with ourselves, our sketches, our colleagues, and our clients. From my perspective, such conversations are the essence of design thinking and the design process.

However, listing ingredients is always risky. Just because I give you flour, milk, yeast, eggs, and an oven does not mean that you know how to make bread at all, much less the multitude of variations that you might find in a good bakery. Nor does it mean that these are the only ingredients that contribute to the final product. So it is with design—except that the recipe (to the extent that there may be such a thing) is likely far more complex and varied than that for bread. And, as with bread, the importance lies as much in the interplay among the ingredients as in the ingredients themselves.

We can explore this by looking a little deeper into the last element in our list, the critique, or crit. This is the one element in the list about which you will be hard pressed to find anything in the literature, despite the key role that I have observed it playing in both design education and practice.

I have frequently asked myself and others why nothing is written about this. The best that I can come up with is this: Critique is so fundamental to design that there is no more need to talk about it in a book than there is to mention that designers also eat breakfast or breathe air.

Actually, I did find one book on critique called *Design Juries* (Anthony 1991). However, its focus was on a very specific type of critique—what can best be described as a portfolio examination. This is frequently an adversarial type of review where a victim (frequently a student) risks having their work torn apart in public by some famous professor or designer. Anthony gave a pretty colourful and accurate portrait of the angst, stress, sleepless nights, devastation, and humiliation that all too frequently accompany this type of criticism.

Although every designer goes through such a thing from time to time, and it may not be a bad thing to give the student some experience in handling it before going out into the world, this is decidedly *not* the nature of critique that I am referring to in this book.

Rather, what I am describing is something that may well be a daily or weekly practice, and something that may be spontaneous rather than scheduled. It is something that is among peers as much as it is between peer and supervisor or peer and subordinate. But perhaps most of all, it is decidedly not where you present the project, but rather a critical component of how you do the project. It may just be between two students who demonstrate that the true mark of a friend and colleague is to be honest and respectful when commenting about each other's work. It is a testament to my axiom:

> **It is better to have your preliminary work critiqued by your colleagues while there is still time to do something about it—no matter how difficult the criticism might be—than to have the finished project torn apart by strangers in public.**

But of course, this works only if the conversation is fair and respectful, with the purpose of finding truth (or the best design). That, rather than scoring points, is the only true way to win, and the best part is, all participants get to win—if they play fair.

Here is how I see this form of critique fitting in with the other elements in the list at the start of this section. The essence derives from a statement made to me by the Canadian print-maker and teacher, Richard Sewell

> I can't critique just one thing.

In employing sketching to come up with multiple solutions, the designer is not just offering variety. They are also making the following important, albeit tacit, declarations: "I have not committed to any of these ideas. I don't know which is best or if any is what we want. It is too early to decide."

It says that the person who created the sketches is not invested in any of the alternatives being evaluated. And when I say "invested" I mean that emotionally, as well as in terms of time. As sketches, the concepts are disposable, and so therefore may be the concepts that they represent. And all this opens the door that enables the conversation that follows to be about the work, the ideas that it represents, and the objectives being pursued, rather than about the person.

Without sketches the process won't work. Without multiple solutions to any question, the process is highly vulnerable. Without the ability to see all the work at once, spread out, relationships will be missed, and the conversation and subsequent designs will suffer.

And, like any contact sport, if the players don't know, understand, respect, and follow the rules of the game, all bets are off. The thing that designers need to keep in mind is that most nondesigners don't even know that the game exists. Furthermore, many (most?) designers are not all that aware that this game, which is part and parcel of their daily life—part of their whole ethos—is largely foreign to those outside the profession, and it is one of the things that most distinguishes them as a culture and a profession.

In playing the game, it is important to keep in mind that the intended outcome is to make a decision or choice to move the design forward. This means that the resulting conversation involves making an assessment. This brings up its own set of issues, which also have cultural implications, something discussed by Snodgrass and Coyne (2006):

> A common lament is that the assessment of design is insufficiently objective, that rigorous standards of assessment are lacking in studio critiques. Interpretation theory, however, shows objectivity to be an impossibility. On the other hand, if we concede that assessment is subjective, then it presents as mere opinion, subject to the vagaries and whims of the assessor. If objectivity is a chimera, and subjectivity anarchic, how is fair assessment possible? The assessment process seems to be mysterious and inexplicable. (p. 119)

A bit further on, they add:

> Design evaluation, therefore, is not free or haphazard; it is limited by the tacit understandings of the hermeneutic community of designers; and it is not haphazard because the assessor has acquired a tacit understanding of design value and how it is assessed, a complex set of tacit norms, processes, criteria and procedural rules, forming part of a practical know-how. From the time of their first 'crit', design students are absorbing design values and learning how the assessment process works; by the time they graduate, this learning has become tacit understanding, something that every practitioner implicitly understands more or less well.
>
> An absence of defined criteria and procedural rules does not, therefore, give free rein to merely individual responses, since these have already been structured within the framework of what is taken as significant and valid by the design community. An absence of objectivity does not result in uncontrolled license, since the assessor is conforming to unspoken rules that, more or less unconsciously, constrain interpretation and evaluation. If not so constrained, the assessor would not be a member of the hermeneutical community, and would therefore have no authority to act as an assessor. (p.123)

These excerpts do a good job of reflecting the cultural foundation of design practice and cultivation of values. In the sense that I can see more than a few of my friends from the engineering and financial professions cringing when they read these passages (seeing them as vague, soft, fuzzy, wishy-washy, etc.), they also illustrate some of the cultural differences that distinguish the design profession: something that is good for all participants in the product development process to be aware of, especially since so much of it is tacit and vague. These differences are easy to miss, and therefore can easily lead to otherwise avoidable misunderstandings.

I believe that we can come up with an approach to design that meets the new demands imposed by the emerging types of products and technologies that we are being asked to design. Furthermore, I believe that we can do so in a way that respects and takes advantage of the unique skills and cultures of all the disciplines that are required to succeed in this. My topic is design, so that is where my focus is. But if we ever lose sight of the interdependence of the various skills making up the whole, we will most likely fail.

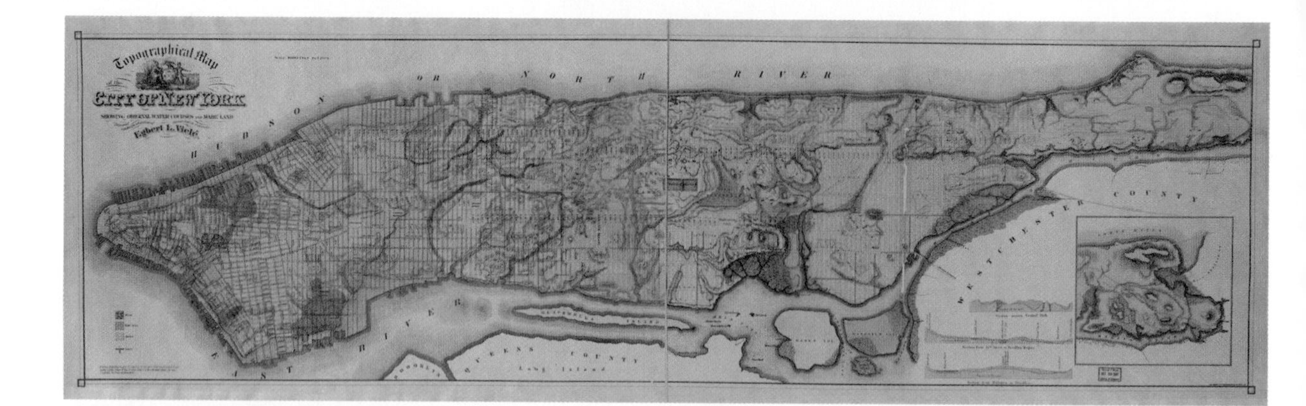

Figure 72: Long Term Thinking

The Commissioner's Grid of Manhattan Island was laid out in 1807. Officials at the time envisioned an adaptable design that is capable of evolving over time while maintaining a basic organizational structure. Today, we see Manhattan as the iconic metropolis. In designing cities, urban planners have to develop scenarios and strategies for designs that last hundreds of years.

Sources: Top - Library of Congress

Bottom - © 2007 Microsoft Corporation / Virtual Earth™

The Second Worst Thing That Can Happen

...failure, though painful, is better than frustrated longing.
—Earl Denman

I give a lot of talks. One of my motivations is that it gives me a way to test ideas that I am working on. In a way, talks are the preliminary sketches for what I eventually write.

In 2003 I tried out some of the material intended for this book. The talk was at a usability conference. I thought that it went pretty well, and this was reinforced by the audience reaction as well as the comments that I received afterward.

That is, until the next day. Someone came up to me; after expressing how much he enjoyed the talk, he went on to tell me how lucky I was. Now, although I do consider myself quite lucky, I couldn't help but ask why, in particular, he thought so. His response led me to the bleak realization that my talk had been a failure.

What he said was, "You are lucky because you get to work on such great projects, and to think so far ahead into the future. In my job, I have to focus on my current project, so I can't afford the luxury of going so far out."

What this said to me was that I had completely failed to communicate the relevance of what I was saying to his day-to-day activities. As interesting as my examples may have been, they were irrelevant to him, and therefore little better than entertainment.

The good news is that the sketch performed its function—I learned something. And so I dedicate this section of this book to that audience, in the hope that it makes up for what I failed to deliver in the talk: namely, why it is important to look five years down the road, even when your immediate concern has to do with the next quarter or two.

We are about to go through some examples of sketches, prototypes, and models. Some, perhaps even most of them, may appear to be kind of "far out" in concept or time. But I want to argue that, if anything, they are too conservative. Not only are they not too far out, but responsible design demands that you think that far into the future! It is not only critical that the designer understand this, but that he or she be able to explain this to management. Here is why.

I assume that if you are reading this, you are in some way involved in the design or production of interactive technologies. In light of this, consider the following:

The worst thing that can happen with a new product is that it is a failure. The second worst thing that can happen is that it is a huge success.

Product	Company	Year	no. of Years
1-2-3 (VisiCalc)	IBM (Lotus) (Personal Software / VisiCorp)	1979	28
Director (VideoWorks)	Adobe (Macromedia) (Macromind)	1985	22
Studio (Alias/1)	Autodesk (Alias Research)	1985	22
Illustrator	Adobe	1987	20
Photoshop	Adobe	1990	17
Maya	Autodesk (Alias|Wavefront)	1998	9

Table 1: Successful Products Breed Legacy Code

The table lists a number of products and the year of their release. All of these products are still on the market. Where the name of the product or the company selling it has changed, the original is in parentheses. The decisions that the original developers made still have impact on the programmers who must maintain and add features to the code, on the customers, and consequently on the companies that sell them. The dates do not tell the full story, as these are the ship dates. Work on Maya, for example, began in 1994. Product designers cannot afford to think only 1-2 years out. Their product may not even be shipping by then.

What? How can success be a bad thing? The answer is easy. The more successful a product, the more of them are out there and the longer your company, clients, and successors have to live with the design and architectural decisions that you made. Furthermore, this is true for all kinds of products, from automobiles to buildings, not just computer-based systems.

If you did not do your best to anticipate the technical, social, and commercial ecology within which your product must live throughout its entire life, you have not done your job.

Without appropriate design, yesterday's success is tomorrow's straightjacket, since today's great applications are tomorrow's legacy systems.

But if we have to take the future into account as part of the design process, how far do we have to go? Well, the answer depends on how successful you want your product to be. Table 1 helps shed some light on this.

It lists a number of products along with their original release dates and the name of the company that released them, and, if different, sells them today. The oldest was released in 1979, and the newest in 1998. Since they are all still on the market, it is reasonable to consider them all successful.

Now look at Figure 73. It shows the data from Table 1 in a different way. It is intended to represent a "mirror" that reflects the release date of the products as far into the future as the release was in the past, relative to 2007. For example, relative to 2007, the 1998 release date of Maya is as far in the past as 2016 is in the future.

So, if Maya were released 2007, and had at least the same product life, the decisions that we made in designing it now would still have to be current in 2016. Yet, how many people designing a product today seriously analyse what the product and technological ecology will be nine years into the future? To emphasize the point, keep in mind that while Maya was released in 1998, work started four years earlier, which brings our nine-year look-ahead up to 13 years. And Maya is the newest product in our examples. What if the product you are designing had the legs of Lotus 1-2-3? Now most people say that you cannot predict the future, much less five years out. They use this as an excuse for not making the effort, or even contemplating it. I believe that this reflects a lack of training, technique, or responsibility on the part of design or management.

I need to qualify that. Of course we can't accurately predict the future.

In all of this, I am not ignoring the fact that products get updated and that often the new versions incorporate changes that could not have been anticipated in the original design. There is always room for improvement.

What I am trying to say is that bad decisions made in the initial product release often pose a significant impediment to making such inevitable changes and improvements later on. Furthermore, many, if not most, of these bad decisions could be avoided if an appropriate process had been adopted in the first place.

So, no, we can't predict the future and we don't have to get it completely right. But we sure do need to make our best effort. The good news is that we are not flying

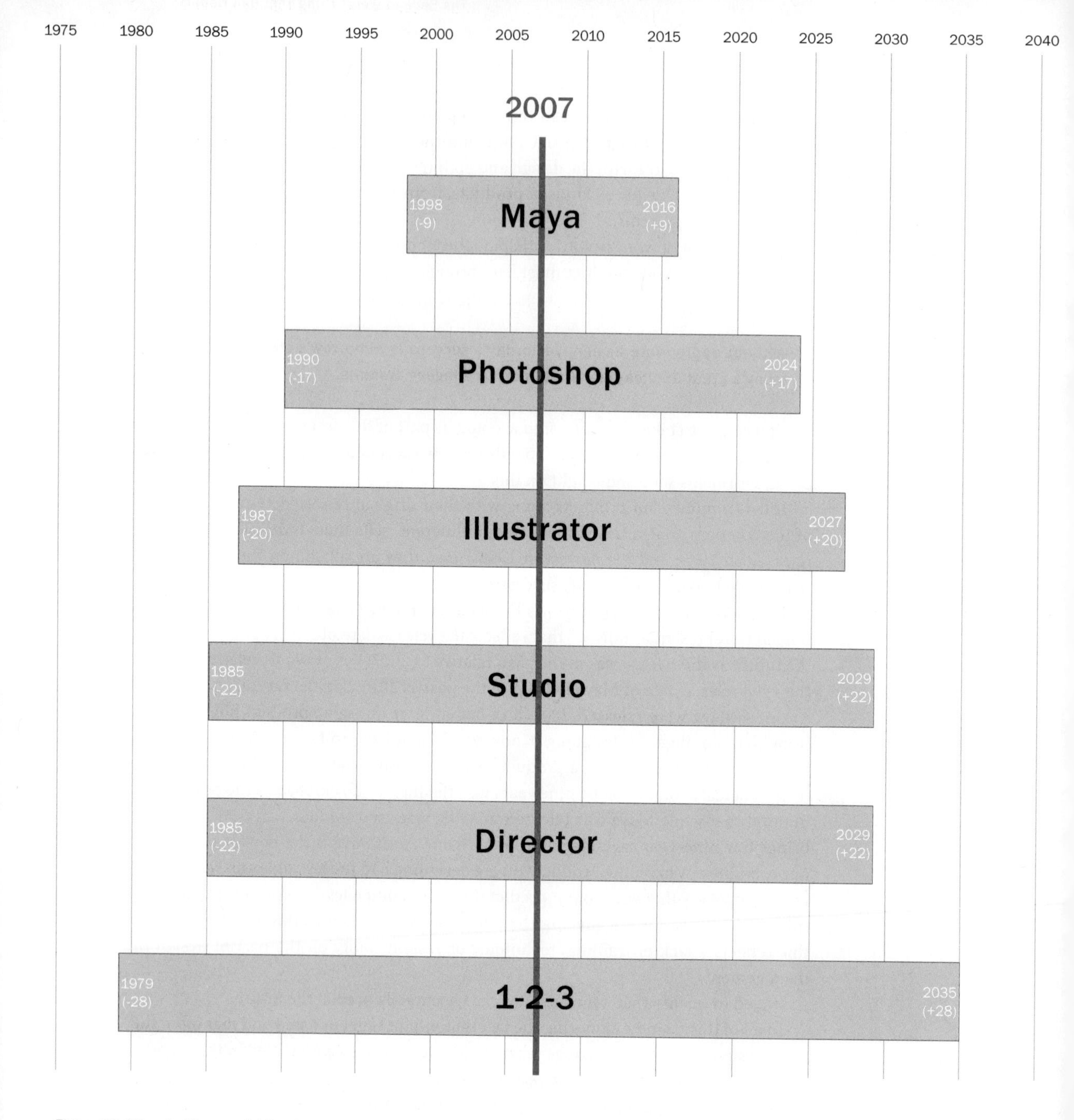

Figure 73: Mirroring Successful Products

The image is intended to represent a "mirror" which reflects the release date of the products listed in Table 3 as far into the future as the release was in the past, relative to 2007.

entirely in the dark. There are techniques that we can use and basic data that we can apply. The following examples relate to the computer industry, but I am sure that one could come up with a comparable list for almost any type of product:

- Moore's law tells us, more or less, that computational power for a given price will double every 18 months.
- A variant on this tells us that the amount of network bandwidth that we can get for a given price is doubling at twice the rate of Moore's Law (i.e., about every 9 months).
- We know that large format displays are becoming less expensive, and will increasingly be embedded into the ecology of home, work, and school.
- We know that there will be a parallel growth in the power and ubiquity of portable wireless devices.
- We have techniques (many to be discussed later) that enable us to take these considerations into account during the design process.

Perhaps most of all:

- We should not count on any *deus ex machina*. We should not expect any magic bullets. It is highly unlikely that there will be any technology that we don't know about today that will have a major impact on things over the next 10 to 20 years.

This last point is perhaps the most controversial, so I will spend a bit of time on it. The fastest way to summarize it is by means of a quote of William Gibson from an NPR interview, Nov. 30, 1999:

> ... the future is already here. It's just not very evenly distributed..

This is a variation of Ecclesiastes 1, 7-9, "And there is nothing new under the sun." Both have an element that rings true for me. For example, I can think of nothing that is shaping our current technological environment that I was not personally familiar with 10 to 20 years ago. Furthermore, I suspect that that is true for anyone else who has been in the industry this long.

Let's just take two devices that we are all familiar with as examples: the mouse and the CD.

The first mouse was built by Doug Engelbart and William English in 1964. I saw my first mouse seven years later in 1971. Mice were in use at Xerox PARC by 1973 with the first Alto computer. The first commercially available computer that came with a mouse was, I believe, the 3 Rivers Systems PERQ, which was released in 1980. The Macintosh was released with a mouse in 1984, making them more widely known, but it was not until the release of Microsoft's Windows 95 operating system that they became ubiquitous. That is to say, it took 30 years for something whose benefits were plainly visible to make the transition from first demonstration to broad usage!

The idea for the CD came to its inventor, James Russell, around 1965. He patented

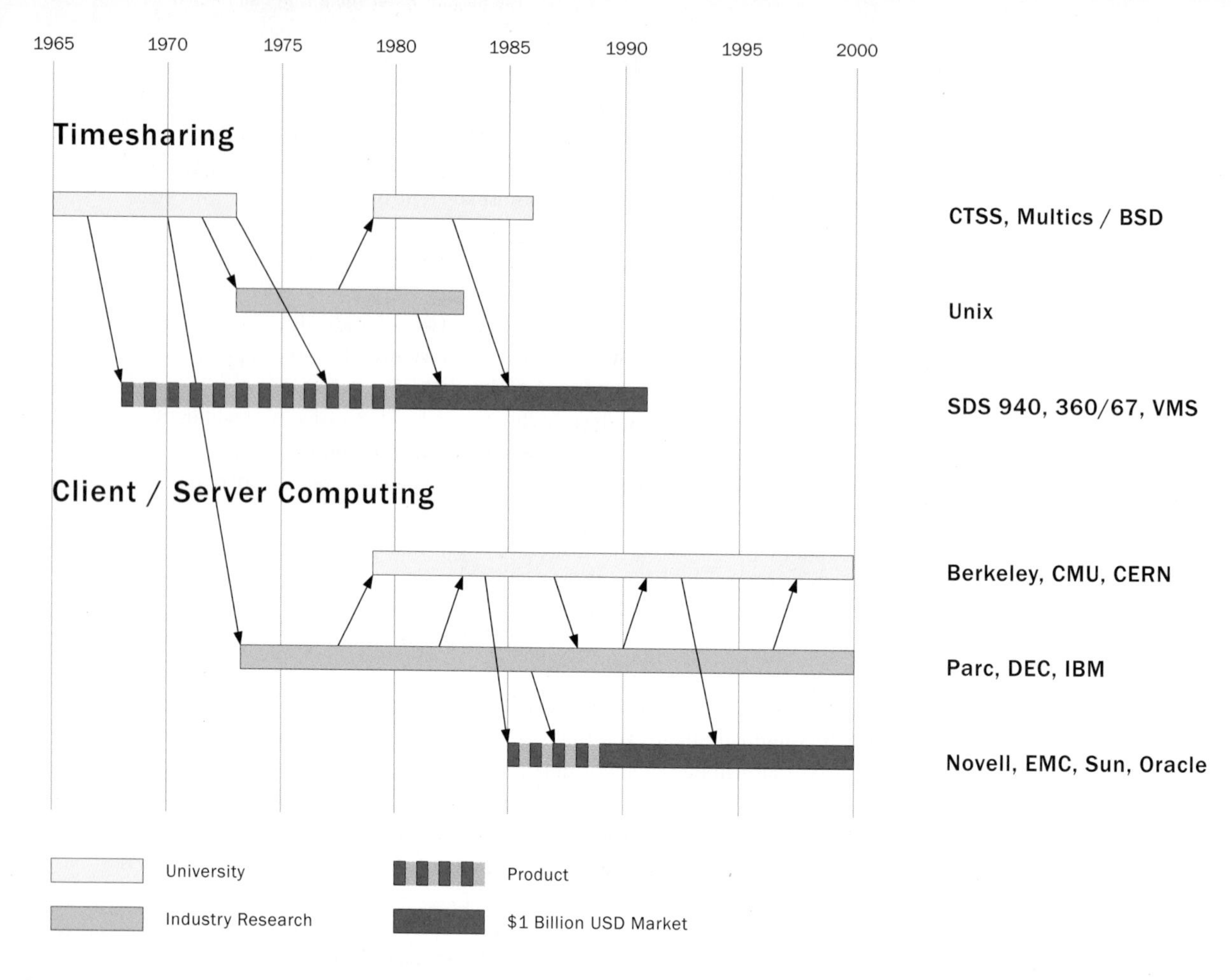

Figure 74: The Evolution of Client-Server Technology
The evolution of client-server technology is shown from its inception in the laboratory to the point where it became a $1B industry.
Figure: Butler Lampson based on data from the Computer Science and Telecommunications Board of the National Research Council, 2003

the basic concepts in 1970. Philips released its first production CD players in April 1982, and it took until 1990 for the industry to reach $1 billion—that is, 25 years after the initial concept and 20 years after the first prototype.

Now if you think that this 20-year period is surprisingly long, then just think about Alexander Bain and Frederick Collier Bakewell. Bain was a Scotsman who invented the first fax device. Bakewell, a British physicist, made the world's first successful fax transmission—between Seymore Street in London and Slough. The catch is that Bain filed his patent in 1843, and Bakewell made his transmission in 1847, about 140 years before facsimile machines took off in the market place (Huurdeman 2003).

Alternatively, you might look at the case of Friedrich Reinitzer. He is an Austrian chemist who discovered liquid crystals. What will surprise most, besides the fact that he did so in the process of trying to understand cholesterol, is that he did so in 1888—about a hundred years before LCDs became ubiquitous on our wrist watches, laptops, televisions, and so on.

Compared to the examples of the fax machine and the liquid crystal, a 20-year interval between invention and widespread commercial exploitation seems almost instant. But nevertheless, in absolute terms, 20 years still seems like a long time, and certainly, there can be huge market advantages to those who can figure out how to cut this number down. In fact, a great deal of this book is about processes and techniques that can help you do so while minimizing the risk. One take-away lesson in all this is the following:

> **Innovation is not primarily about alchemy. Rather than trying to make gold, it has far more to do with learning how to find it, mine it, refine it, and then work it into something of value. If Gibson is right, then the innovator is likely best to trade in his or her alchemist's chemistry set for some prospecting tools, and learn about geology, mining, smelting, design, goldsmithing, sales, and marketing, so to speak.**

The other has to do with the goldsmithing part:

> **Hearken back to our earlier discussion of the iPod. It is generally not the underlying technology itself, but its deployment and associated value proposition, that brings us surprise and delight, as well as generates wealth for those who executed well on their insights.**

But I suspect that these arguments are still too ad hoc, anecdotal, or "touchy-feely" for many readers. They play well to the already converted, and make for entertaining conversation, but they will never convince the skeptic. So let's look at some harder data that make the same point.

In September of 2003, I was invited to speak at a meeting of the Computer Science and Telecommunications Board of the National Research Council at the National Academy of Science in Washington. While I was there, I had the pleasure to attend a presentation given by Butler Lampson on a report that the committee had just completed, called *Innovation in Information Technology* (Computer Science and Telecom-

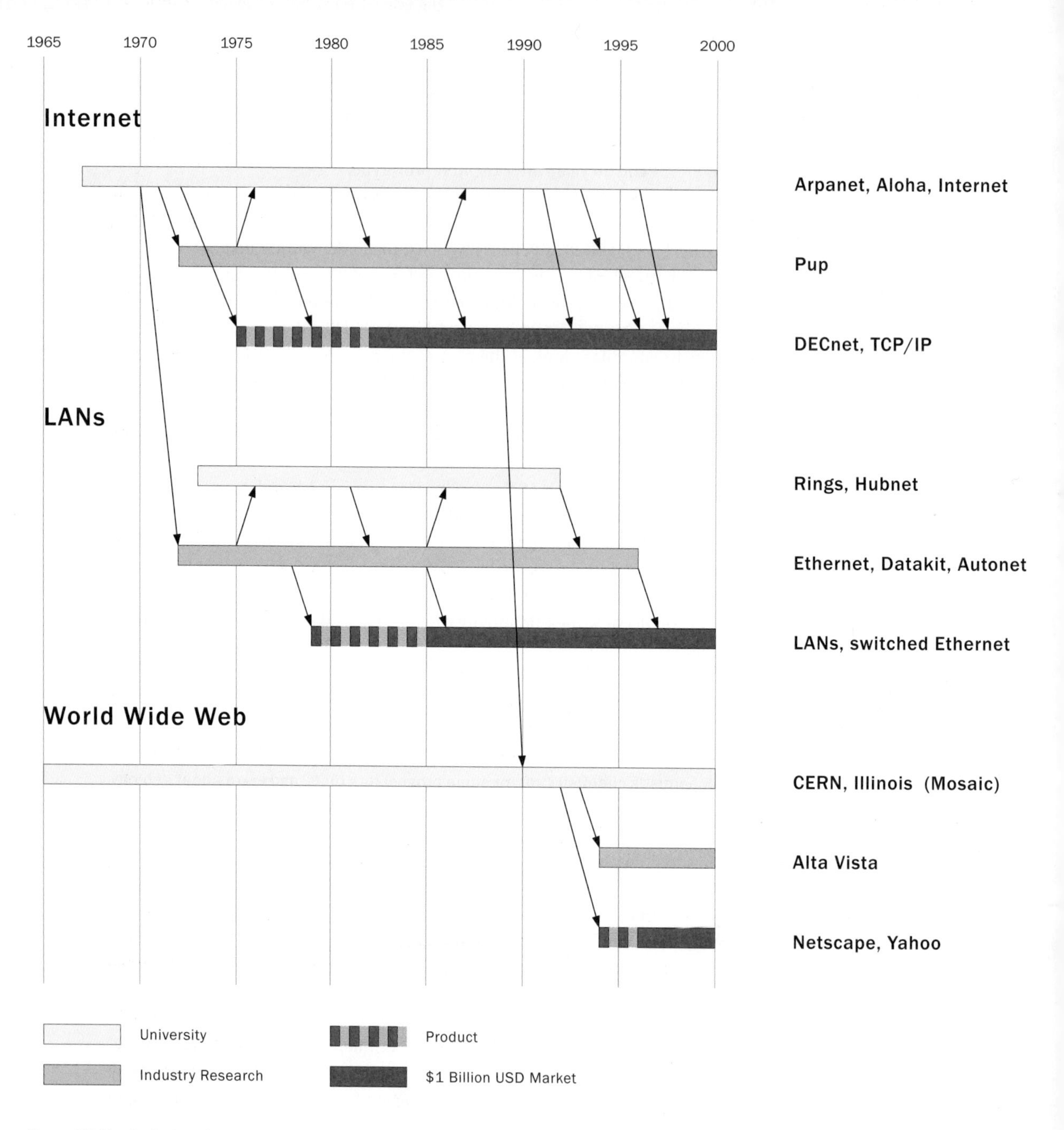

Figure 75: The Evolution of Network Technology

The evolution of network technology is shown from its inception in the laboratory to the point where it became a $1B industry. For key to the figure, see Figure 66.

Source: Butler Lampson based on data from, Computer Science and Telecommunications Board of the National Research Council, 2003 munications Board of the National Research Council, 2003).

The report studied the genesis of a large number of the telecommunications and computer technologies that are having such a large impact on our lives today. These included graphical user interfaces, portable communications, relational databases, and so on. For each technology examined, the study looked at the duration and path that the technology took between initial discovery to the point where it reached being a $1B industry.

Two such examples that come from Butler's talk are shown in Figures 74 and 75. These show the evolution of client–server and network technologies, respectively. If you read the report, what you will see is that the data clearly support the case that I made. That is, the time for all these technologies to go from inception to maturity was about 15 to 20 years.

Just to make the point even stronger, my friend and one-time student, Brad Myers, did a comparable study where he looked at the evolution of some of the key user interface technologies that we are using today (Myers 1998). A modified representation of these is reproduced in Figure 76.

Here we see the same thing. The period from concept to product is about 20 years in the industry in general, and in user interface technologies, specifically.

So much for fast-changing technology!

There is nothing that I can see to suggest that this is going to change. The bad news is that it takes so long for things to take hold. The good news is, we have an "early warning system" in the scientific literature that helps us make educated guesses about scenarios to consider for the future.

If history is any indication, we should assume that any technology that is going to have a significant impact over the next 10 years is already 10 years old!

I really feel that it is an obligation, not a luxury, to consider the future. And in the coming pages I hope to convince you that there are reasonable tools, science, and techniques to enable you to do so.

Likewise, I hope that having been confronted with these data, we might take a bit more care in rationalizing the balance between our investment in inventing new technologies, compared with innovation around accelerating the successful commercial adoption of technologies and concepts that are already known.

Finally, recent research (Della Vigna, Stefan, & Pollet, Joshua Matthew 2005) suggests that even when the trends are known, investors do not tend to look beyond a five-year window into the future. Understanding this aspect of behaviour gives us one more insight into how one can break from the norm and innovate: break with the status quo and look further ahead.

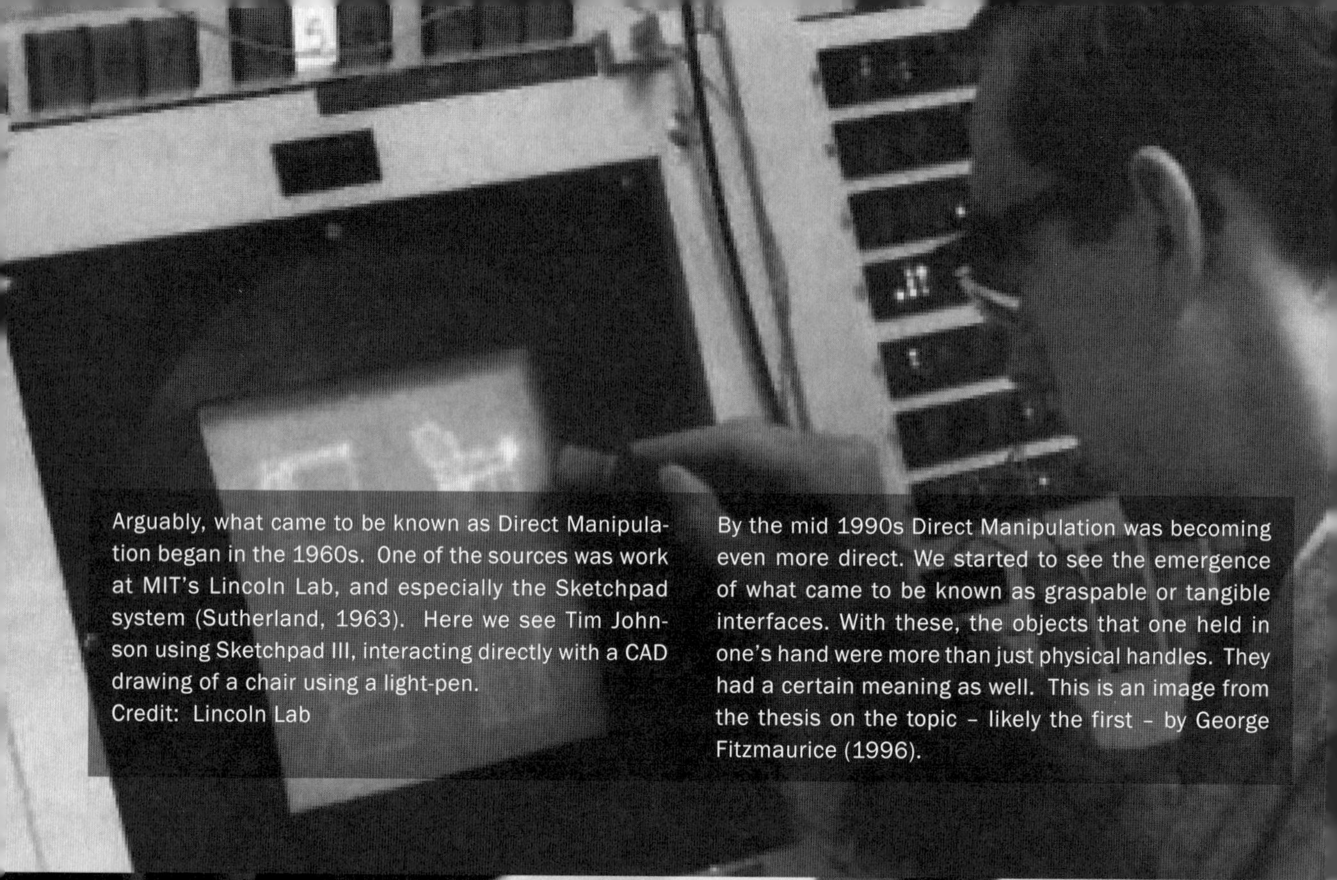

Arguably, what came to be known as Direct Manipulation began in the 1960s. One of the sources was work at MIT's Lincoln Lab, and especially the Sketchpad system (Sutherland, 1963). Here we see Tim Johnson using Sketchpad III, interacting directly with a CAD drawing of a chair using a light-pen.
Credit: Lincoln Lab

By the mid 1990s Direct Manipulation was becoming even more direct. We started to see the emergence of what came to be known as graspable or tangible interfaces. With these, the objects that one held in one's hand were more than just physical handles. They had a certain meaning as well. This is an image from the thesis on the topic – likely the first – by George Fitzmaurice (1996).

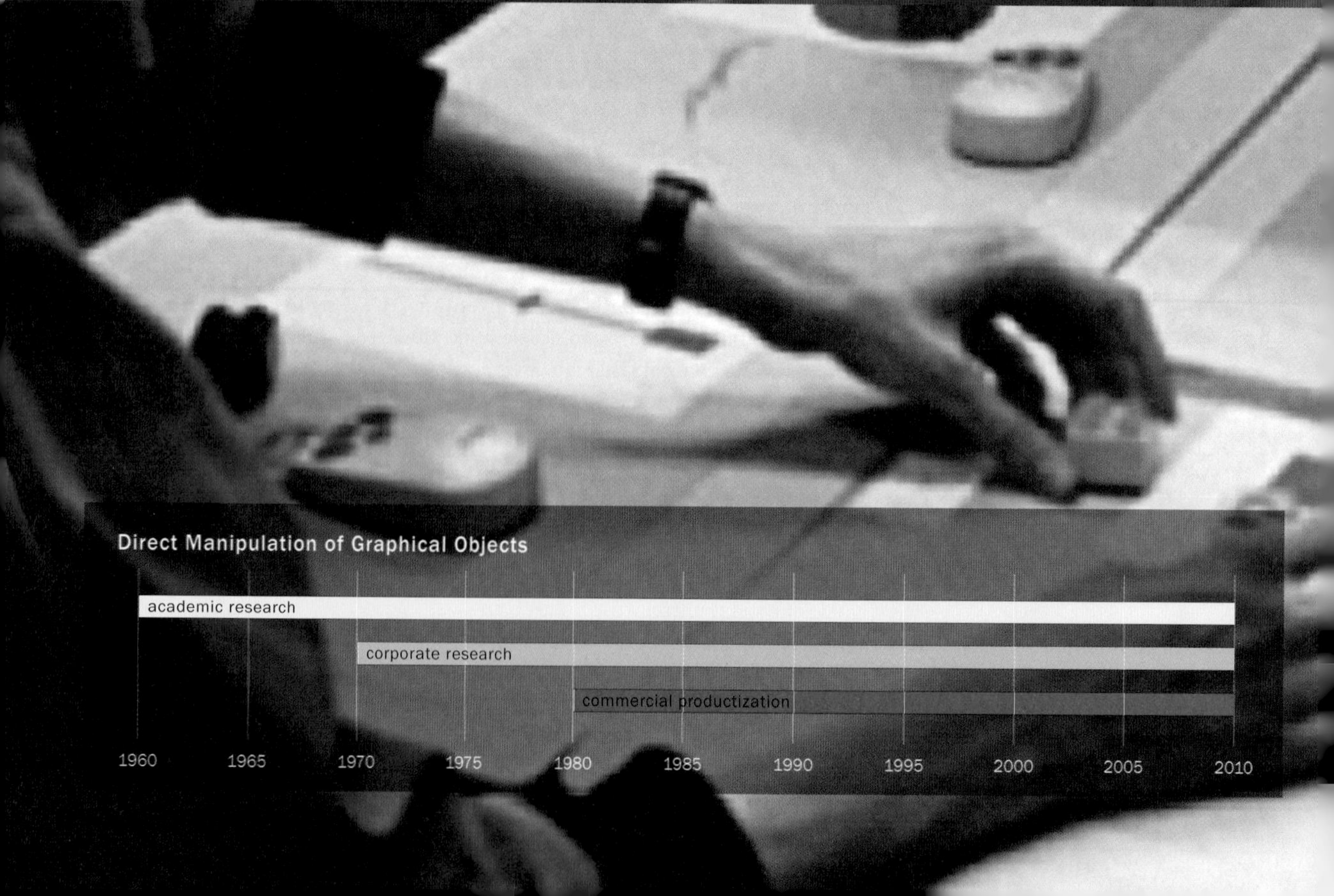

Direct Manipulation of Graphical Objects

academic research

corporate research

commercial productization

1960 1965 1970 1975 1980 1985 1990 1995 2000 2005 2010

The first mouse that I ever saw or used. It was a clone of Engelbart and English's original mouse. It was used in two systems, both studies in human computer interaction. One was a very early system for computer animation, and the other for computer music. It was while composing a film sound-track using the latter, in 1971, that I encountered the device.
Photo: National Research Council of Canada

A small sample of the various mice and would-be mice that are part of my collection of input devices. Those illustrated range from an early mouse from Xerox PARC, to one of the first mice from an Apple Macintosh, to a 3D mouse with a 3D trackball mounted on its back. There is also a "Head Mouse" that tracked head movement, as well as a "Foot mouse."
Photos: Ilene Solomon

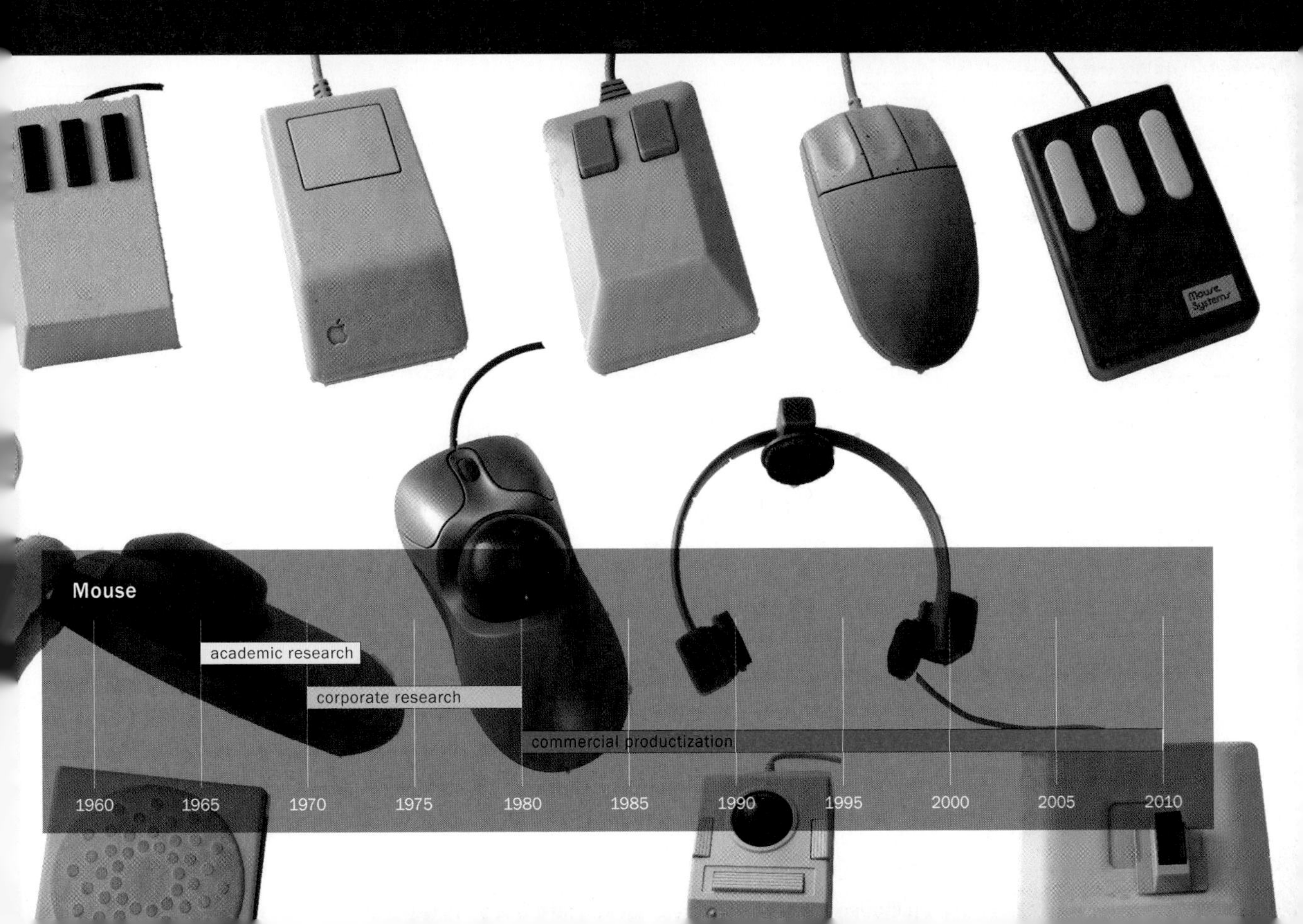

TJ-2: Type Justifying Program

General Description

TJ-2 accepts English text from typewriter or reader, and reproduces it at any line length via typewriter and/or punch. So much as possible both left and right margins are aligned in the output. To accomplish this the program doubles some of the spaces in the output line, and may hyphenate words, getting hyphenation data from its dictionary or from the operator via the display.

immediately precede a tab. Any number of adjacent spaces is treated as a single space. Backspace and the unused Concise codes are illegal. The program simulates tabs by spacing to

This is the first page of the description of arguably the first page layout / text editing application, TJ-2. It was developed by Peter Sampson on the PDP1 computer at MIT in 1963. It supported right justification, word-wrapping, automatic hyphenation, centring, indentation and page breaks. This document, written by Sampson, was created using the program.
Image: Daniel P. B. Smit

From one extreme to the other. This is an example of one of the chapters of this book being revised in Microsoft's Word 2007. It supports different fonts and type-faces, as one would expect. It also supports rich annotation, mark-up, correction, and versioning. Perhaps the most interesting thing about this new version of word is that the company revamped the user interface.

Sketching1.2_Dec10 - Microsoft Word

File Edit View Insert Format Tools Table Window Help Adobe PDF Acrobat Comments Type a question for help

this is not sustainable in the long term. As a rule, companies will eventually have to release truly new products, or see their business shrink to the point where it is likely no longer of interest to investors, and perhaps not even viable.

To understand why, let's narrow our discussion and focus on software companies.

With software products, each n+1 release must contain sufficient incremental improvement to contribute to two objectives. First, it should help motivate those who have not already purchased the product to do so. Second, and perhaps more

The problem with relying on n+1 products is that the cost of achieving an improvement that is greater than or equal to this Value Threshold increases with each release. Furthermore, my experience suggests that the higher the release number, n, the higher

Text Editing

academic research

corporate research

commercial productization

1960 1965 1970 1975 1980 1985 1990 1995 2000 2005 2010

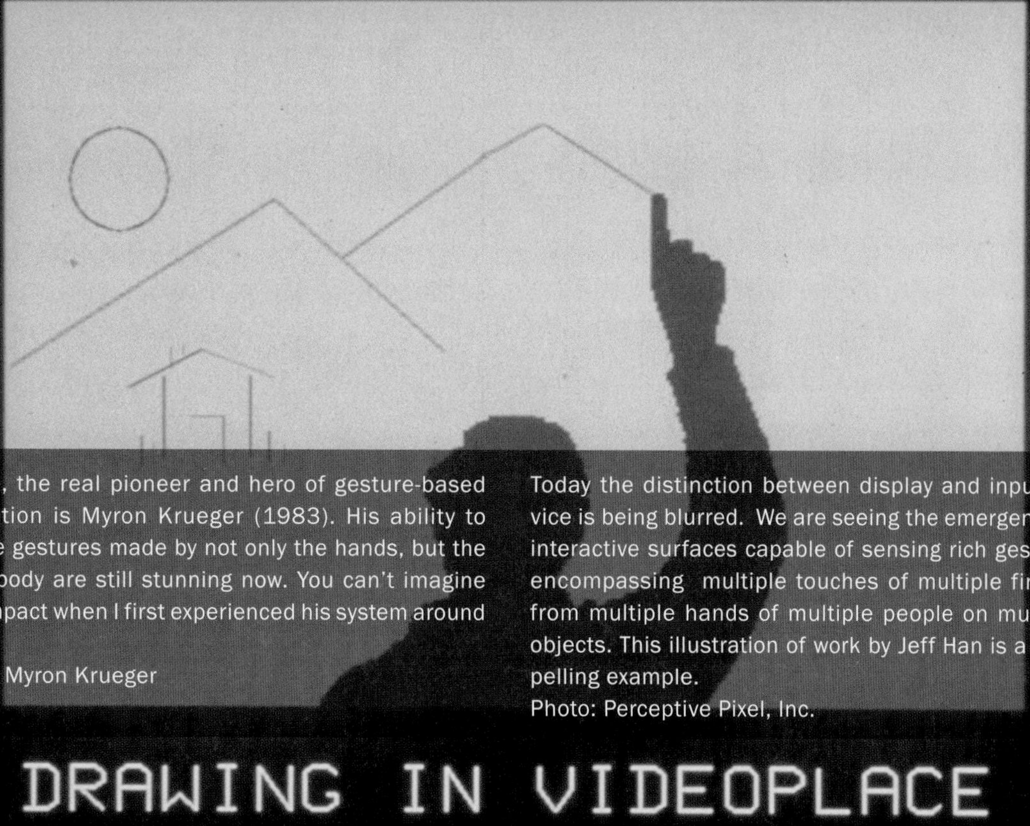

For me, the real pioneer and hero of gesture-based interaction is Myron Krueger (1983). His ability to capture gestures made by not only the hands, but the whole body are still stunning now. You can't imagine their impact when I first experienced his system around 1984!
Photo: Myron Krueger

Today the distinction between display and input device is being blurred. We are seeing the emergence of interactive surfaces capable of sensing rich gestures encompassing multiple touches of multiple fingers from multiple hands of multiple people on multiple objects. This illustration of work by Jeff Han is a compelling example.
Photo: Perceptive Pixel, Inc.

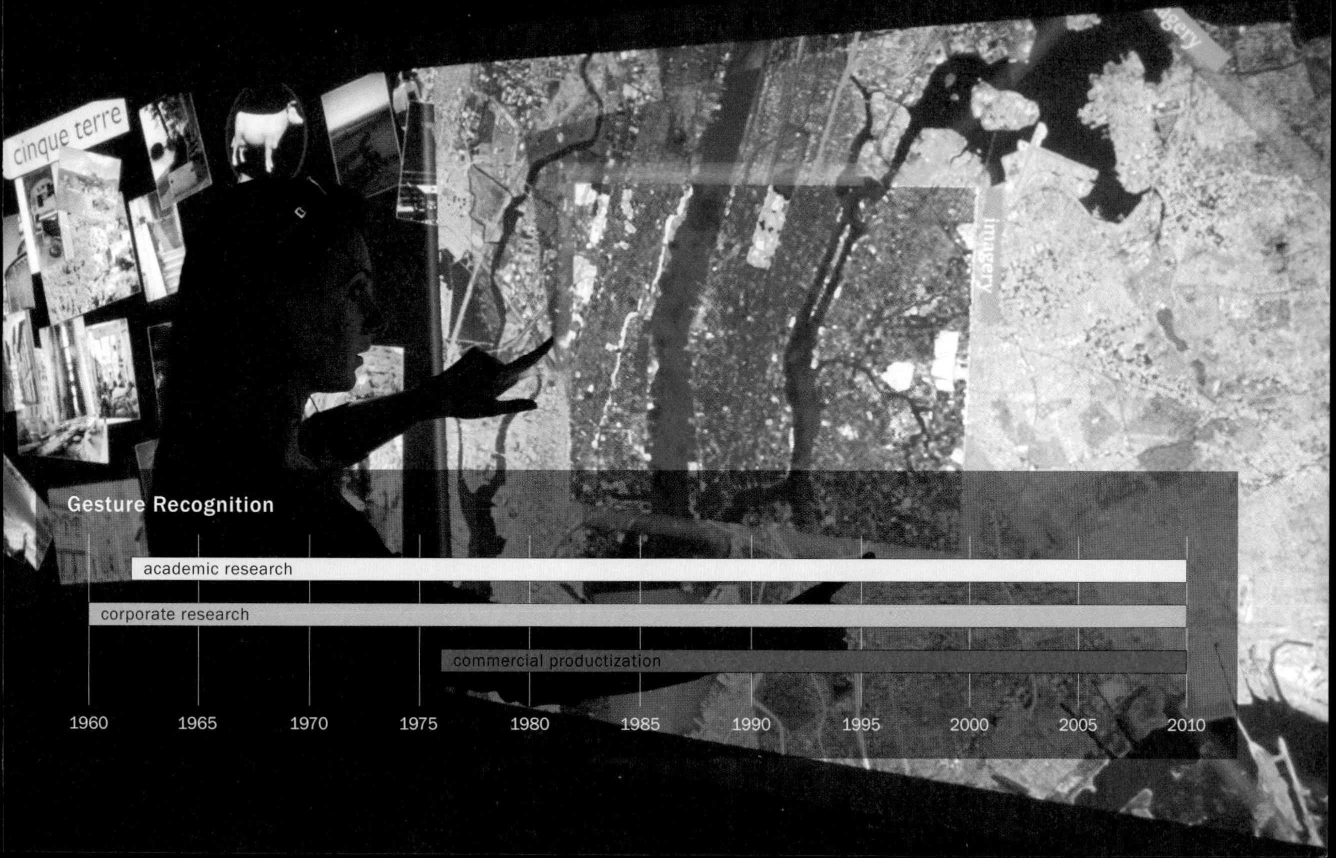

DRAWING IN VIDEOPLACE

cinque terre

imagery

Gesture Recognition

academic research

corporate research

commercial productization

1960 1965 1970 1975 1980 1985 1990 1995 2000 2005 2010

Figure 77: A River Runs Through It

In fly fishing, how do you know where to cast if you have no sense of where the fish might be? Likewise, how does a company know where to allocate resources, or do its employees know how to direct their creativity, if it has no shared view of the future?

Source: Sony Picture Classic

A River Runs Through It

The best way to predict the future is to invent it.
—Alan Kay

It is interesting to note the reluctance of product designers to deal with a future that extends much further than the company's current revenue forecast. Design is about the future. I want to close this section by taking some of the burden off of the designer, and sharing it among the rest of the organization.

Innovation and creativity cannot be delegated to some particular part of the organization, such as the design department. Doing so can be the death knell of a company, regardless how creative that particular department may be. In a talk in San Jose in 2002, Alan Kay said something that really struck me:

> It takes almost as much creativity to understand a good idea, as to have it in the first place.

From the perspective of management, the take-away lesson is that you must foster an overall culture of creativity within your organization—one that not only has good ideas, but also understands them, is receptive to them, and knows what to do with them. Otherwise, you will lose both the benefit of the ideas that you paid for as well as your most creative people. Good ideas are not sufficient and innovation and creativity cannot be compartmentalized.

The significance of this is further emphasized in the following, which is my corollary to Alan's statement:

> **It takes even more creativity to productize a good idea than it does to have the idea in the first place.**

I make these points now because I want to stress the importance of leadership and vision in actually realizing any of the ideas that I have been talking about so far. Yet, I am no management guru. I am not going to pretend that I know how to help you find the right vision for your company or organization. And I have wasted enough time in meetings where high-priced consultants tried to coach companies that I have worked for in the process of finding one. So I know it is hard. But nevertheless, vision—as amorphous as it is—is really important. The best that I can add is that in my experience, it comes as much from instinct as analysis. And, for some reason, whenever I think of corporate vision, I think of Norman Maclean's book, *A River Runs Through It* (Maclean 1976), and the 1992 film by the same name, directed by Robert Redford.

I guess the reason has to do with fly fishing, around which the story was framed. Knowing where to cast implies some assumptions about where the fish are, and involves the same type of experience-based gut feeling, coupled with rational analysis, as plotting strategic vision. The first thing is, you have to cast. Second, if you don't want to go home empty handed, you had better cast where you think the fish are. By analogy, that is the future. As we reel in our line, it cuts a path in the water, right to our feet. Where the hook emerges is the present. Where we are now.

I know that the metaphor is pretty simple. But I prefer models that are simpler than the thing they are modeling, and this one helps me conceptualize what I am trying to get at.

One thing that emerges from seeing a path from here to there, rather than just some point off in the future, is the recognition that I am already on the path—just at one end of it. The second thing it helps me recognize is that what I do in the short term may not need to be a radical departure from the present, it just needs to be along that path.

We have already seen an example of this in a previous case study. Yes, mountain bikes brought about a radical change in the bicycle industry. But on the other hand, think of how conventional Trek's first mountain bike looks with today's eyes (see Figure 27).

As time goes by, we may revise where we cast. That is normal. The fish may move, or we may have learned something along the way. But even if we see a fish jump in a pool way off in another direction, we should question the long-term benefit of pursuing it. Any moves off of our chosen path are potentially expensive diversions, regardless of short-term benefits.

Shared vision is important because it provides a coherent guide for innovation and execution. It provides the appropriate landmarks to guide behaviour so that people "know where to go" rather than have "nowhere to go" with their ideas and skills.

> One reason Paul caught more fish than anyone else was that he had his flies in the water more than anyone else. "Brother," he would say, "there are no flying fish in Montana. Out here, you can't catch fish with your flies in the air." (Maclean 1976)

Likewise, your ideas will never be realized if you leave them floating in the clouds—no matter how good they are.

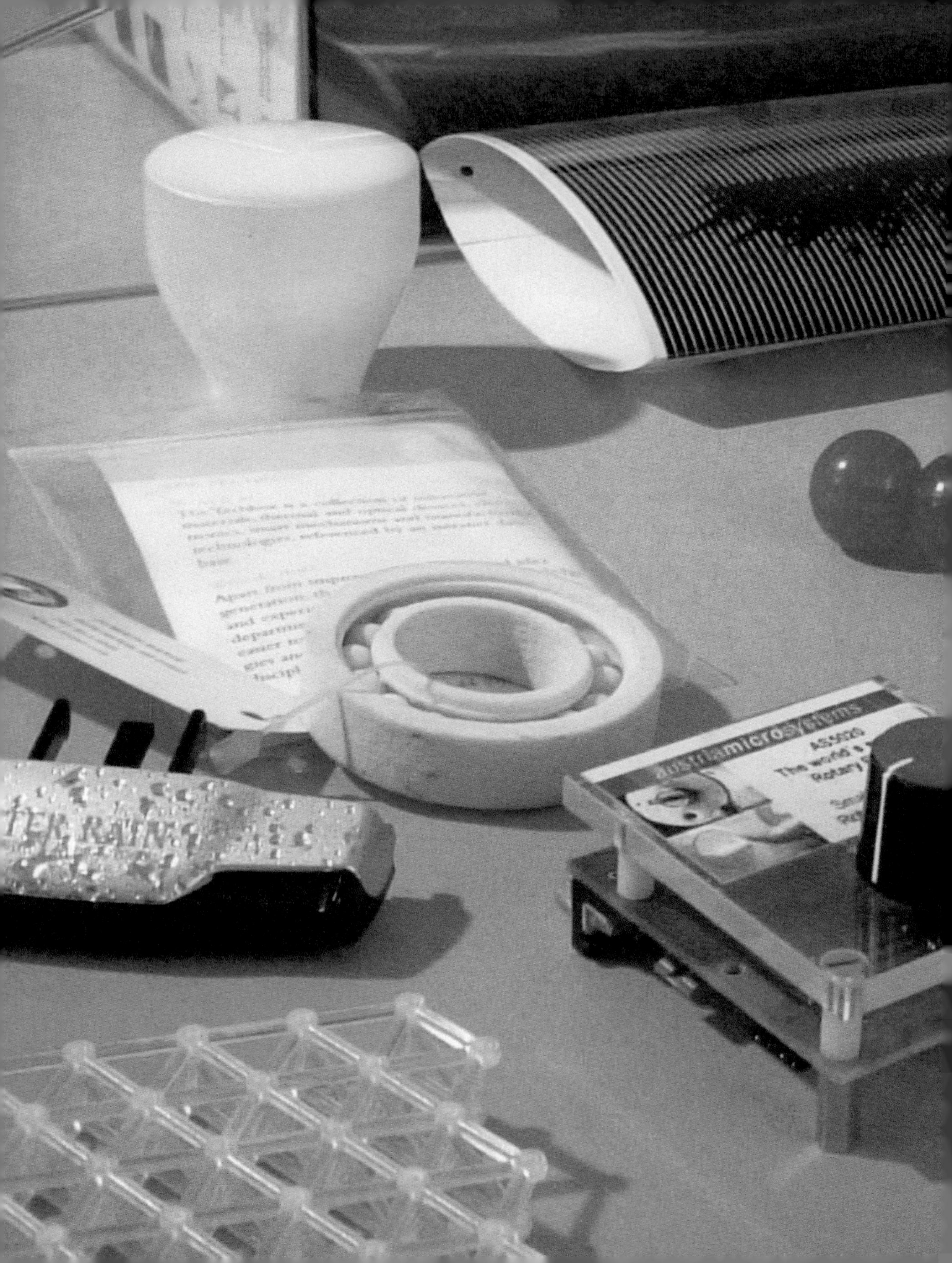
austriamicrosystems
AS5020

Part II: Stories of Methods and Madness

Figure 78: Workshopping Ideas

One of the best ways to draw out the best from people, designers and users alike.

Photo: Brooks Stevens Design

From Thinking On to Acting On

...we are in danger of surrendering to a mathematically extrapolated future which at best can be nothing more than an extension of what existed before. Thus we are in danger of losing one of the most important concepts of mankind, that the future is what we make it.
—Edmund Bacon

Now we change gears.

What we are going to do in this part of the book is look at the work—and more particularly, the working methods—of very good designers, from established professionals to talented students. This approach serves five important functions:

To illuminate what I perceive as best practices.
To help those who work with the design team (including managers and the executive team) to understand these practices and their output.
To foster a shared literacy among the design team of some of the relevant "classic" examples from our diverse traditions.
To show exemplary student work side by side with that of those who pioneered the field in order to show that what I am advocating is attainable.
To give a sense of some of the basic competencies that I would expect in an interaction/experience design team, and hence in the educational programs that train them.

When I speak of "best practices," I am referring to a repertoire of techniques and methods with which I would expect any Experience Design team to have a reasonable degree of fluency. This is not a "How to design a great product" manual, or a treatise on "How to be creative," but it does stake out part of that turf, namely a subset of design primarily relating to ideation and sketching. There is a good chance that someone who reads this section will be familiar with some of what I discuss, but I suspect that there will be few for whom there is not something new. And, even with familiar material, I hope that I am able to bring a sufficiently fresh perspective to contribute new insights.

As for the second point, before product managers or executives dismiss the material in this section as being irrelevant to them, they might want to recall Alan Kay's quote that I mentioned earlier:

> It takes almost as much creativity to understand a good idea, as to have it in the first place.

One of the best steps toward fostering a common culture of creativity among a diverse team is to become as fluent as possible in each other's languages. I have tried to make this book as accessible to the businessperson as the designer because I think that the designer's efforts will be for naught if the executive and product manager don't understand the how and what of the designer's potential contribution to the organization. Just think back to the case of Jonathan Ive at Apple. Do you want to squander the potential of your design team, as was largely the case until Steve Jobs came back to Apple, or do you want to improve your ability to exploit it the way that Steve did? Simply stated, the sooner you understand a great idea, the more lead time you have to do your part in executing it. That is why you need to read this section. Not to become a designer, but so that together with your design team (which you are paying for anyhow!), *and with the rest of your organization,* you can make design a more effective differentiator in your company.

As to the third point, I confess to being captivated by history— of my profession and of almost everything I am interested in. To me, history is both interesting and part of basic literacy. I think that it is important to the effective practice of our craft. The problem is, the experience design team of today involves people from many different traditions, each with its own history. I would hope that those from each tradition would know their own history, but I would never assume that they know each other's. For example, industrial designers will likely know about Christopher Dresser (Whiteway 1991), Norman Bel Geddes (Bel Geddes 1932), Henry Dreyfus (1955), or Raymond Loewy (Tretiack 1999), and why they are important. But more often than not, these names will draw a blank when given to a user interface designer who has a computer science or psychology background. By the same token, names such as Doug Engelbart (Bardini 2000), Ivan Sutherland (Sutherland 1963), and J.C.R. Licklider (Waldrop 2001), which should be familiar to the user interface designer, are most likely unknown to those from the tradition of industrial design.

Yet, the histories of each of our various disciplines, including marketing, have the potential to lead to more informed design. Knowing each other's histories lays the foundation for shared references and the common ground that that creates. So, whenever appropriate, I have chosen to mix key historical examples from various traditions into what follows. Although it is not a history lesson per se, hopefully it will make some contribution toward building a shared literacy and tradition among the emerging culture of experience design.

Fourth, while familiarity with some of the classic examples from our history is important, it can also be intimidating. By relying on such examples, am I setting the bar too high? Is this standard attainable by a student, or is this too much to ask from someone that you are thinking of hiring? I think not. I have consciously also incorporated examples from the work of students from around the world to convince you of this point. For me, meeting these students and being exposed to their work was one of the most encouraging and enjoyable parts of researching this book.

Finally, a new approach to design implies a new approach to design education. Let's say that what I talk about makes sense and that by some miracle executives all over the world say, "Yes! Let's incorporate something like this in our company." Who are they going to hire? Where are the people going to come from? What kind of skills and experience should one be looking for? This section provides the basis for a partial answer. But I need to add, yet again, a cautionary note: This is not a comprehensive manual on product design. I am only trying to fill a gap in the literature, not cover the whole space. There are other books on topics such as Participatory Design, User Centred

Design, Usability, Industrial Design, Ethnography, Marketing, and so on. We do not have to start from scratch. Second, no individual will or can have equal competence in all the requisite skills. So the second thing to keep in mind is that we need coverage of the larger skill set distributed among a heterogeneous team, not the individual. But, and this is the important "but," for that team to function well, the players must have at least basic literacy in each other's specialties, if not a high level of competence.

Is this section going to be technical? On the one hand, yes, we are going to dive into the design funnel and talk about what goes on inside. On the other hand, it is not going to be any harder to follow than what we have already discussed. And I certainly hope that it is as interesting and relevant. It is definitely not going to take the form of some academic analysis of formal design theory or methodology. Why bother? As Chris Jones says in his book, *Design Methods*:

> There seem to be as many kinds of design process as there are writers about it. [There is] little support to the idea that designing is the same under all circumstances, and ... the methods proposed by design theorists are just as diverse as are their descriptions of the design process. (Jones 1992; p. 4)

In many ways, we wouldn't be in our current situation if formal design theories and methodologies worked as advertised, with their many boxes and arrows that map out the process. Gedenryd (1998) makes this argument pretty well. Speaking about architecture, Snodgrass and Coyne (2006) say:

> Contemporary architecture theory now largely ignore the vast literature on systems theory and design methods ... (p.24)

And in his book, How Designer's Think, Bryan Lawson remarks:

> Well, unfortunately none of the writers ... offer any evidence that designers actually follow their maps, so we need to be cautious.
>
> These maps, then, tend to be both theoretical and prescriptive. They seem to have been derived more by thinking about design than by experimentally observing it, and characteristically they are logical and systematic. There is a danger with this approach, since writers on design methodology do not necessarily always make the best designers. It seems reasonable to suppose that our best designers are more likely to spend their time designing than writing about methodology. If this is true then it would be much more interesting to know how very good designers actually work than to know what a design methodologist thinks they should do! (Lawson 1997; p. 39)

Whenever possible I have video clips that compliment what I say with words and pictures. These can be accessed from the companion website to this book: www.mkp.com/sketching.

I have structured this section in a kind of musical "E-A-B-C-D" form. Perhaps this is my earlier life as a composer coming out. I am going to start with a few rich examples that foreshadow where we are going, then pull back to a simpler world. From there I will build back up toward where

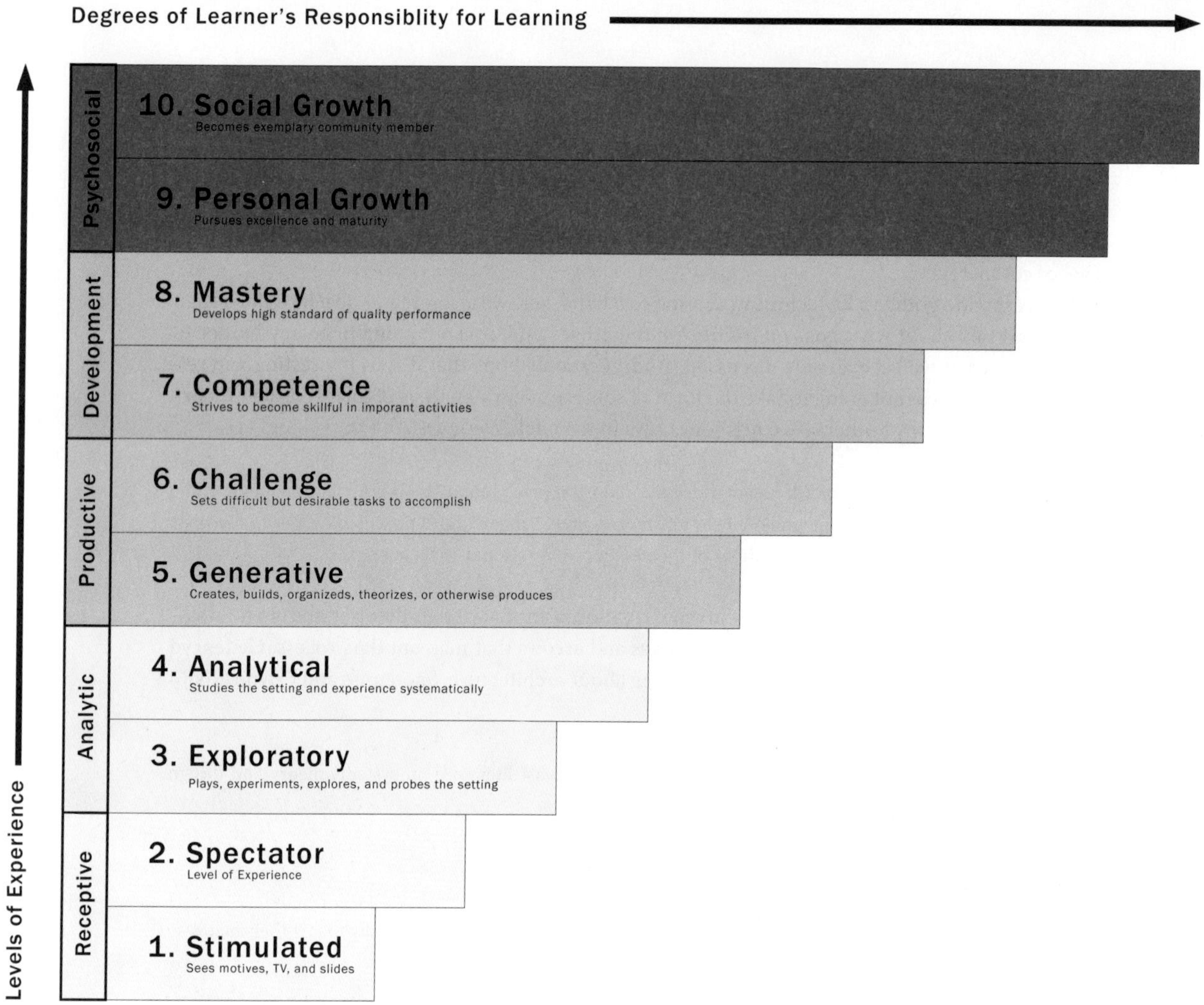

Degrees of Learner's Responsiblity for Learning

Levels of Experience

Psychosocial

10. Social Growth
Becomes exemplary community member

9. Personal Growth
Pursues excellence and maturity

Development

8. Mastery
Develops high standard of quality performance

7. Competence
Strives to become skillful in imporant activities

Productive

6. Challenge
Sets difficult but desirable tasks to accomplish

5. Generative
Creates, builds, organizeds, theorizes, or otherwise produces

Analytic

4. Analytical
Studies the setting and experience systematically

3. Exploratory
Plays, experiments, explores, and probes the setting

Receptive

2. Spectator
Level of Experience

1. Stimulated
Sees motives, TV, and slides

Figure 79: Scale of Experience in Learning

Ten levels of increasing experience in learning are shown. As the level increases, the learner takes on additional responsibility.

we started, laying more of a foundation in the process. And just as a warning, somewhere in the middle, I am going to insert an interlude where I can add some meta-comments and examples.

But when I talk about richness or space, what is the scale on which my A, B, C, and so on lie? I am going to draw on a tangentially related field, experiential learning (see Kolb 1984, for example). In this literature, Gibbons & Hopkins (1980) developed a *Scale of Experience*, illustrated in Figure 79. With it, they attempt to establish a kind of taxonomy of levels of experience. Although a legitimate target for debate, it can serve our purpose.

At the lower levels are things where one is at the receptive end of experience. The notion is that although you can experience seeing a train or a bear in a movie, there is a higher level of experience seeing it live. Likewise, there is a deeper level still if you get to play with the train or (hopefully Teddy) bear, rather than just see it. The argument made is that as one goes up the scale, one moves through different modes, from Receptive through Analytic and eventually through to what they call Psychosocial mode.

If we push too hard on this its relevance to our work diminishes. After all, the scale was developed for a different purpose—education rather than design. There are really only three things that I want to draw out of it.

First, when I say that I am going to organize this section on an E-A-B-C-D structure, I am going to start with a few examples from the high end of a scale analogous to that of Gibbons & Hopkins. I will then drop back to examples and techniques that are at the lower, receptive, level of the scale, and work my way back up.

Second, Gibbons and Hopkins argue that higher levels of experiential learning imply a higher level of responsibility on the part of the learner for what they learn (the auto-didact). This is represented by the horizontal axis in Figure 79. Likewise, from the design perspective, our renderings (be they sketches or prototypes) afford richer and richer experience as we go up the scale. However, reaping the potential benefit of the design knowledge, or learning, that can be extracted from these renderings also depends on assuming the responsibility for using them appropriately.

Third, going a step further from the previous point, keep in mind that the level or type of experience that one can get out of renderings at the lower levels should not necessarily be considered impoverished. Seeing something live is not necessarily better than seeing it in a movie—it is just different. There are different types and levels of experience. Knowing how to use them appropriately in design is where the artistry and technique come in.

Finally, before proceeding, I want to point out that I did notice the "Those who can, design, and those who can't, write about design" aspect of the earlier quote by Lawson. The irony of including it, much less Lawson's writing it in the first place, is not lost on me. I have tried to keep its message in mind in what I write. Second, I think that there are times that design goes through transitions due to new demands that are made on it. In such times, thought and writing about design can provide a useful role in helping us get through those transitions with minimal problems. I view us as being in the midst of just such a transition, hence my sticking my neck out and taking up my proverbial pen.

The Wonderful Wizard of Oz

The purpose of this sidebar is to introduce the story of *The Wonderful Wizard of Oz*. For many readers, this may seem rather strange, perhaps even bizarre. After all, this is one of the most popular stories in American culture. Yet, although the so-called *"Wizard of Oz Technique"* is known by virtually everyone who read drafts of this book, not one of the non-native English speakers were familiar with the story after which it was named. This is a brief attempt to remedy this. For those who know the story and initially thought it strange that I was going to tell it, let this be a lesson to you (as it was to me) about the importance of not making any assumptions when designing something that is intended to span cultures.

The Wonderful Wizard of Oz is a children's book that appeared in 1900. It was written by L. Frank Baum, and highly illustrated by W.W. Denslow. It tells the tale of a young girl named Dorothy who, along with her dog Toto, is swept away by a cyclone from the farm in Kansas where she lives with her uncle and aunt. They are taken to the land of Munchkins, where they are very popular since the farmhouse (which blew away with them) killed the Wicked Witch of the East by landing on top of her.

Here begins Dorothy and Toto's odyssey to return home to Kansas. Graced by a kiss from the Good Witch of the North, and a pair of silver shoes, Dorothy and Toto head off down the Yellow Brick Road to the Emerald City, where they hope that the Wizard of Oz will help them get home.

Along the way they have a number of adventures, which include picking up an entourage of fellow-travelers, including a Lion who is in search of courage, a Scarecrow on a quest for a brain, and a Tin Woodman, who is seeking a heart.

On reaching the Emerald City, one by one they each see the Wizard in order to plead for his help in their individual quests. "Oz, the great and terrible" manifests himself differently for each, but gives them all the same answer: he will only grant their wishes if

they go off and kill the Wicked Witch of the West, who rules the Winkie Country. Having no other hope, they reluctantly go off on the quest, which they accomplish inadvertently when Dorothy throws a bucket of water on her that causes her to melt.

Surprised by their success, the troupe heads back to the Emerald City to collect their reward. Together they have an audience with the great Oz, who this time manifests himself as a great voice, but not visually. The other thing that he doesn't do is grant their wishes as promised.

So now we come to the part of the story that is relevant to our purpose. Here is what happens, as stated in the book:

> The Lion thought it might be as well to frighten the Wizard, so he gave a large, loud roar, which was so fierce and dreadful that Toto jumped away from him in alarm and tipped over the screen that stood in a corner. As it fell with a crash they looked that way, and the next moment all of them were filled with wonder. For they saw, standing in just the spot the screen had hidden, a little old man, with a bald head and a wrinkled face, who seemed to be as much surprised as they were. The Tin Woodman, raising his axe, rushed toward the little man and cried out, "Who are you?"
>
> "I am Oz, the Great and Terrible," said the little man, in a trembling voice. "But don't strike me— please don't—and I'll do anything you want me to." (Baum 1900; p. 183)

So, the reality behind this great and terrible wizard was a timid little man who was so convincing in his ruse that everyone had thought he was something that he was not.

And here I leave my telling of the story, other than to reassure you that Dorothy and Toto get home, the Lion finds his courage, the Scarecrow a brain, and the Tin Woodman a heart. Everyone lived happily ever after.

Figure 80: The Wonderful Wizard of Oz

The cover of the first edition of one of the most unexpectedly influential books on the topic of interaction design.

The Wonderful Wizard of Oz

I can't leave the future behind.
—Suze Woolf

Perhaps the most important book in terms of informing our endeavor is *The Wonderful Wizard of Oz* (Baum 1900). To some this might seem a strange statement. But think about our purpose: it is to experience interactive systems, before they are real, even before we have arrived at their final design, much less their implementation. Keeping that in mind, think again about Baum's book, or the film version of it, and consider this:

> **Up to the point where Toto tipped over the screen and revealed the Wizard to be a fraud, all of Dorothy's reactions were valid psychologically, anthropologically, and sociologically. To her the Wizard was real, and therefore so were all her experiences.**

In the Introduction to this book I said something that read like a paradox:

> **The only way to engineer the future tomorrow is to have lived in it yesterday.**

What Baum's book teaches us is that if we do an effective job of following the example of the Wizard, we too can conjure up systems that will let users have real and valid experiences, before the system exists in any normal sense of the word. In all of this, the four most important things to glean and carry forward are:

- It is fidelity of the experience, not the fidelity of the prototype, sketch, or technology that is important from the perspective of ideation and early design.
- We can use anything that we want to conjure up such experiences.
- The earlier that we do so the more valuable it generally is.
- It is much easier, cheaper, faster, and more reliable to find a little old man, a microphone, and some loud speakers than it is to find a real wizard. So it is with most systems. Fake it before you build it.

I'm not sure who first used the term "Wizard of Oz" in the context of interaction design. I first heard it from one of my early influences, John Gould, of IBM. But regardless of who coined the term, the meaning is pretty well understood internationally (unlike the story): the *Wizard of Oz Technique* involves making a working system, where the person using it is unaware that some or all of the system's functions are actually being performed by a human operator, hidden somewhere "behind the screen."

The objective is not to make the actual system, but to mock up something that users can actually experience, thereby enabling us to explore design concepts in action and as experienced far earlier in the process than would otherwise be possible. Such a system should be cheap, quick to realize, disposable, not the real thing, and only have sufficient fidelity to serve its intended purpose. That is, it should have all the attributes that characterize a sketch.

Inherent in all this is the following rule:

> **Generally the last thing that you should do when beginning to design an interactive system is write code.**

Now I know that a very large proportion of interactive products are software driven, and that the status quo is that software engineers play the primary role in their development. Having gone through this myself, I also know the amount of time and effort that these same software engineers have expended in acquiring their skills. I even know that they or their parents probably spent a fortune paying for that education. But I also know the following rule:

> If the only tool you have is a hammer, you tend to see every problem as a nail. (Abraham Maslow)

Having gone through all the effort and expense to develop these skills, the natural inclination is to damn well use them. And the better you are at it, the more likely it is that this will be the case. After all, that excellence probably came at the expense of not developing other skills (since there are only so many hours in the day and nothing comes for nothing). So the point here is that all the biases in the status quo stack up against the Wizard, and in favour of programming and engineering. Not just from the individual perspective, but also from that of management, who have made a huge investment in the best programming talent available. ("What? I hire all these great computer scientists and engineers and now you are telling me not to use them?")

It is precisely because of the magnitude of this investment that we must make a conscious effort to manage things along a different path. These resources are too expensive to squander on a product that is almost certain to fail—and fail it will most likely do, if one leaps into engineering prematurely.

I hope that it is clear that I am not trying to downplay the importance of programming and engineering skills. I am simply saying that the front-end of the interactive design process is generally not the appropriate place to apply them. Furthermore, the need to balance the biases of the status quo is yet another reason that design, and not engineering, should own this front part of the process. As I have said previously, engineering is culturally as unsuited for the task of managing design as designers are for running engineering. With product engineering in charge of the front end of the process, the temptation to fall back into old habits will be just too great to resist.

So if we are not going to start by programming, and we do want to follow the Wizard, then some examples may help shed some light on how to do so.

Airline Ticket Kiosk

Historically, the first instance of using the *Wizard of Oz Technique* that I am familiar with was in the field-testing of a self-service airline ticket kiosk concept (Erdmann & Neal 1971). In the past few years, these kiosks have appeared in most airports, but in 1971 that was not the case. In fact, computerized public kiosks were not yet familiar to the general public, much less ubiquitous—this was the same year that the first automated teller machine was deployed. Compared with today, even credit cards were not that common. Consequently, there were few, if any, precedent technologies on which design decisions could be based. So they built a prototype, and after testing it in the lab, field-tested it at Chicago's O'Hare Airport.

Since it was being used by real customers, using real money to buy real tickets on actual flights, it had to work from both the user-interaction perspective, as well as that of the back-end ticketing and financial system. But the system wasn't finished. That was the whole point: they wanted to test it before deciding to proceed with the product, much less commit to a final design.

So they went straight down the Yellow Brick Road and faked it, like any good wizard. To the user, everything looked real. But behind the curtain, there was a human being, acting as The Great Oz, who keyed in the pertinent information using a conventional terminal, and then issued the ticket. Although expensive and inefficient from the short-term perspective of actually delivering the service to customers, the potential value in terms of informing the design, usability, and acceptability of a potential future system was immeasurable.

Figure 81: The Wizard's Listening Typewriter

A perfectly functional listening typewriter is implemented simply by having a fast typist, hidden behind the screen, who would enter the text captured from the microphone.

The Listening Typewriter

Practically everyone involved in user interface design has had at least one person come up to them and explain how, “If we could just talk to computers like we talk to people, and if they could understand what we are saying, then the user interface problem would be solved.”

However, just because people have been saying such things for a long time does not make their assertions true. On the other hand, it doesn’t make them false either.

To shed some light on the matter, John Gould and his colleagues at IBM decided to perform a study to investigate the potential benefits of a listening typewriter (Gould, Conti, & Hovanvecz 1983).

At the time, the late 1970s and early 1980s, IBM had a lot at stake. They had a major position in office information systems, and they wanted that position to consolidate and grow. If there were potential benefits to be derived from a listening typewriter, then they wanted to be the ones to reap them. The problem confronting them was how to understand what such benefits might be, especially when the underlying technologies did not yet exist. They could spend a fortune on R&D and wait years for it to work sufficiently well to support a realistic test. But what if the expected benefit wasn’t there?

How could they make an informed funding decision that was more of a calculated investment than a gamble? Here is where Gould and his colleagues stepped in, and engaged the Wizard in order to test the system before it existed.

The way that they did this is illustrated in Figure 81. Their approach was as ingenious as it was simple and effective. Playing the role of the Wizard was a speed typist, who really did hide behind the curtain. Whatever the user dictated into the microphone, this “typing Wizard” entered into the computer such that it appeared on the screen of the user. Not surprisingly, the system worked fantastically well. Words were properly spelt, and paragraphing and punctuation matched the speaker’s intention. In fact, that system probably worked better 20 years ago than any real system does today. Using the *Wizard of Oz Technique*, they were able to leap more than 20 years ahead in the technology curve, and collect real user experience data! And, they were able to do so in a matter of weeks, without writing any substantive code. This paper is a classic, and should be studied by students of interaction design the same way that art students study the classics of the renaissance.

Figure 82: An Example of a Display That Changes According to Location

A camera's viewfinder is an example of a display whose content changes depending on what you point it at. Even pointing at the dog, what you see changes as you move the camera left-right, up-down, or in-out.

Chameleon: From Wizardry to Smoke-and-Mirrors

Only a god can tell successes from failures without making a mistake.
—Anton Pavlovich Chekhov

One of the people who learned from Gould and his colleagues was one of my PhD students, George Fitzmaurice. Like others, he realized that one could create the illusion of a working system using techniques other than having a human Wizard perform some of the functions of the system. In some cases, such as the example that we are going to look at, the magic, or illusion, comes by means of the clever use of technologies and techniques on the part of the designer. Rather than a human Wizard, "smoke-and-mirrors" technologies are applied to realize an interactive sketch of a concept in a form that users can actually experience.

In the early 1990s, we were investigating

> ... how palmtop computers designed with a high-fidelity monitor can become spatially aware of their location and orientation and serve as bridges or portholes between computer-synthesized information spaces and physical objects. (Fitzmaurice 1993).

I know that that was a mouthful, so here is a simple version of the question he was asking: What if the contents on the screen of a handheld device could be determined by what the device was near, what it was pointed at, or how it was moved up-down, left-right, or in-out?

Imagine looking through the viewfinder of a camera, such as that illustrated in Figure 82. It has all these properties. If you point it at a dog, you see the dog. If you move it closer, the image of the dog gets larger on the screen, but you see less of it. If you pan the camera left, you see the dog's tail in the viewfinder. If you pan it to the right it shows his head, and so on.

With the camera there is no magic. The screen just reflects what is in front of the lens. But what if the device in your hand was not a camera, but a PDA, for example? What if instead of panning across a dog, as in our example, you were panning over a large spreadsheet or some other document? What if you were doing so simply by moving the PDA left-to-right as if you were operating a camera? Likewise, what if when you came up to an object in a museum and the PDA could "know" what was in front of it, and display additional information on the artifact depending on where you pointed it?

George called this type of position and motion sensing display, *Chameleon*. He wanted to gain some sense of what it would be like to use such a system. How would it feel? Would people be comfortable using it? What would the experience be compared to that of using a desktop computer?

These are simple questions to ask (at least once you have come up with the idea). In order to answer them, "all" that he needed was a handheld computer that sensed where it was in space, and the computational and graphics power of the most powerful workstations of the day.

Figure 83: A Small LCD TV Masquerading as a Position-Sensitive PDA

All we could afford was this little LCD TV display. It worked, so it was all that was needed. Anything more would have been superfluous.
Source: Fitzmaurice (1993)

Figure 84: Faking a Position-Sensitive PDA

The scenario is much like that in Figure 72, except this time the Wizard is replaced by a video camera and other miscellaneous technologies that we could borrow or scrounge.
Source: Fitzmaurice (1993)

But there was this minor problem. He was doing all this in 1992, a year before the Apple *Newton Message Pad* was announced and four years before the release of the original *PalmPilot*. Nevertheless, he was able to implement a system that let one personally experience this type of interaction. Here's how.

To begin, he bought the small $100 Casio LCD TV shown in Figure 83. This served as a surrogate for the "handheld computer" with which future users might interact.

So far, so good. But this was just a TV and had no computer power. However, what it did have was a video input. So, George hooked it up to a (borrowed) video camera and pointed the camera at the screen of a (borrowed) state-of-the-art SGI graphics workstation. That caused a video image of the computer display to appear on the Casio TV. *Voila!* He now had the makings of palmtop computer with the computational and graphics power of an SGI workstation.

He then had to figure out how to sense the position and motion of the LCD display as it was moved around by the user. To do this, he borrowed a device called a Bird, made by Ascension Technologies. This is what is known as a motion-capture device, originally designed for computer animation. It consisted of a small cube that George unobtrusively attached to the back of the Casio TV, and it then connected to the SGI, where it provided the spatial information required to drive the interactions seen on the handheld display. The overall configuration is shown in Figure 84.

Admittedly the device was tethered, and therefore did not have the mobility that we envisioned in the future. Nor did it have the graphics resolution that we anticipated would become available in the future. But by using it with some 3D software that he borrowed from another project, it did let George get a good first approximation of what it would be like to interact with such a device, and more to the point, it let him observe the experience of others doing so.

It also enabled him to explore different types of applications and transactions. Figure 85, for example, illustrates panning across a display. Unlike a desktop computer, where you use the scroll bars and scroll arrows to move the document on a stationary display, here you move the display over the surface of a stationary virtual document in a manner analogous to our camcorder example.

Likewise, Figure 86 illustrates how the behaviour of the *Chameleon* can be driven by its position over an object in the physical world, rather than some virtual document. In this case, the object is a map. But when coupled with the *Chameleon*, the paper map becomes a guide for browsing for more detail about its surface. So, for example, if you move the display over Winnipeg, Manitoba (where George is holding it), it might give you details about the city's important role in the fur trade, or why it is the mosquito capital of Canada.

CANADA

Figure 85: Panning Across the Display
The user navigates across the information space by moving the handheld device much the same way that you would pan across a scene using a camcorder.
Source: Fitzmaurice (1993)

Figure 86: Chameleon in Context to the Physical Environment
The photo illustrates how something in the physical world, in this case a map of Canada, can work with Chameleon. For example, by bringing the handheld device into the vicinity of southern Manitoba, as illustrated, the device would provide information about Winnipeg.
Source: Fitzmaurice (1993)

In short, by employing smoke-and-mirrors techniques, George was able to initiate research into a new class of interaction, in a way that enabled him and his users to gain direct personal experience, at least 10 years before such interfaces were commercially viable.

So, is George's *Chameleon* example a sketch? I want to spend a bit of time on this question since it provokes some issues that run through this part of the book.

First, I'm not sure. Given the tools and resources available at the time that it was done, it probably lies somewhere just to the left of centre on the sketch-to-prototype continuum.

Second, perhaps the real value in drawing a marked distinction between sketching and prototyping lies not in the end points, but in recognizing that there *is* a continuum between them. An awareness of it, its properties, and its implications, may help guide us in how and when we use different tools and techniques.

Third, as we shall see later in our discussion of "paper prototypes," how a technique is used is the ultimate determinant of whether one is sketching or prototyping. At the time that it was done, the *Chameleon* project would most accurately be described as "rapid prototyping." Yet later, I give a more recent example where I explored the concept in a context that I would call sketching.

Finally, regardless of label, I think that our repertoire of techniques needs to incorporate the facility and mindset reflected in George's example.

Having said that, I also want to address one thing missing in his work. As agile as his approach was, he could not take the interactive result (rendering) out into the field to test. Its size and complexity meant that in order to use it, people had to come to the lab.

There are trade-offs and consequences of the decisions that we make, and we have to be aware of them. But if we are aware of them, we can incorporate techniques into our plan that help compensate for any shortcomings. For example, notice how George dealt with this particular issue. Through the use of staged photographs taken in situ, such as that shown in Figure 86, he was able to connect the dots between what one could experience live in the lab, and various visions of its application "in the wild." (See his original paper for further photographic and diagrammatic examples.) This is a nice example of how various media (including the accompanying video) complement each other in order to tell the larger story.

The lesson here is that it is rare that only one form of rendering will suffice, which is one of the reasons that the design team must have command over a range of techniques. I am not going to give any magic rules or guidelines when or where to use which technique. Far more effective is for you to get practical fluency with the techniques that we discuss. The experience gained in the process will provide far more effective insights about their relative strengths and weaknesses

Figure 87: A 2D Chameleon Sketch

By sticking a piece of paper under a Tablet-PC to reduce friction and attaching a mouse to the side, one can quickly make a working sketch of what it is like to navigate around a virtual map by sliding the tablet around on the surface of a desktop, like a moving window.

than anything that I could write. Words are not sufficient to provide such understanding. It must be based on experience.

Finally, I want to revisit George's rendering of *Chameleon* with today's eyes, and in particular, the tools and technologies that are now at our disposal. Recently I was involved in a project to explore new concepts for lightweight slate (tablet) type computers. Since we were talking about ways to navigate over maps and other documents that are too large to entirely fit on the screen, I thought of George's example.

What I was able to do, right in the meeting, is render a working interactive sketch of what I was talking about: a desktop version of *Chameleon* that worked in 2D. First I placed a piece of paper under my Tablet-PC so that when I slid it over the table surface, there was minimal friction. I then loaded a map program on the computer and scaled it to the full screen size. Finally, I stuck a miniature wireless mouse to the side of the PC. I was then able to navigate over the surface of the map by moving the tablet on the desktop, thereby realizing a version of George's *Chameleon* idea. This is illustrated in Figure 87 and the associated video.

Yes, I had to hold the left mouse button down, and yes, I had to have the mouse-tracking symbol over the map surface for it to work. And so as not to hide anything, I also had to adjust things like the mouse control:display ratio and orientation to get the feel that I wanted. In a matter of minutes everyone in the room understood the basics of the concept, as well as some of the issues underlying its implementation. Furthermore, despite some in the room already having heard or read about similar ideas, everyone left with a renewed appreciation for the difference between cognitive versus experiential understanding.

Not only was this a sketch (meeting all the criteria that we identified), it was the experience that was sketched. Furthermore, it was done on the spot in the wild, so to speak.

One quick final comment to emphasize the notion that design has as much to do with attitude as technique: What is significant, but not obvious, is that I shot the video clip that illustrates this sketch right in the meeting. I always carry around a compact digital camera with a really big memory card with which I can collect reference images of things that interest me, as well as shoot videos, such as this one. Hence, I can share material with others on the team, even if they are not there to experience it first-hand. This "hunting and gathering" of reference material is a habit that borders on mania with every good designer that I have met. We shall return to this theme again.

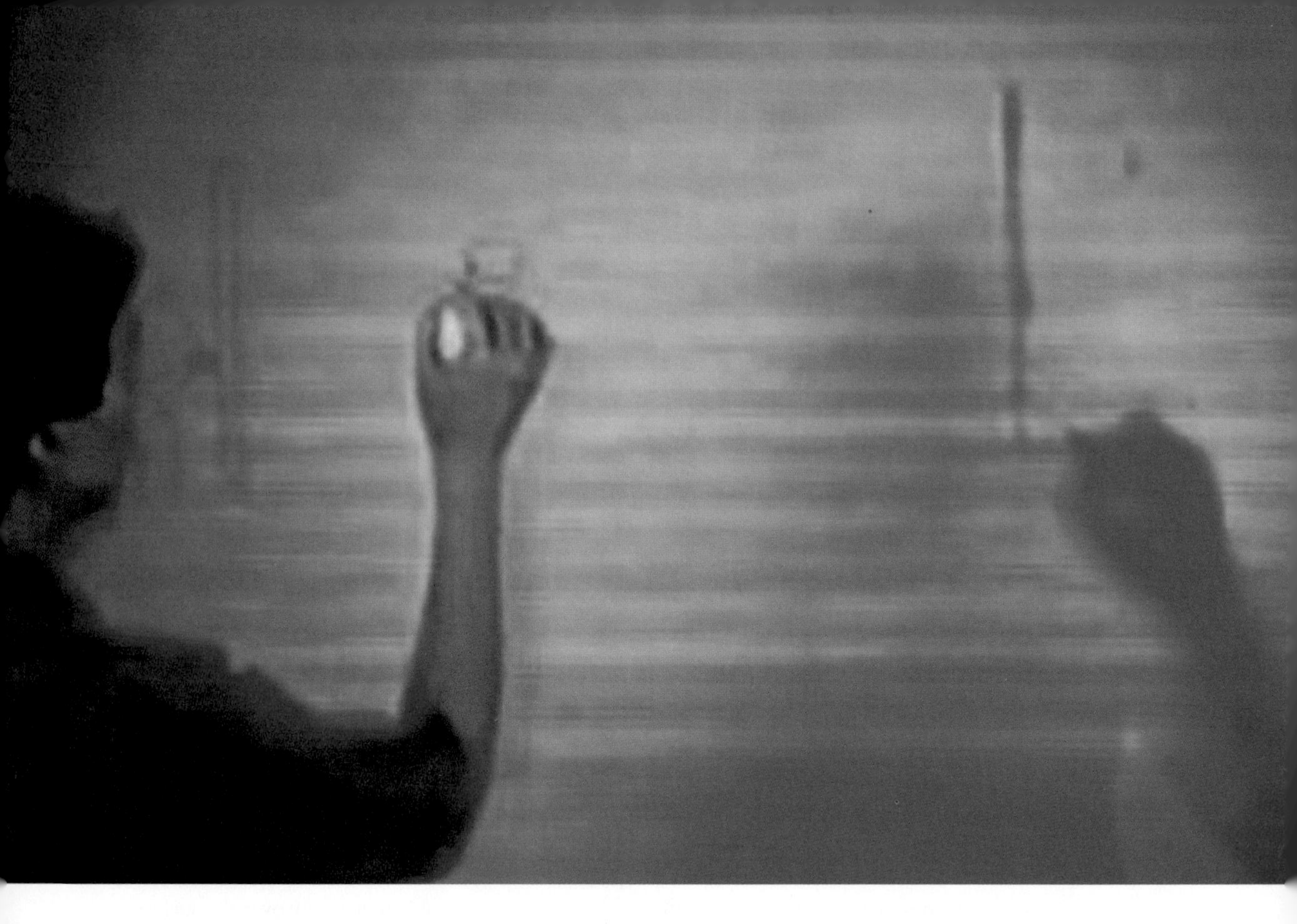

Figure 88(a): Video Whiteboard

This is a shared drawing surface on which the marks created by both the local and remote person are combined. Furthermore, the remote person has a presence by virtue of a life-size shadow that appears to be just on the other side of the surface, regardless of how many thousand miles away they might be.

Source: Tang & Minneman (1991)

Figure 88(b) Winky Dink and Remote Drawing

Refer back to Figure 66 of the TV show Winky Dink and You. This is a shot of what was going on in the TV studio. It illustrates how the host and child could both draw together. Other than the minor fact (!) that the host couldn't see the child or what was being drawn at the other end, it is a wonderful example of how popular media can contain the seeds of future ideas, and be rich terrain for relevant reference material

Photo: CBS.

Le Bricolage: Cobbling Things Together

Are you experienced?
—Jimi Hendrix

I now want to complement the previous examples with one that exudes the elegance, simplicity, and effectiveness of the best of sketching. What I love about it is that it is highly interactive, provides a very high fidelity experience, is a surrogate for a futuristic technology, yet involves no computer at all in its implementation. It was made using materials found in most modern offices or universities.

The Video Whiteboard

The project is *The Video Whiteboard* and it was done by John Tang and Scott Minneman at Xerox PARC (Tang & Minneman 1991). They were interested in collaborative design, where the collaborators were separated by geographic distance. The scenario that they envisioned was a large electronic whiteboard at each location, which provided each participant a surface on which to draw, as well as to see what the other person was drawing.

To this point, the idea is interesting, but not that special. Many people have imagined and even built surfaces for shared remote drawing. What was special with Scott and John's vision was the idea that you could see the shadow of your collaborator on the surface, as well as what they drew, and that it would be life size, appearing as if the person was just on the other side of the surface. This is illustrated in Figure 88.

The reason that this concept is so compelling is that it integrates the space of the task (the shared drawing) with the space of the person (the shadows).

With almost all other shared drawing programs, the only presence that a person has on a remote screen—besides the marks that they make—is the tracking symbol of their mouse or stylus. Thus, they are restricted to the gestural vocabulary of a fruit fly! Thus, they cannot refer to the shared drawing with any of the manual gestures that they would employ if physically present.

Not so with *The Video Whiteboard* , where a user can see the remote person approaching, such as when they want to point at something, or when they are about to work in a particular region, or change something. Furthermore, as can be seen in the figure, the closer that the remote person comes to the board, the sharper the focus and the darker their shadow. All this gives an unparalleled sense of presence. The result is that the partners have the visual cues to enable them to anticipate the other person's actions and say, for example, "No, don't change that!" or "Okay, you work there and I'll work over here."

So, what I tried to convey in the preceding paragraph is that this was a pretty cool idea that beautifully captured a sense of shared presence in collaborative work. The fact that I like the idea so much (and 15+ years later am still waiting for a product that even begins to deliver its

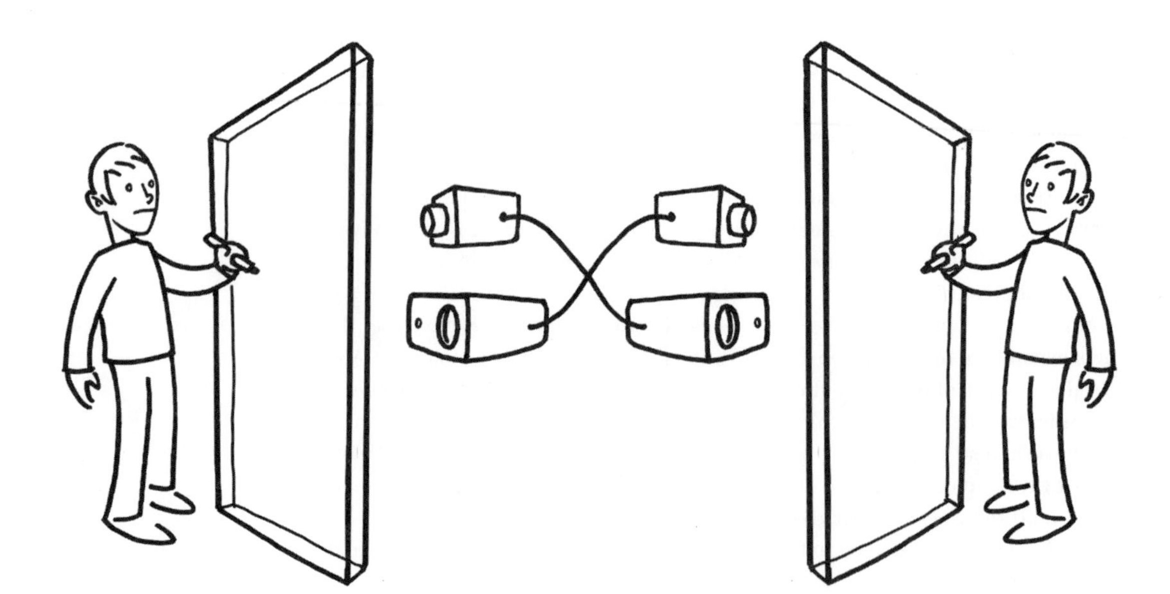

potential) only heightens my reason for including this example. For, as interesting as the idea is, how they sketched it out is just as good or even better.

Now if you had this idea yourself, perhaps the fastest way to "build" a sketch of it that you could experience might be to do what I have illustrated in Figure 89. That is, just get two people to stand on either side of a window or glass door and have them draw on it using conventional whiteboard markers.

This is a good start (and something that I encourage you to do, rather than just read about). But, it only partially captures the concept. What one gets is a "full fidelity" image of the remote person, not the shadows that John and Scott wanted.

The next step would be to do the same thing, but this time using glass that is fogged. That way you can get a very fast sense of the trade-offs of using shadows versus full-fidelity representations of the remote person. (For an account of a parallel project that uses the full fidelity approach, see Ishii & Kobayashi 1992.)

After doing these quick proof-of-concept sketches, you can explore the case where the participants are separated by more than just a sheet of glass. Here is how Scott and John did it, and in so doing, won my undying respect.

Their solution, illustrated in Figure 90, involved each person having their own rear projection screen on which (like in the previous implementations) they drew using conventional whiteboard markers. Pointed at the back of each screen was a video camera that captured what was drawn on the screen, as well as the shadow of the person doing the drawing (by virtue of the ambient back-lighting). The captured video was then projected onto the back of the screen of the other user.

Augmented by a bidirectional audio link, the two screens (and therefore the two people) can be as far away from each other as you like, and still appear to be just on the other side of the glass. And, in the implementation, there is even automatic control over permissions: each user can erase only what they themselves drew.

The reason that I like this example so much is its simplicity, and the way that one can quickly implement and iterate through the experience sketches that I have described. It is a great idea, as relevant today as when it was first conceived. It is *very* relevant to collaborative computing, and yet there is not a bit of computer or digital technology involved in its implementation. In fact, I would argue that doing so would most likely distract one from the basic idea, rather than help.

For me, the *Video Whiteboard* is one of the classics that should be studied and replicated by any serious interaction designer.

Figure 89: First Iteration: Nothing but Glass

The first iteration in a video whiteboard is simply having people drawing on either side of a sheet of glass or Plexiglas. All that is needed is something like a window (such as a glass door) and whiteboard markers.

Figure 90: Video Whiteboard Schematic

What is drawn by the person on the left, as well as their shadow, is captured by the video camera on the left. This is then projected onto the back of the other screen, where it is visible to the person on the right. Likewise, the person on the right and what they have drawn is captured by the right camera, and projected on the screen on the left.

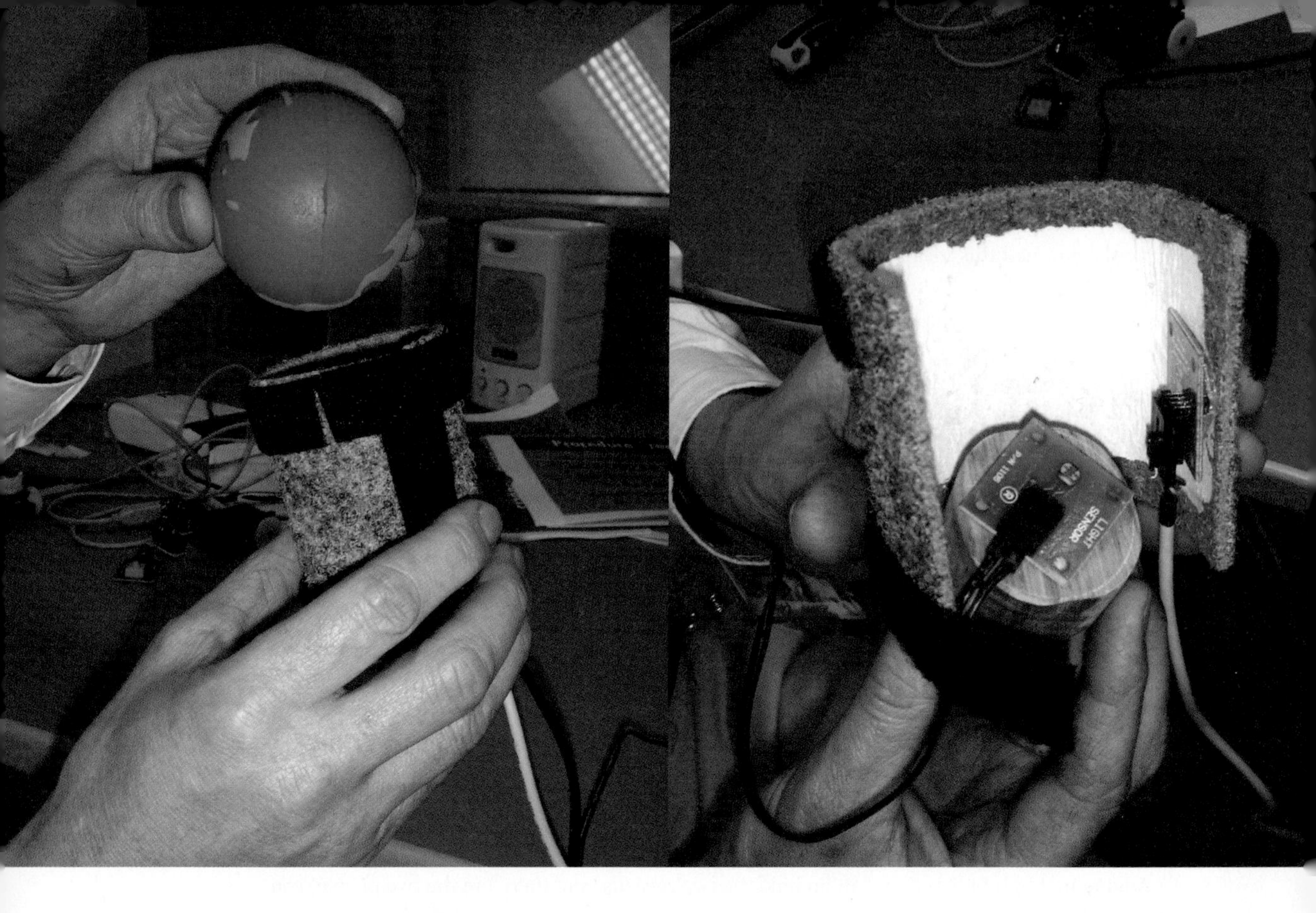

Figure 91: A Working Concept Sketch of a Gesture-Controlled MP3 Player

Gesture-Controlled MP3 Player

I stated that, "the tools and technologies that are available to us today are opening up new ways of approaching things." One such example is the emergence of tool-kits that let us rapidly assemble hardware and software components to build working devices, much in the way that we assembled LEGO bricks to make things when we were kids. Of course, one always could build things with miscellaneous hardware (electronic and otherwise) and software. The difference today is that doing so is within the reach of normal people, and does not require the specialized skills of an electrical engineer and computer scientist.

Here is an example that illustrates the kind of thing that is now within the reach of the design team.

In April of 2006 Abi Sellen and I invited Caroline Hummels and Aadjan van der Helm of the ID-Studiolab in Delft to conduct a workshop on sketching using physical prototypes for the group that we are part of at Microsoft Research Cambridge. Aadjan and Caroline arrived with multiple bins of materials and tools, as well as a broad assortment of *phidgets* (www.phidgets.com).

The task was to build a working MP3 player. We split into four teams of three or four, and had to design and implement a player that was appropriate for a particular persona—all in one day—including learning to hook up the hardware, program it, and iterate through the design. Oh, and one rule: the programmers were not allowed to program, and hardware people were not allowed to do hardware. The idea was to demonstrate that in short order, using the right tools, making such things was within the capability of social scientists and industrial designers, for example.

Figure 91 shows the player made by my team, which included the social scientists Richard Harper, Andrew Fogg, and Martin Hicks. As is illustrated in the first image, one opened up the world of music, so to speak, by lifting the ball off the top of the device, which was made of fabric in a cylindrical form. The volume was controlled by how much the top was uncovered. To change tracks, one punched in a new track by thrusting the device forward. To change genres, one shook the device up and down. Not something that is going to make anyone a million dollars, but it worked.

The second image shows the guts of the device, a photosensor on the bottom, and a 2D accelerometer on the side. The whole thing was programmed using *MAX/XSP* and Jitter (www.cycling74.com), and an MP3 module supplied by Caroline and Aadjan.

The *MAX* program that controlled our player is shown in Figure 92. It is divided into three parts, each represented by a horizontal region in a distinct shade of blue. The middle section, labelled "Product Functionality," is the heart of the MP3 player—the part that actually plays music. This module was provided to each team by Caroline and Aadjan. It does the work, but has no external user interface.

The bottom section, labelled "Product Feedback," was also provided by them. It provides the kind of visual information that you would expect to see on the display of a CD player or your car radio.

Finally, the top layer, labelled "Sensor Mapping," is the part that we "wrote" to implement our particular gesture-controlled player. The top-left box in this section, labelled *r ai2,* is an icon that represents the source of the numerical data that *MAX* receives from the light-sensor that we want to control volume. By drawing lines out of the bottom of this box and connecting them to the top of another, we can make the current value "flow" from one box to another. In a perfect world, to control volume using this sensor, we would just have to draw a line from the bottom of this box to the top of the box labelled *Volume* in the *Product Functionality* section. But we don't

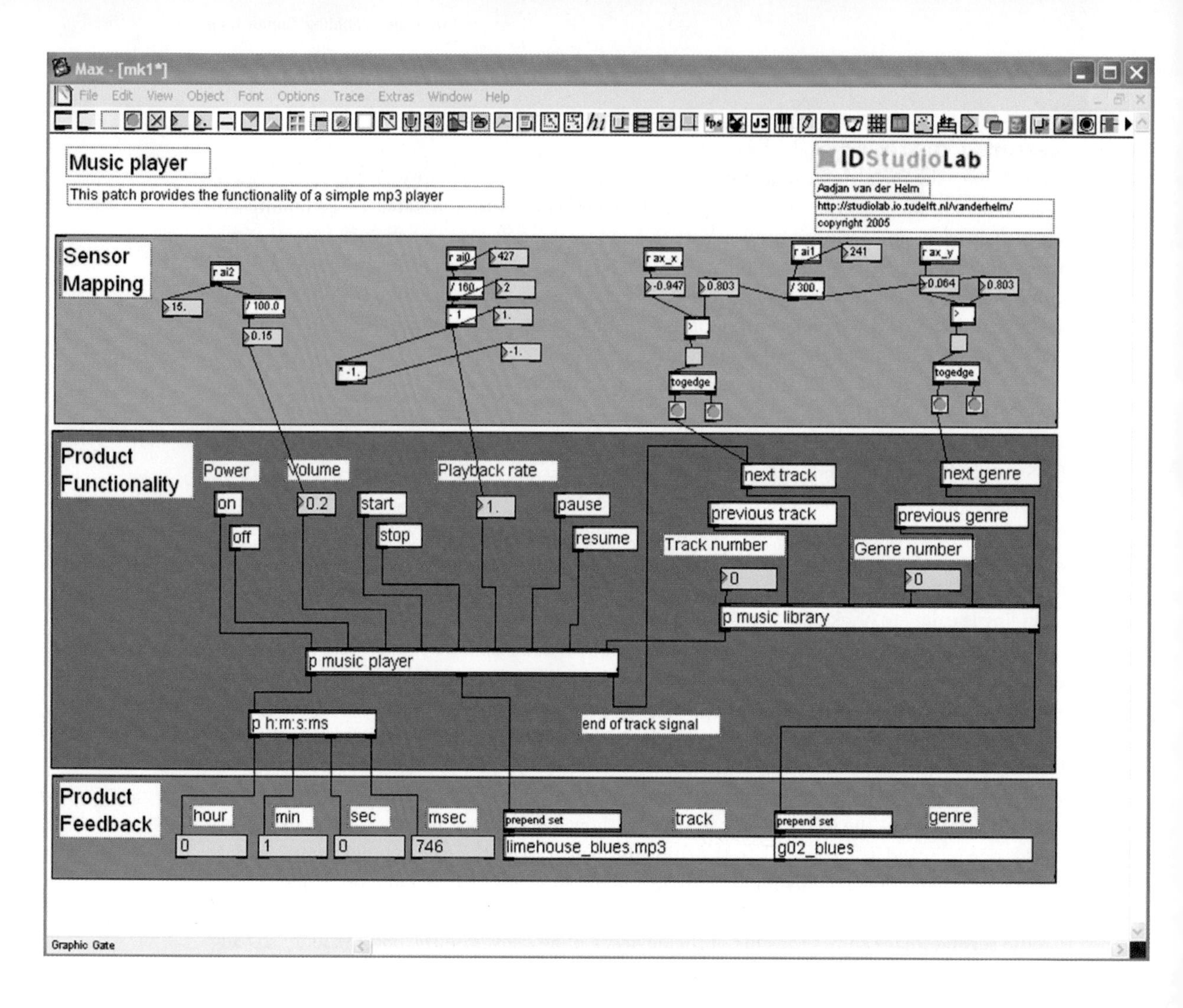

Figure 92: Program to Control the Gesture-Controlled MP3 Player

live in a perfect world and being true to life, our exercise was not that simple. The range of values output by the sensor were far too large for the MP3's volume control. For example, if the volume control went only from 1 to 10 but the range of our sensor produced numbers from 1 to 100, we would be able to use only 10% of the full dark-to-light continuum.

We knew that the range of the volume control was 0 to 10. So our first step was to see what the minimum and maximum values coming out of the light sensor were. To determine this, we added a numerical display box that would show us the instantaneous value coming out of the sensor. This is the box to the lower left of the r ai2 box. It shows us that at the time that this figure was made the output was 15.

Experimentation told us that the range of values from the light sensor was 0 to 1000. Consequently, we passed the sensor's output through an arithmetic box, which divided the incoming value by 100, thereby enabling us to use the full dynamic range of the sensor, and still remain within the proper range of the volume control.

That this operation worked can be seen in the numerical display connected below the division, which shows that the value has been changed to 15. This is the value that is then piped to the input of the *Volume* control.

We then used similar techniques to map the outputs of the other Phidget controllers to the Playback Rate, Next Track, and Next Genre controls of our MP3 player.

Stepping back and looking at the *Wizard of Oz*, *Smoke-and-Mirrors*, and *Bricolage* examples that we have discussed, I have to concede that their roots are mainly in research rather than product development. The fact is, this is simply where the techniques originated (at least in terms of the publicly available literature).

Being sensitive to the not infrequent gap between research and product cultures in organizations, I want to emphasize that these roots in no way disqualify the use or relevance of such techniques in product design. In fact, many innovative designers will likely yawn at some of my examples as old hat, having used them for years.

There are two general take-away lessons here. The first is the way of thinking about and exploring the design ideas that these examples represent. The second is hinted at by my modern return to *Chameleon*. It gives us our first clue that the tools and technologies that are available to us today are opening up new ways of approaching things. Not only is the barrier-of-entry to using these techniques dropping really fast, but those same changes mean that we are frequently able to take these things out of the lab and into the wild, whether for user testing or participatory design. (For more on this approach to physical interfaces, see Greenberg & Fitchett 2001, Holmquist, Mazé & Ljungblad 2003, Holmquist, Gellersen, Kortuem, Schmidt, Strohbach, Antifakos et al. 2004, Greenberg 2005 and Villar, Lindsay & Gellersen 2005.)

There may be as many approaches to sketching interaction as there are products to design. In this volume we will scratch only the surface. But we have to start somewhere.

These have been my opening "E" examples. Now we are going to drop down to the "A" level on the experience scale, and work our way back up.

It was a Dark and Stormy Night...

If you really want to see something just close your eyes.
—Alex Manu

Throughout the second half of 2003, I seemed to be visiting a lot of design studios, schools, and conferences. Design—especially interaction and experience design—was front and centre in my consciousness. That is what I talked about in my presentations, and that is what dominated the conversations I had with the people I met. One such conversation was with my friend Alex Manu, who teaches at the Ontario College of Art and Design in Toronto. We had a long lunch where I did my best to explain to him how I was thinking about sketching in the context of experience design. I was hoping to extract some kind of reality check from him. Did any of what I was thinking about make sense to someone from a more traditional industrial design background? (Of course, I may be the first person in the world to call Alex "traditional," but that is just another way to break tradition.) Anyhow, the next day he sent me the following e-mail:

> I keep thinking about the sketch gesture versus thought and know what bothers me with going visual too soon (in the case of the "literal" sketch): it is the commitment to form. Too many designers commit too soon to form through sketching and then they just reduce the field of design to what is visible. Mind play sketches allow you to discover and design functions and relationships not apparent to the senses. (Alex Manu, personal communication)

Reading this was a revelation to me. I had seen the parallels between film-making and interaction design in terms of the need for preproduction and the use of video to capture demonstrations, and such, but I had missed the more subtle (and perhaps more important) parallels between storytelling in film and interaction design. I even lost sight of the fact that most films start as words, not pictures.

Alex's e-mail reminded me of the importance of doing exactly what I have been doing throughout this book—telling stories.

If interactive sketches need to be timely, cheap, and quick to produce, then words are one of the most efficient weapons in our arsenal. Even my old college dictionary supports the notion of sketching with words. In fact, it does so in terms of both prose and theatre:

> **Sketch: 3 a:** a short literary composition somewhat resembling the short story and the essay but intentionally slight in treatment, discursive in style, and familiar in tone ....c: a slight theatrical piece having a single scene. (Webster's New Collegiate Dictionary)

In my experience, it is important for the well-equipped designer to have a facility with sketching in terms of both nuances quoted in this definition. However, anyone who has worked closely with me will be forgiven if they shake their head in surprise when they read this. The reason is that they have heard me repeatedly groan, or worse, when I have been subjected to tedious and contrived scenarios ("Bob is a ..." or "Mary works as a ...") that are so favoured by some people in marketing and usability engineering.

It's not that I don't like stories or don't think that they are important. Anyone who has heard me speak knows that I work very hard at being good at telling a well-crafted story. In the user interface community there is a solid body of literature that discusses the value of using scenarios and persona as part of the design process. A couple of examples from the academic perspective are Carroll (1995) and Carroll (2000). From the practitioner's literature, one example is Cooper (1999), and for me, the most useful is Jenson (2002). From the perspective of using personae, perhaps the most thorough treatment is that of Pruitt and Adin (2006). Lest we inadvertently assume that this is a recent approach to design, or something that began within the human-computer interaction community, it is worth noting that its genesis is in the 1930s, and is reflected in Joe, Josephine and their children—creations of the industrial designer Henry Dreyfuss (Dreyfuss 1955). Finally, there is also a small but relevant literature emerging on the relevance of storytelling on organizational change (Denning 2001; Brown, Denning, Groh & Prusak 2005).

> A [good] story is worth a thousand pictures. (Gershon & Page 2001)

Yes, the right story told in the right way at the right time by the right person can be extremely helpful. What concerns me is that although most people have a pretty good sense of their ability to draw or sketch with a pencil, my impression is that they are far less adept at assessing their story-telling or acting ability. Like any other tool in our arsenal, this one takes practice to develop a reasonable level of skill. With bad sketches, you can just look away. Stories take time—especially bad ones. So they are much harder to escape, and therefore have the potential to be far more painful and embarrassing than a bad drawing.

Perhaps the problem with many of the "scenario" stories that I find so trying is that they violate the spirit of design. Rather than invite, suggest, and question—three of the key properties of sketches that we have identified—they tend to tell, show, explain, and try to convince. The key to stories, as Alex describes them, is to help discovery. Taking our lead from the previous definition, it is the discursive element—the back and forth playing with the story among the design team or the audience—wherein the insights are found and where the value lies.

There is one key word in Alex's e-mail that I can't ignore and still do him justice. It is *play*. Play is not something that I would have talked about had he not written to me, but it is central to understanding his approach to design. It is also something that resonates with me, especially when I try to look objectively at my own behaviour.

In short, I seem to be someone who is constantly reminded by others that I don't take serious meetings seriously enough. What they really mean is that I'm constantly playing with what is being discussed, and throwing in seeming non sequiturs, bad puns, or simply comments that I, at least, think are curious, interesting, relevant, funny, or all of the above. Without Alex's prompting, I would not have even mentioned my firm belief that my best ideas consistently come from verbal playing around with thoughts triggered during conversations with others. This leads me to share two of Alex's mantras around this topic:

> Without play imagination dies.
>
> Challenges to imagination are the keys to creativity. The skill of retrieving imagination resides in the mastery of play. The ecology of play is the ecology of the possible. Possibility incubates creativity.

Or one of my own personal mantras:

> **These things are far too important to take seriously.**

Stories, and more importantly, story-telling and play, are a critical part of design. And like other forms of sketching, we can capture their essence in a persistent form that we can "stick" on our digital corkboards as reference material to share with others, be it in audio recordings, video, snapshots, key words such as the title, or even just in the name of one of the characters.

For example, if you have read this far, then you are part of my "community of shared references," and now all that any of us has to do is say the words "mountain bike" or "orange juicer" and the rest of us will know the story. Those brief words will conjure up a whole belief system around innovation and experience design, one that is special to our specific community. Conversely, those same words will be meaningless, or invoke perhaps other, contrary images, to those from outside our culture. Therein we see both the potential for developing our own community, but also the risk of isolating ourselves from others.

Our stories give us shared references that constitute a shorthand for key landmarks that aid us in navigating within the otherwise amorphous space of design. And sometimes shorthand means really short, and the whole story itself may only be a few key words—as long as they conjure up the right images and memories on the part of the listener, as in the following anecdote:

> When British Rail wanted to develop a new design for their InterCity trains they invited a number of leading designers to submit proposals. The winners were in fact Seymour/Powell who, at that time had no previous experience with train design. The Seymour/Powell submission was not based on drawings or traditional design documents. They simply explained to British Rail that their design would be 'heroic' in the manner of the British Airways Concorde and that it would once again make children want to become train drivers as in early times. We can only imagine that such a description must have triggered childhood memories in the minds of some senior British Rail executives, and that they carried with them their own image of such a train. (Lawson 1997; p. 255)

Like Alex Manu, Thomas Erickson (1995; 1996) has been an articulate voice in advocating the use of storytelling in interaction design. To me, he makes a really important point in the following words:

> Stories are particularly useful for communication within the organization for two reasons. First, stories are memorable. People will remember—and retell—the ... story long after they have forgotten the more formal principles ... Second, stories have an informality that is well suited to the lack of certainty that characterizes much design-related knowledge. (Erickson 1996; p.35)

To be sure, a good story helps communication. But this is not just about you communicating to a colleague or a customer. Good stories are retold. Hence, they not only help you explain your ideas and make converts. Through their memorability and retelling, they provide a means whereby your audience itself becomes an effective conduit for spreading the debate and understanding reflected in your tale. In short, they are a form of "viral marketing" for design ideas.

Two Sides of Role Playing

And as my old college dictionary suggests, our stories can be theatrical as easily as not—another kind of play. So contrary to what you were told in grade school, you can act up. It is not only encouraged, it is an essential part of experience design. Interaction is about roles and their changing relationships.

These roles can be of two very different types: most commonly, it is the role of the user that is undertaken; less commonly, it is the role of the product. Since it is the exception, I will talk about the latter first.

In 2003 I visited the Interactive Institute in Stockholm. This is a research centre for interaction design that has studios in a number of locations in Sweden. While reviewing some of the projects, I distinctly remember one that struck me as really novel. It involved looking into new types of devices and services that could help people find their way around town, both in terms of getting from point A to point B, and in terms of accomplishing a particular task, such as renewing a driver's license.

In order to better understand the problem and get a sense of the types of dialogues one might have with such a device, they adopted a methodology that I had never heard of before. They simply had members of the design team play the role of the device, and provide the service to sample users as they went about their day. That is, their role was to be a surrogate for the device and to respond to requests in a manner that was consistent with the proposed device.

Despite having a human undertake the role of the system, the approach is distinct from the *Wizard-of-Oz* technique since there is absolutely no attempt to fool the user into believing that the system is real. Nevertheless, the designer-surrogates were not there as a companion or friend of the user, and remained in character throughout. The conversations were recorded, the "computers" and users debriefed, and all this contributed to the data that was used to inform the design process.

At least, that is how I remember it. But now I am starting to doubt my own powers of recollection. The example stuck in my mind and made enough of an impression that when I returned to Toronto I wrote to the institute to get more documentation on the project, since I wanted to document it for this book. To my surprise, nobody knew what I was talking about, and subsequent efforts to sort this out have come to naught.

So either I misunderstood what I was being told, I didn't explain what I was asking for properly, or I simply imagined the whole thing. The fact is, I can provide no reference to where to find out more about this specific project. But the technique itself is somewhat like what John Chris Jones (1992) calls Personal Analogies, which he describes like this:

> The designer imagines what it would be like to use one's body to produce the effect that is being sought, e.g. what would it feel like to be a helicopter blade, what forces would act on me from the air and from the hub; what would it feel like to be a bed? (Jones 1992; p. 279)

In my example, the designer went beyond imagining, and actually assumed the role of the device. Does it matter if I imagined the whole thing? No. The important thing is that the technique is interesting and can be useful in certain contexts, and the story gives you something to remember it by—which is what stories do.

So now let's look at the case where the role assumed is that of the user, rather than the product. Here, the intent is not to test a product, but to use role playing to understand the experience that it engenders. The hope is to gain insights that might guide future actions and/or cultivate understanding.

I am going to discuss two examples. Both are pretty extreme cases of "putting yourself in someone else's shoes." But then, as my climbing partners are prone to say, there is nothing like a little extremity to help bring on a healthy dose of clarity.

The first example is a pretty famous book from the early 1960s, *Black Like Me*, by John Howard Griffin (1961). The second is the less known, *Disguised: A True Story* (Moore & Conn 1985).

In case you are not familiar with it, *Black Like Me* documents the author's experience in trying to understand what it was like to be black in the southern United States. The two key things to know are first, that Griffin was white, and second, this was in 1959—a period of high racial tension in the south, right at the front-edge of the civil rights movement that came into its own during that decade.

What Griffin did was change his appearance, including the pigmentation of his skin, so that he could pass for black. He then spent a little over a month traveling through Mississippi, Alabama, Louisiana, and Georgia. He did not want to observe the black condition, he wanted to *experience* it, and there is no question in the mind of anyone who has read the book that he did it "in the wild." Now it would be naïve to confuse Griffin's experience with actually being black in the south. The economic and educational circumstances of his upbringing, the fact that he was only black for a month, and his knowledge that he could and would return to his white middle-class background were insurmountable barriers to his being so. But on the other hand, his experience was visceral, real, and far beyond what he could have gotten from interviews, observation, or reading the literature.

Black Like Me did not come from the design perspective—other than in the sense of trying to bring about a redesign of American social and cultural values. My second example did. It comes from another book, *Disguised*, which could just as easily be called *Gray Like Me*. It is the story of an industrial designer in her mid-20s, Patricia Moore. The question that motivated her work was:

> How could I expect to understand their [elderly people's] needs, to help design better products, develop new concepts, without a sharper sense of their daily experience? (Moore & Conn 1985; p. 39)

Between 1979 and 1981, with the assistance of a professional make-up artist, she regularly and convincingly assumed the role of an 85-year-old woman. Actually, in order to experience how apparent economic circumstances affected how the elderly were treated, she variously assumed the role of one of three such women: a wealthy elderly matron, a middle-income lady, and a bag lady.

However, she did not just want to assume the appearance of being old. She wanted to experience the feeling of being old, and having to function in the physical world. So, she also adopted what I would call negative prosthetics. For example, she put baby oil drops on her eyes to blur her vision and wax plugs in her ears to impede her hearing. She wrapped her legs in bandages, taped a wooden splint behind her knee, and wore an elastic belt below her hips—all to limit her physical agility to be comparable to someone of that age. She also wore gloves to conceal that her fingers were taped, so as to restrict their motion much as would be the case if she had arthritis.

Now some of this might seem extreme, but take the taped fingers, for example. Can you think of a better way to give one a deep understanding of the difficulties and frustration of dealing with things like latches on your purse, or fumbling with change or keys? Such negative prosthetics expanded the range of her experience from that of how people treated her to what it was like to navigate in the world with the mobility, dexterity, and sensory facilities of a senior.

Of course, she was not old, and only assumed her role about once a week, and only for a day at a time. And even easier than Griffin, she was able—at will—to switch back into her younger self. But I think that it takes little imagination to understand the difference in the depth between understanding that is founded on experience, compared to that which is purely intellectual.

Regardless, as interesting as these two examples might be, in some ways it is fair to ask how they relate to sketching, or our overall theme. On the one hand, Griffin's case in *Black Like Me* doesn't even conform to some of our basic criteria. It sure wasn't cheap or quick. There was only one of him, and in order to work, the rendering of the character had to be perfect, not sketch-like. Nor was the role easily disposed of. Unlike Moore, for example, he was not able to take off his make-up, costume, and prosthetics in the evening, and go back to his regular life.

The case of Moore is perhaps more helpful in establishing the link. For example, you don't have to go to the full extreme that she did. Let's assume that you are designing a new change purse. Reading her book will not only remind you to take elderly users into account in your design. It points out alternative ways of doing so that you may not have thought of otherwise. For instance, following her example, you could quickly tape your fingers, as she did, or hold some ice cubes for five minutes to freeze them, in order to approximate the experience of an elderly person with arthritis.

So, yes, my examples are extreme, but in a way, that is where their value lies: in helping focus on the role-playing. What is most important about both is that Griffin was not black and Moore was not old, so the experience of each was a taste, or approximation, not the real thing. But it was visceral, and in situ, rather than in the lab or some artificial environment, and the experience of each brought forward insights that shed light on the questions or issues that they were investigating—insights that were almost certainly more meaningful and helpful than they would have been had they gotten them in any other way.

These are properties that are absolutely in keeping with our topic. In a way, all that we need to do to make this type of role playing fit our purpose is to reel them in a bit, and at the same time, recognize their limitations. Concerning the former, it should be pretty clear that there are often real advantages to walking in the shoes of your target user, and to do so in context, in the field. Likewise, it should also be quite obvious that if you are trying to address some problem that a brain surgeon is having, that there might be a few minor limitations on the extent that you can assume their role.

One final point about these examples. I stated earlier that I wanted to use this part of the book to help build a deeper awareness of our collective traditions and history. The case of Moore and *Disguised* is a good example of why. As I mentioned, she comes from industrial design, and wrote a book that is very relevant to the use of role playing in design. Yet, in the dozens of publications that I read on role-playing and theatre in experience design, only one article—by the designer Bill Moggridge (1993)—mentions her pioneering work. Too bad. We are not good enough at our craft to be able to afford to lose such lessons. Did this happen because of bad scholarship or incompetent people? No, I don't believe so. My assumption is that the literature is just really vast, and it is hard enough to stay on top of our own area, much less that of others.

This sounds like one more reason to share our stories across cultures.

Theatrical Sketches of the Less Extreme Kind

As I just have hinted at, there is a fair amount of literature on the use of role playing, scenarios, and other theatrical techniques as an aid to the design process. Our interest is mainly around using scenarios and theatrical techniques as an alternative approach to identifying or working through problems, or to communicate or validate ideas from one stake-holder to another. As a kind of modern cousin of Brecht's explicitly didactic use of theatre, the term *informance* has been used to describe one approach to theatrical performance that informs (Burns, Dishman, Verplank & Lassiter 1994; Burns, Dishman, Johnson & Verplank 1995). Likewise, the term *bodystorming* (Burns et al. 1995; Oulasvirta, Kurvinen & Kankainen 2003) has been used to describe the use of acting out as a means to generate new ideas and insights about a particular question or problem. In this case, the term is an explicit reference back to brainstorming, which Osborn (1953) first wrote about over 50 years ago.

Other examples of using scenario- and acting-type approaches in ideation include Ehn, P. (1988), Brandt & Grunnet (2000), Strömberg, Pirttilä & Ikonen (2004), Iacucci & Kuutti (2002) and Iacucci, Kuutti & Ranta (2000).

Rather than dive into a deep analysis of these and related works, I think that the most effective thing to do is try and give some overview of the issues that I have extracted from reading them. For me, the easiest way to do so is in theatrical terms, literally. Hence, the types of variables that one can mix and match to expand one's repertoire of techniques include:

Script: What is the level of scripting? Is there a general scenario on which the actors improvise, or is it more scripted. If the latter, by whom?

Director: Is there a director or coach, and if so, who? In some cases it is the designer or the people building the product. In others, someone with professional experience in theatre is brought in.

Figure 93: Simple Acting Out "On TV"
There are simple techniques that one can adopt in order to reduce people's inhibitions when acting out scenarios or telling stories. Here, the device used is a cartoon-like TV bezel. Something as simple as this can be enough to enable people to come out of themselves.
Source: Van Rijn, Bahk, Stappers & Lee (2006)

Actors: Who is doing the acting? If we are designing a system intended for use by nurses or hairdressers, are representatives from those professions doing the acting, or the designers, or some third party, such as professional actors? Although there can be benefit in having the designer "walk a mile in the customer's shoes," there are limits. As we have seen, one can play the role of an old person to significant effect. On the other hand, there may be little benefit in having designers act out the role of a neurosurgeon.

Audience: For whom is this piece of theatre? For example, designers may act out scenarios for the intended users of a particular technology in order to get some feedback. Or, the designers may be the observers while a scenario is acted out by the users. It may be that designers and users watch professional actors, or that the designers are the audience for the same performance in which they are acting.

Setting: The scenario may be acted out "in the wild" (on location, so to speak), in some mocked-up simulation thereof, or in some generic space such as a conference room at the designer's office. The actual location may help build up the design team's understanding of the eventual context in which a product will be used. On the other hand, on-site explorations may be disruptive, inappropriate, or may overly bias ideas by the status quo.

Performance or rehearsal? Does the director or audience let the actors go right through the scenario without interruption, as in a performance, or can it be stopped, mid-stream, in order to ask questions, give notes, make suggestions, or change the script?

Props: Props can have a large impact on what scenarios you can do and how they are played out. This is not just a question of whether props are used. If so, which ones? Who designed them? Who built them? When were they introduced? In many ways, much of what follows (as well as our previous discussion of the *Wizard of Oz* and *Smoke-and-Mirrors*) is about prop building, and so can be read in this light.

Taken together, this defines a pretty rich space. My sense is that the most effective approach is to keep things simple, and build up your skill sets gradually. Svanæs & Seland (2004) is a good example of how one can build up experience and technique through a series of smaller studies. It also speaks to some of the issues in involving outside professionals in the process.

Perhaps the most important overall lesson here is to lose your inhibitions, and find your senses of humour and invention.

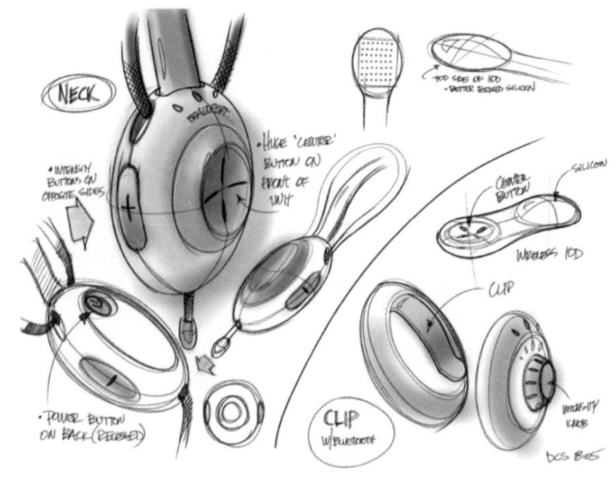
NECK
•HUGE 'CENTER' BUTTON ON FRONT OF UNIT
•INTENSITY BUTTONS ON OPPOSITE SIDES
TOP SIDE OF IOD
•BETTER FORMED SILICON
CENTER BUTTON
SILICON
WIRELESS IOD
CLIP
•POWER BUTTON ON BACK (RECESSED)
CLIP W/BLUETOOTH
INTENSITY KNOB
DCS 8-05

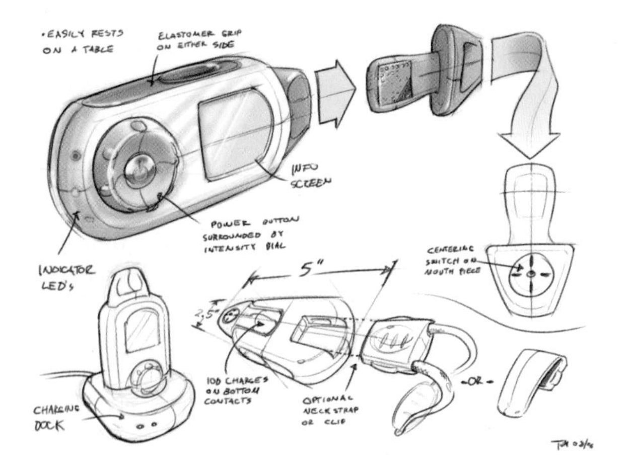
•EASILY RESTS ON A TABLE
ELASTOMER GRIP ON EITHER SIDE
INFO SCREEN
POWER BUTTON SURROUNDED BY INTENSITY DIAL
INDICATOR LED's
CENTERING SWITCH ON MOUTH PIECE
5"
2.5"
IOD CHARGES ON BOTTOM CONTACTS
OPTIONAL NECK STRAP OR CLIP
-OR-
CHARGING DOCK

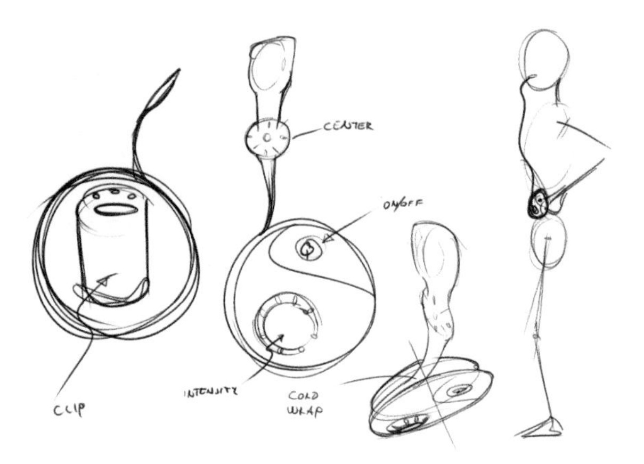

Figure 94: Acting Out with Sketched Props

One can workshop ideas where groups come up with design concepts. But their intent and value may only become evident when coupled with some theatre that explains their use in context.

Images: Brooks Stevens Design

Asleep on the Job

In the winter of 2004 I was working with a team at Microsoft Research in Cambridge England. We were looking into technologies specifically designed for the home. In so doing, of particular concern was showing respect for the moral order of the home. That is, the home has certain social, cultural and behavioural conventions that distinguish it from school, the supermarket, or the office.

Working with the sociologist Richard Harper, who was part of the team, gave me a whole new level of appreciation for such things, as well as a new term to add to my vocabulary. Having bought into the importance of the concept, my next thought was to try and find an effective way to capture its importance in a way that could easily be communicated to others.

My solution was a small piece of theatre coupled with my digital camera—the result of which is captured in the photograph on the facing page.

It shows Richard Harper in a meeting, at the office, wearing his pajamas and dressing gown! The reason that the image is unusual is because his being thus garbed violates the moral order of the workplace—at least at Microsoft Research. Therein lies the key to the point that I was trying to draw out: why is populating the home with technology designed for the office any more acceptable than wearing pajamas (which are perfectly acceptable at home) at the office? Through this piece of theatre, we gained a common reference object (the photo) that we could all use as a prop to facilitate future conversations about the moral order of both home and office. And we got to laugh at Richard.

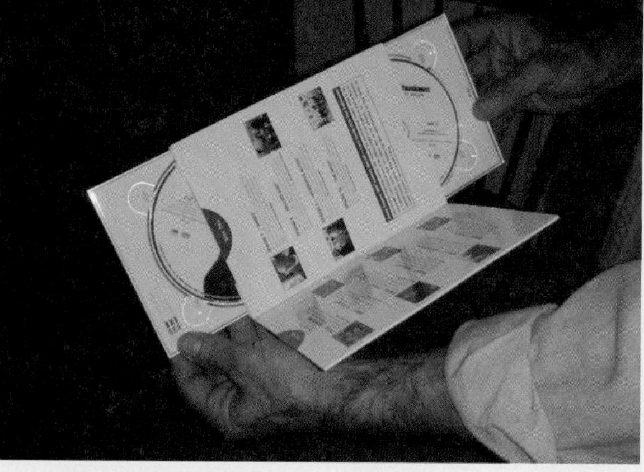

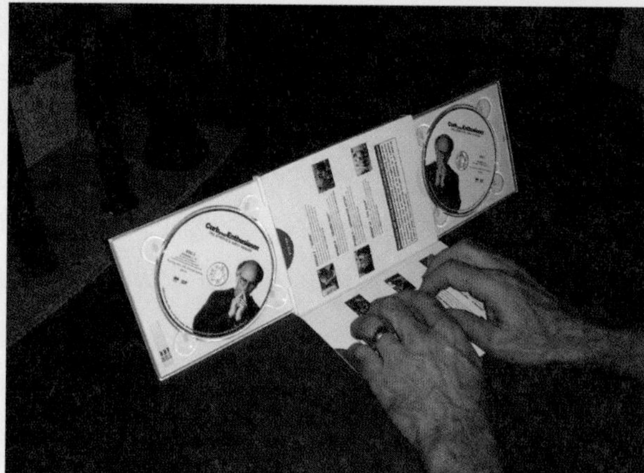

Curb Your Enthusiasm

These images show how a found object, here the packaging for a DVD, can serve as a concept model for a collapsible, portable computer. In the concept, the package opens like a regular clam-shell laptop. However, one can expand the screen into a panorama format by pulling sidewing displays out from the central panel.

To fully appreciate the example, however, you really need to experience it yourself. I actually mean it. Read what I have written. Look at the photos. Then go out and buy the package at your local DVD store.

For the cost of a lunch, you get something that will serve as an effective and compelling reference for the whole studio. By investing about five minutes more, you can photograph the scenario so that it can be shared even further. Finally, if you like Larry David, you get to watch his entire first season as a bonus.

Such things are the norm, not the exception, in design. There is a reason that earlier I described designers as hunters and gatherers. This DVD cover is just one of the treasures that came from this process. And anyhow, we didn't have a refrigerator box to play with!

Photos: Elizabeth Russ

Storytelling with Found Objects:

A Concept for a Collapsible Display for Ultra-Portable Computers

Have you ever watched a child unwrap a gift and then play with the box rather than the toy that was inside? As a child, when your parents got a new refrigerator, did you not take the box and transform it into a fort or a spaceship? We have all seen and done such things–made free associations between objects and their meaning and purpose. The key observation here is that such transformations are as fundamental to design thinking as they are to childhood imagination and discovery. Let me give you an example.

I was in a meeting in Cincinnati with Larry Barbera, Director of Industrial Design, at the design firm, Kaleidoscope. The purpose was to talk about new kinds of display technologies and how they might be packaged.

Larry had scattered a number of everyday objects on the meeting room table. One of them was the DVD package for the first season of the HBO Video series, *Curb Your Enthusiasm–The Many Moods of Larry David.*

I had never heard of the show. Nor had I ever seen a DVD packaged the way that this one was. It opened much like a book. Then the two disks pulled out symmetrically from the top and the bottom of the package. What was subtle about all of this was how the two were linked: you couldn't pull one disk out without the other. Somehow the mechanisms were linked together.

I immediately got the point. The object itself was its own script. All that it needed was for me to act it out. What the plot revealed was that this seemingly unrelated object was really a novel representation of a new concept for an ultra-portable computer. But I can't adequately explain this to you in words. I have to show you–hence the accompanying photographs.

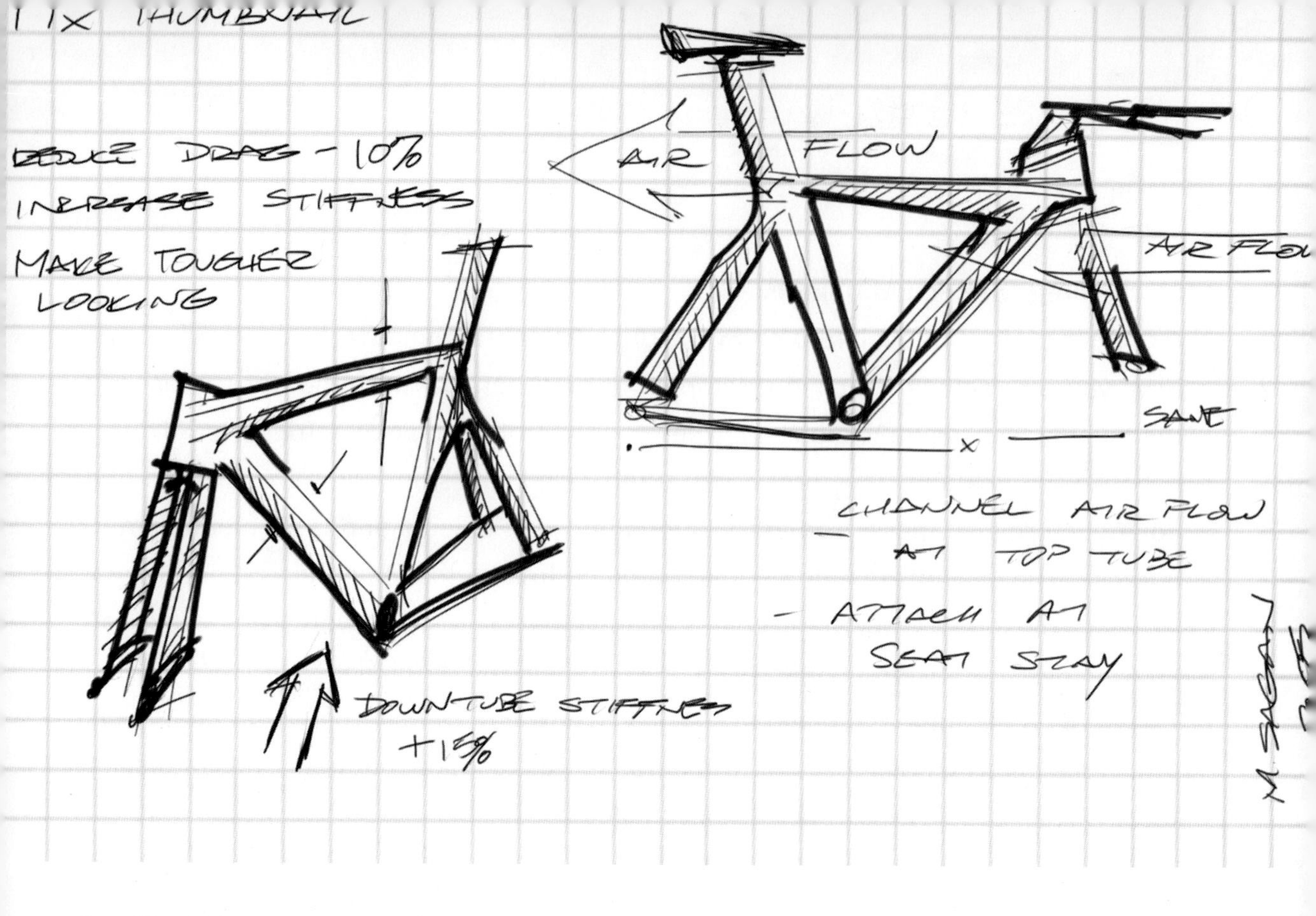

Figure 95: Words Meet Images

Here a combination of drawings and words give information that neither could do on their own. Besides understanding just the form, we gain insights about the rationale for the design, including issues such as aerodynamics and strength.

Figure: Michael Sagan.

Visual Storytelling

Every Picture Tells a Story.
— Rod Stewart & Ron Wood

From words we now move on to images, and visual story-telling. Of course, in so doing, we don't abandon the word altogether. We just add another tool in our sketching arsenal. In fact, the bicycle concept shown in Figure 95 is one example of how words can be incorporated into a sketch.

The foundation for this section was laid back when we were discussing the sketches of the Trek time trial bike. These provided a means of introducing the role of traditional sketching in design. I bring them up again here as a reminder that traditional sketching has an important place in experience design—they are just not sufficient.

So, having people on the team that can draw and produce effective traditional sketches is really valuable. But then, that leaves me out. My wife, Liz, is a professional painter and my eldest son, Adam, an illustrator. What that means is that I need no reminder of the limitations of my drawing skills. What my eye doesn't tell me, they will. Consequently, I do my best to work with people whose skills complement mine, and who have the same facility with a pencil as Liz and Adam. But despite my best efforts, reality dictates that sometimes I have to fend for myself. So what is one to do?

The first thing that I would say is, "Get on with it." I take marks off my student's work if their sketches are too good—among other things, it indicates that they spent too long on them. So don't worry if your drawings don't look like Rembrandt's, and remember the *power law of practice*. It confirms that although practice will not necessarily make you perfect, it will certainly help you improve.

To help you with this, there are lots of books and courses. Nanks & Belliston (1990) is just one. If you are quite accomplished already, then look at books like Powell (1985). No matter what, practice!

In terms of improving my own limited skills, one of the things that gave me great encouragement took place in New York City in 1997. My wife and I were going through a retrospective exhibition of one of my heroes, Robert Rauschenberg, at the Guggenheim Museum. About three-quarters of the way through the exhibit, I had an epiphany: There was nothing at all in the exhibition that suggested that Rauschenberg could draw!

What the exhibition told me was that one of the most influential visual artists of the second half of the twentieth century achieved that stature with only limited drawing skills. As an aside, a follow-on speculation was that those same limitations may well have been a key catalyst to his creativity.

The take-away lesson from this story—even if the art historians in the crowd discount it totally—is that Rauschenberg points to a way for people like me to augment their meager talent: just find another way that you can manage, and get your ideas down. To help get you started, let me give you a few hints that work well for me.

The first is to do what we used to do in kindergarten—trace. What I do is take digital photos, or

Figure 96: Sketch as "Manual Photocopy"

Since my talent at drawing is about the same as yodeling (don't ask), the fastest way for me to get a sketch of a phone is to trace a photo of the real thing. I load it into a sketching program as the background layer (top left), add a second layer on top, and trace the outline in red (top right). I then make the background layer invisible, and end up with what looks like a reasonably well-proportioned and recognizable sketch of a mobile phone.

Figure 97: Phone Graffiti

Here I have used the same phone as the background. This time I have added a hand-drawn screen on the display.

scans of images, and load them into a drawing program. As Figure 96 illustrates, I add a transparent layer on top of the image, such as the mobile phone in the top-left image, and trace over it. (In filmmaking, the fancy name for this is rotoscoping, so if you feel too old for tracing, then you can use the fancy term.) Regardless of what you call it, once you have finished tracing the object or person (top right panel in figure), you can hide the background photo. This leaves you with only the hand-drawn version, which you can save and then use in other drawings—sort of like your own sketch clip-art. You have something that you can do quickly to function as a sketch, and which has proper proportion, perspective, and such. So it is a start. I did the example in Figure 96, including capturing the screen snaps, in less than five minutes.

The second technique is a slight variation on the former. In this case, you just draw on top of the photographed object. As is shown in Figure 97, I have added a (clearly) hand-drawn interface onto the screen of the photographed phone. The resulting composite has a sketch-like feel, despite the bulk of the image being a photograph.

The potential impact and use of this kind of technique is shown to really good advantage in the two images in Figure 98. In both cases parts of the original photos were traced over and transformed into cardboard-type sketched characters. By tracing, the position, pose, and proportion are correct. Achieving this effect is technically simple in terms of technique and technology. The key ingredient is imagination.

The result is really effective. The photographs were clearly taken by someone with professional level skills, and would never be considered to be "sketch-like" on their own. That is precisely why the traced figures stand out so much and work so well—they are just so incongruous in that visual context. They catch your eye, and despite the detail being in the photo, your eye is drawn to the sketch. As Andrea del Sarto says in Browning's poem, "Less is more."

In these examples, be clear that I am not trying to give a comprehensive or even short course on image making. I am just trying to make a few points:

Conventional sketching has an important place in interaction design.
Even those without a great deal of natural talent can improve their drawing skills with practice.
There are a range of techniques and technologies that can be used to create images that serve sketch-like purposes (we have only touched the surface).
Remember the Rauschenberg Effect: The limiting factor is your imagination, not technology or technique. There is always a way to express an idea appropriately within your means.

That being said, if traditional sketching were sufficient to handle the types of things that we want to design, there would be little need for this book. As I have said previously, the main drawback of conventional sketching has to do with its limitations in capturing time, dynamics, phrasing—the temporal things that lie at the heart of experience.

Figure 98: Hybrid Photo-Graphic Composition

A bit of tracing and drawing can enable a well-crafted photograph to serve as a sketch.

Photos: Aldo Hoeben ID-Studiolab, TU Delft

Sequencing Images: Scott and Ron's Agenda

Let me give you a simple exercise to illustrate what we are up against here. Ideally on paper, but at least in your mind's eye, I want you to make exactly two sketches:

- A sketch that captures a literal representation of the physical nature of your mobile phone (or iPod, or Palm Pilot, etc.)
- A sketch that captures a literal representation of the behaviour of the user interface of your mobile phone (or iPod, or Palm Pilot, etc.)

The first sketch is easy, even if you are terrible at drawing. The second is almost impossible, no matter how good you are at drawing. Pretty interesting, and therein lies the problem. If we are going to exercise anywhere near the level of control over the interface, we need to be able to sketch its essence as fluently as we can sketch the physical form factor. So we need to expand our repertoire of techniques.

The most obvious way to do so is to use more than one image to tell the story. As we already know, this is the foundation for comics as well as the storyboards used in the preproduction of films and video games. This is also a technique that has been used extensively in user interface design.

Another technique is to use something that is called a *state transition diagram*, which can be used to make a kind of map of the displays in an interface. An example of both used together is illustrated in Figure 99.

These drawings were made by Ron Bird, working in collaboration with Scott Jenson. Ron is a pioneer in using all kinds of techniques to make early sketches of interfaces, such as for photocopiers and mobile phones. Scott has a long background in designing interfaces for handheld products, ranging from personal digital assistants (PDAs) to cell phones. The images in the figure represent an early user interface concept that they were exploring for use in the agenda/day-timer of a handheld PDA.

As I said, these sketches represent time in two different ways.

The first technique used in the figure is the familiar convention of a comic strip. That is, Ron has used multiple images to portray the state of the display as one goes through a particular sequence of transactions. In this case, the scenario represented is checking your calendar for the time of a meeting, and sending a message to the person that you are meeting with:

Far-left screen, 1.a: Looking at the agenda and seeing that there is a 10:00 A.M. meeting with someone named "Mary Ford"
Middle-left screen, 2.a: Checking what that meeting is about
Middle-right screen, 2.b: Choosing how to contact Mary
Far-right screen, 4.a: Sending her a text message concerning the "Tour," which was the topic of the meeting

The second way in which time is represented is the state transition diagram technique that I mentioned earlier. Below the sketch of each screen, there is a rough drawing of a navigation "map" that shows the relationship of the various screens to each other. For each of the screens shown, its particular location in that map is highlighted. There are paths connecting some of the screens to each other. These let you know "if you can get there from here." So the screens constitute the "states" and the connecting paths the "transitions" that are possible to or from any particular state (or screen).

To help in reading this map, Ron and Scott have laid out the map in a grid and labeled each screen sketch in a way that identifies its map position: the number, such as in 2.a, indicates in which map column it sits, left to right, and the letter indicates which row, from top to bottom. Hence, we know that we have seen only four of the eight screens of the design.

These types of sketches help the designer explore and communicate the look or character of a particular design approach. They also help explore the dynamics and flow. Although the sketches are static, hanging them up on the wall lets the designer ponder issues such as what action takes you from one screen to another? For example, one might ask, "How do you move to screen 2.a, rather than 2.b. from 1.a?" The representation also helps us contemplate other time-related questions, such as, "What is missing?, Is the flow right?, Are there too many screens?, or Could there be a better way to approach this?"

Because these kinds of sketches are relatively quick and easy to execute, the designer is afforded a way to explore a number of different approaches to the design, in terms of graphical style, functionality, and the flow of the interaction. Far more variations can be explored on a given budget than would be possible if one was implementing the designs in code, for example.

The practice of mocking up screen shots either as a storyboard or in a *PowerPoint* presentation is pretty common in user interface design. So let me tell you why I chose this example. The incorporation of the state-transition diagrams was my first reason. This is something that I think is really useful, but not generally practiced. Second, the biggest and most common mistake that I see is that people start with screen graphics that are so realistic that they could easily be mistaken for the real thing. This is not the case with Ron. The relevance of his hand-drawn artwork to the ultimate product is clear in the example and does not require any great conceptual leap.

High fidelity screen renderings certainly have a place—just much later in the design process, before implementation, but almost never, if ever, at the early ideation phase.

One other comment, following my using the words "storyboard" and "*PowerPoint*" in the same sentence. They are different things, and not just because of the technology. If you presented exactly the same screen shots as a storyboard and as a slide presentation, such as with PowerPoint, the results would be very different. In technology jargon, the storyboard presentation is space multiplexed, but the slide show approach is time multiplexed. In everyday language, with the storyboard, time is distributed in space, and you can see all screens simultaneously. With the slide-show approach, they are seen sequentially, one after another. Both have their place, but they are different, and this is worth keeping in mind.

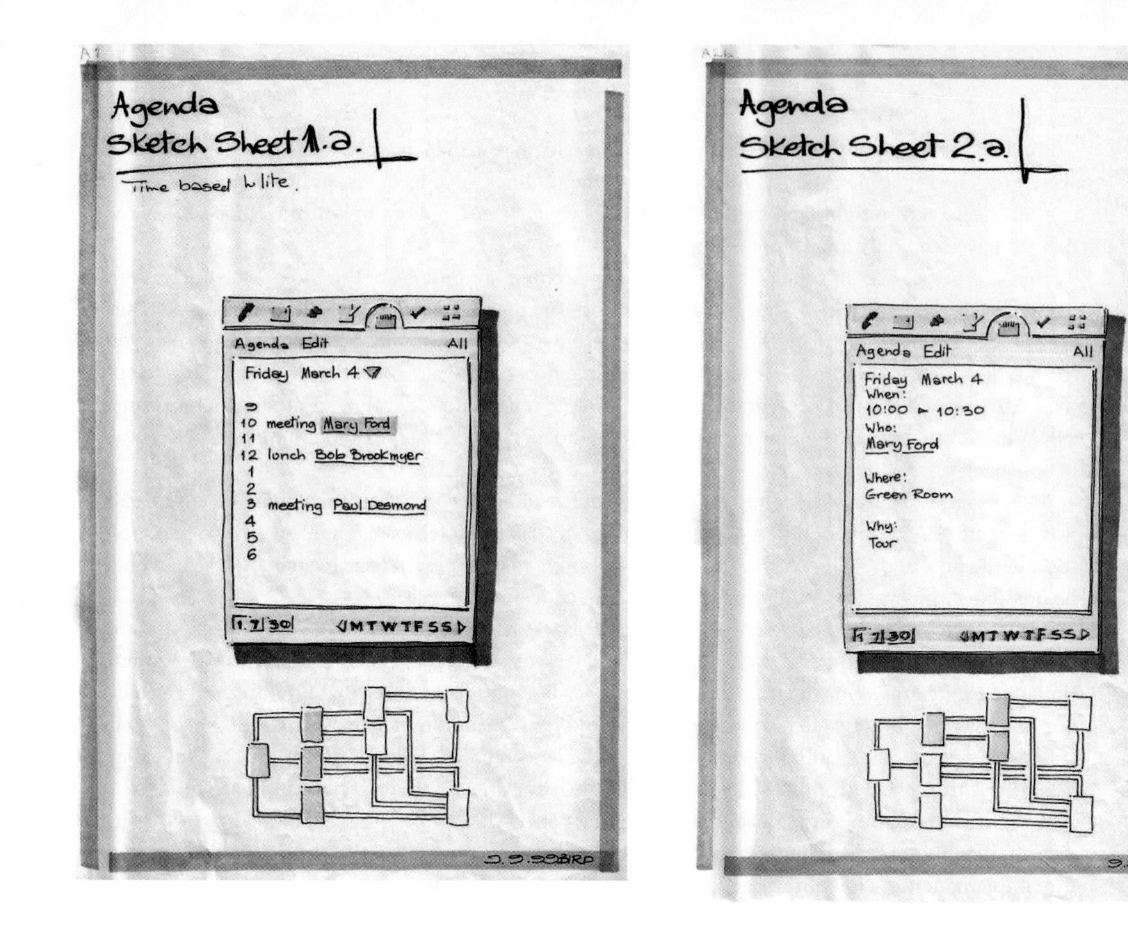

Figure 99: Sketches of PDA Agenda Screens

The sequence of images sketches out a potential design for interacting with a PDA-based agenda. Each image is like a key frame in an animation. Notice the state transition diagram at the bottom of each image, which shows its context relative to the others, as well as the overall page hierarchy. Images: Ron Bird

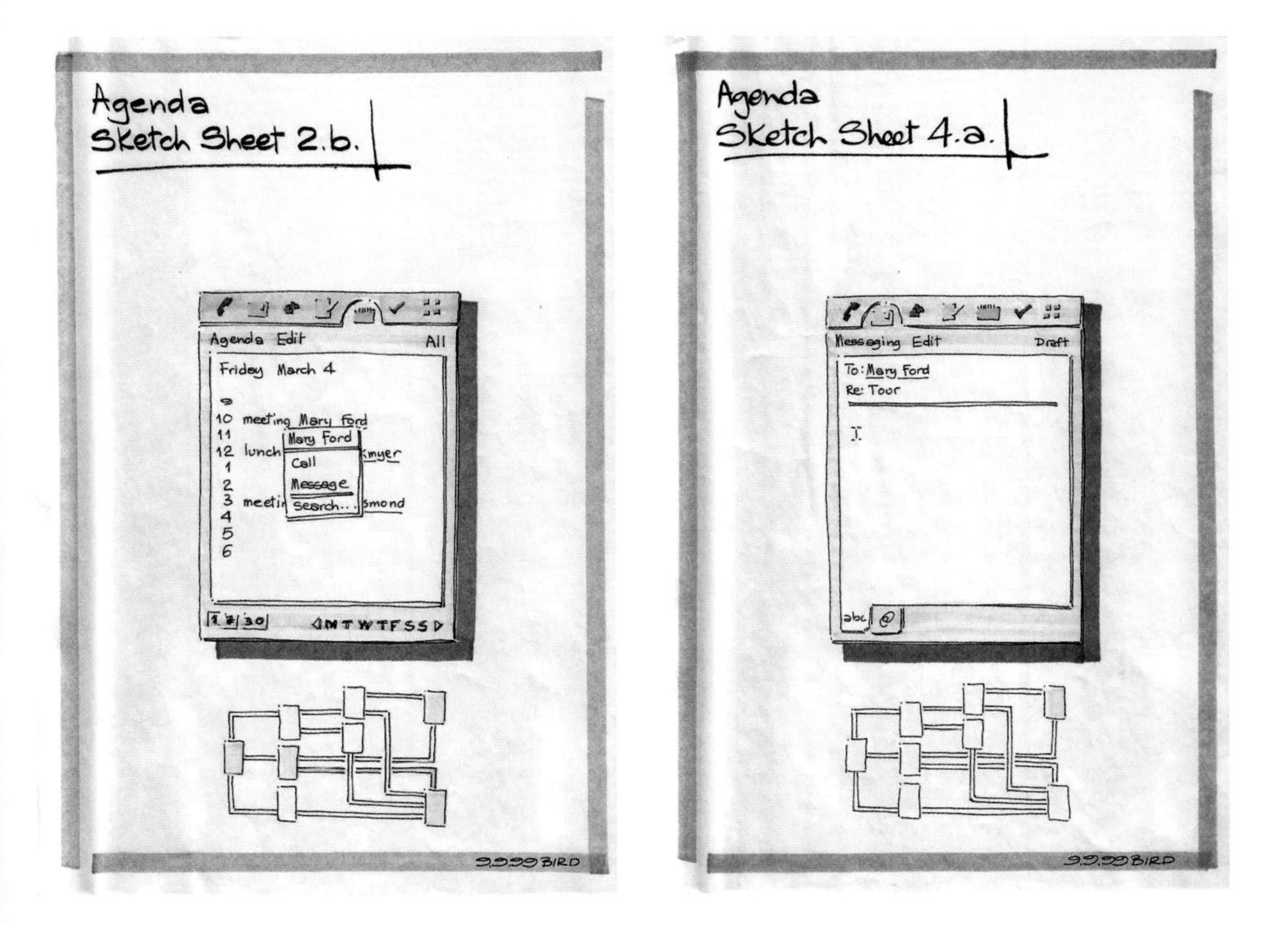
Agenda
Sketch Sheet 2.b.
Agenda Edit
All
Friday March 4
10 meeting Mary Ford
11
12 lunch
1
2
3 meeti
4
5
6
Mary Ford
Call
Message
Search...
smond
9.9.99 BIRD
Agenda
Sketch Sheet 4.a.
Messaging Edit
Draft
To: Mary Ford
Re: Toor
abc
@
9.9.99 BIRD

Figure 100: Zone Tactics

These boards were created to explore the interaction model of a proposed pervasive electronic game. They capture more the spirit of the play than the details of the technology design.

Images: Mark Outhwaite

Boarding a Game to Avoid a Boring Game

There is one other thing that I want you to notice in Ron's drawings. There are no people in them.

This in itself is not a bad thing. But I would argue that before one ever gets to this stage of design, one needs to have spent a lot of time exploring the social, personal, and physical context in which the system is to be used. Almost by definition, this requires representing not only people, but also their emotional state, and where they are physically. How can you get the details right if you haven't looked at the big picture first?

Here again is a place where storyboard techniques can be used to really good advantage as a means to explore, brainstorm, capture, and communicate ideas about use and experience.

To illustrate this, my next example, Figure 100, is from the final year project of Mark Outhwaite, a recent graduate of the Ontario College of Art and Design.

These are structurally similar to the images that Ron Bird made in Figure 99, (minus the state transition diagrams), in that they are a sequence of hand-drawn frames. However, Outhwaite's storyboard is very different in intent. Here the emphasis is clearly on the spirit of the play, and how the lifestyle and the location of the players might impact the design of the physical form factor and the software of the game.

This example also provides an opportunity to reflect on our earlier discussion of the use of theatre and acting "in the wild" to work out concepts. It is easy to imagine how these storyboards could have been preceded and helped by running around outside in different contexts and engaging in various types of collective role-playing.

What I am trying to do here is not only point out the breadth of applicability of any one technique, such as storyboarding, but also that different techniques like story telling, role-playing, and drawing can all complement each other, and in so doing, significantly enhance our ability to explore ideas.

Each technique that we discuss is just going to add to this, and provide us with a veritable arsenal of techniques to bring to the table.

Charmaine:
"Its Tobias, meet me at the top of Tower Hill"
Tobias: Meet me at the top of Tower Hill

"Charmaine, I have class A new Phyzzles!"
Tobias: I have new class A Phyzzles!

"Optic Capture this Tobias."

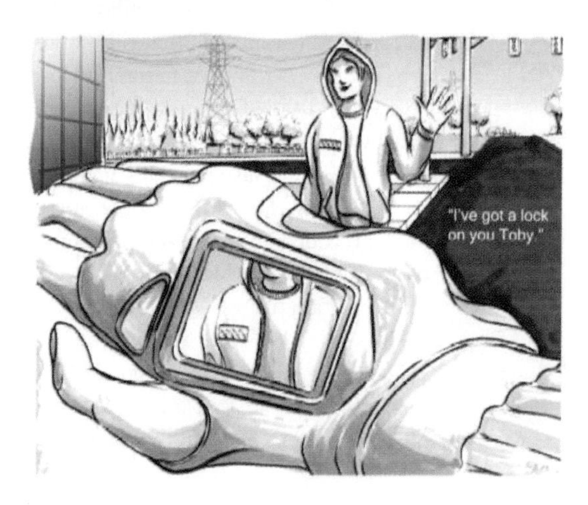
"I've got a lock on you Toby."

Charmaine hiding by the cobalt cube.

Tobias breaking out the Orbit-
A Networked Trans-Reality Toy

"Ready Charmaine?"
"Let it fly!"

"Ready..."

"Set..."

Wait for the patterns to merge...

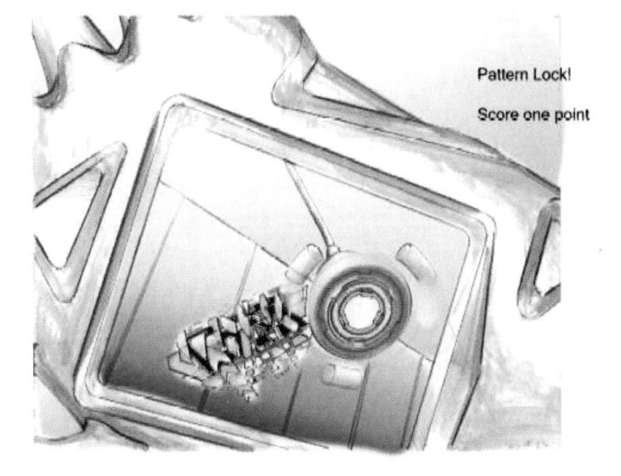
Pattern Lock!
Score one point

Don't drop the Orbit,
or you lose the point.

Right back at you Tobias!

Go get it!

Lets go down to
the quadrant...

Just as bit to the east.

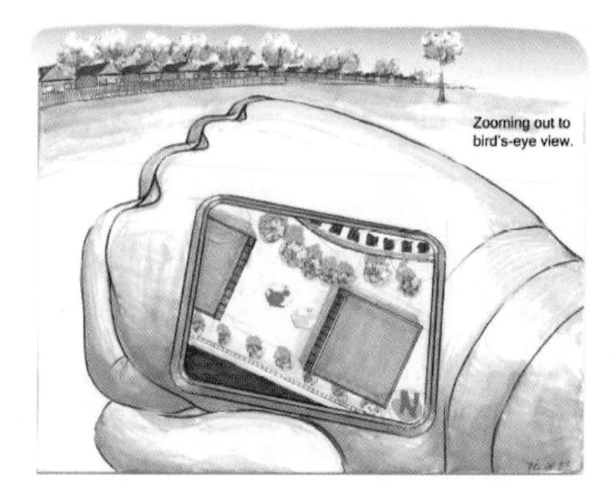
Zooming out to
bird's-eye view.

I'll beat you there!

The Quadrant.

The city's best kept secret.

They're going to be a
couple more minutes...
to be continued...

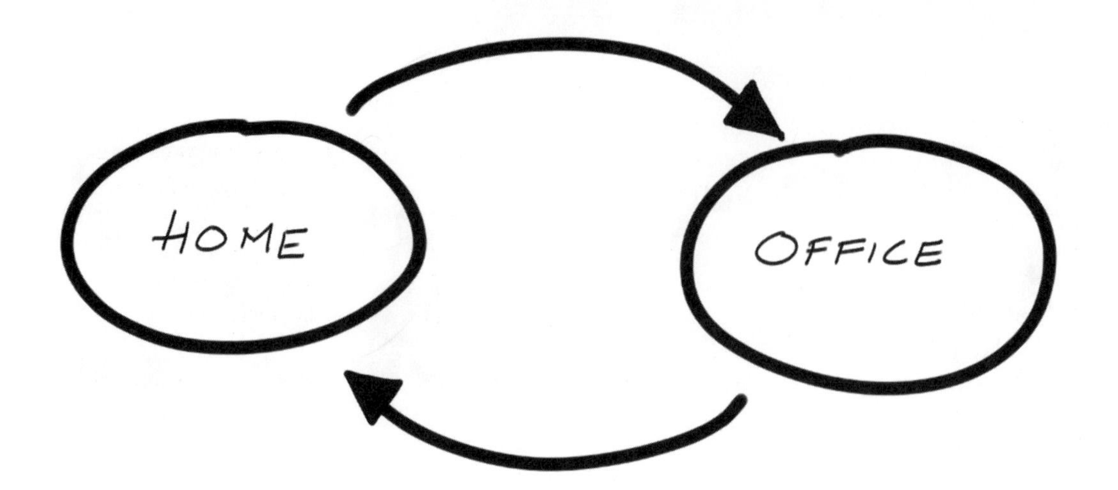

Figure 101: Home-Office State Transition Diagram

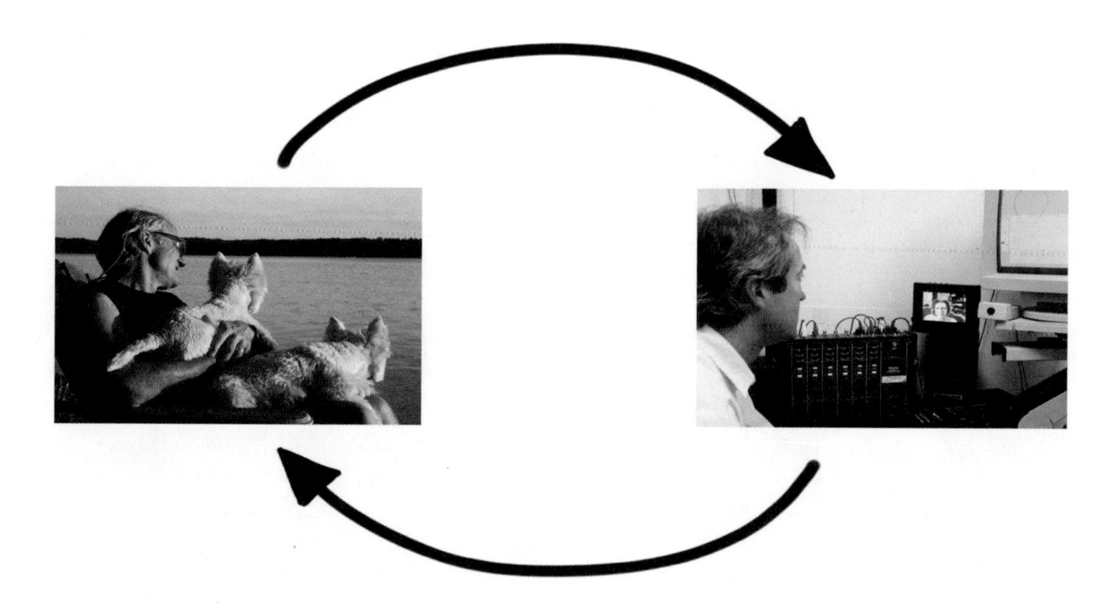

Figure 102: Home-Office Photo-State Transition Diagram

Why Are Transitions Like Canada?

Because they are overshadowed by the states.

Now before you throw something at me, remember what Alex Manu said about the importance of play! And there is a point to my joke besides trying to tie in yet another thing that we have already discussed.

Look again at the storyboards of both Ron Bird and Mark Outhwaite. The focus is on the states, snapshots in time and place, rather than about the transitions. They say, "you can be here" and even "you can get here from there," but they don't tell you how you got there or what the experience was like.

Think about it this way. Let's say that you have two states—"at home" and "at the office"—and that you are always at one or the other, or in transition from one to the other. The resulting state-transition diagram would look like Figure 101.

In this representation there is little more shown about the state (other than the labels Home and Office) than there is about the transitions. Now contrast this with my second Home-Office state-transition diagram. Because of my use of photographs in Figure 102, you know way more about both where I live and where I work (the states) than you did in the previous version of the diagram. But also notice that you know just as little about how I get back and forth between home and work (the transitions).

I include this example to highlight a shortcoming of the storyboards that we have seen thus far. In short, they are like Figure 102. They tell us about the state, but almost nothing about the transitions. This can be fine, but only if we are aware of the missing detail and adequately address it elsewhere, at the appropriate time and with the appropriate tools. The user's experience is shaped as much (if not more) by the transitions as it is by the states. Therefore they must be equally in the forefront during the design process. Yet, in my experience, this is seldom the case. Attention to the transitions nearly always has lagged far behind.

In talking about this, I can't help but think about something the comic book artist and writer Scott McCloud said in his book, *Reinventing Comics*. Of course, I have to quote him in image, rather than words. So look at Figure 103.

To paraphrase Scott, one could also legitimately say that the heart of the interface lies in the transitions. But interfaces are not just sequences of still images. That is, they are more than an interactive slide show. Yes individual screens change, but so do the elements in them. The user's experience largely derives from what moves, and how, when, and where this happens. As people such as Baecker and Small (1990) said in their essay, *Animation at the Interface*, this stuff is really important and we need to pay more attention to it.

So let's step back and see how we might start doing so early in the process.

To me, a good place to start is with Laurie Vertelney (1989). She was the first person that I am aware of who pointed to the film-maker's craft as a source for tools for interaction design. She was a strong advocate of the use of storyboarding, as well as many of the other video techniques that we will discuss as we go along. Her point was that cinematic techniques had been developed precisely to deal with temporal phenomena, such as timing, movement, dynamics, and the like. It made sense to use them in interaction design as well. They were simple and effective. In one way, her message was heard, and the use of video and storyboarding is now commonplace. The problem is that some of the roots were lost along the way.

Figure 103: Life in the Gutter

Image: McCloud (2000) p.1

Figure 104: Story-Boards from The Graduate (1967)

Notice the use of graphic devices to capture the nature of the transitions between frames.

Image: Katz (1991)

As we started to discuss in my home-office state-transition example (see Figures 101 and 102), much of what I see under the name storyboarding is just a sequence of screen shots. Yes, it tells me about sequencing and the design of the various screens. And yes, sometimes this is a legitimate thing to do. But it tells me nothing about timing, movement, or dynamics—something that is at the heart of film storyboards. And, too often, my sense is that practitioners have adopted this limited approach to storyboarding without any awareness that they are leaving out the very thing that they should be focusing on.

We can address this by going back to where Laurie pointed us in the first place: film-making.

There are lots of books on storyboarding for film. Since new titles keep appearing, let me just encourage you to explore them. But if you want my favourite from among those that I have seen, it is *Film Directing: Shot by Shot: Visualizing from Concept to Screen*. It is by the New York film-maker Steven Katz (1991). See also Hart (1999).

Let me give you an example taken from this book that highlights what we need to take into account, and how to do it. The example has three frames of the original storyboards from the classic 1967 film, *The Graduate* which are shown in Figure 104 . At first glance, they seem to be pretty closely related to the *Zone Tactics* storyboards of Mark Outhwaite that we saw in Figure 100. This is good since in many ways they are similar in purpose. But there are a few new elements in the boards from *The Graduate* that are worth noting. First, notice the directions to the cinematographer that accompany each fame. They don't just direct where to move, but *how*, at what speed, and so on. One might wonder if doing so eliminates some of the between-frame magic that Scott McCloud was talking about. I don't think so. First, this is not a comic book, but a sketch of a movie. Second, these directions still leave more than enough room for the cinematographer's imagination to play. Third, think about using this in interaction design. Incorporating alternative directions along with the boards introduces yet another technique that we can use to open up our exploration of the design space.

The other thing that is new and worth mentioning is the use of arrows to describe the motion of the character eventually played by Hoffman. This is potentially a really powerful notation. First, it shows you who or what is moving in the frame and where. But it also has the power to graphically communicate the nature of the movement itself: fast/slow, accelerating or slowing down, smooth or wobbly, and so on. All these types of properties can be captured in how the arrows are drawn, at least when drawn by someone with appropriate technique.

I want to make two final points before wrapping up our discussion of still images and storyboards.

First, even when talking about the early stages of design, I can still hear some people expressing concern about the intentionally "sketchy" nature of rendering that I am promoting. My best response, besides going back to our initial discussion of the nature of sketches, is to defer again to Scott McCloud, who again in Figure 105 says it best in pictures.

I hate the term "low-fidelity" prototype or interface. Why? Because when the techniques referred to are appropriately used, they are not low fidelity; rather, they are at exactly the *right* fidelity for their purpose. I love Scott's phrase, *amplification through simplification*. It is brilliant. It says to me that the fidelity of a sketched rendering can be *higher* than reality—at least in terms of experience (which ultimately is what we really care about). Pretty cool.

And my parents wouldn't let me read comic books! That probably stunted my education.

Finally, you might be asking, or be asked, "Why not just use video or animation to capture the dynamics? After all, are not still images, storyboards, and comic book techniques a poor substitute for this aspect of the interface?"

Figure 105: Learning from Comics

Comics can teach us that so-called "low fidelity" prototypes may be just the opposite; that is, they may amplify our ability to experience or understand the object of design.

Figure: McCloud (1993) p. 30

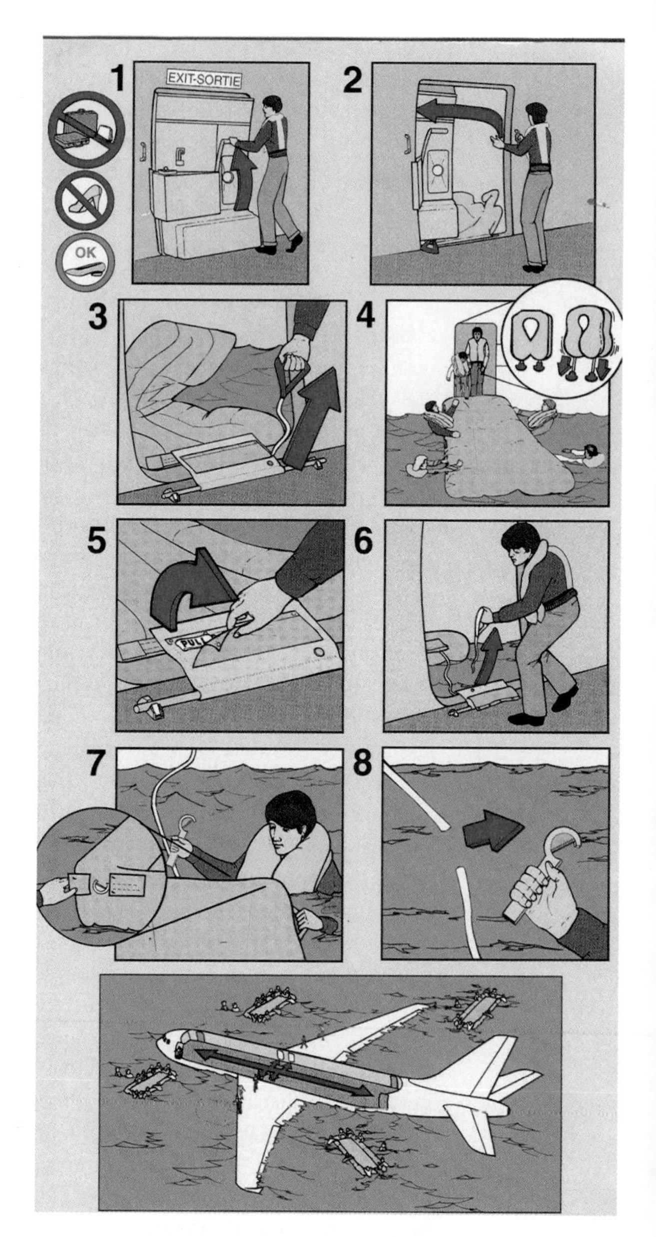

Figure 106: An Airline Safety Card

The technology design and operation are shown in context using a range of graphical techniques, all of which can be exploited in design as well.

Image: Interaction Research Corporation

There are a couple of replies to this. First, for sure animation or other cinematic and interactive forms are really powerful tools in the repertoire of the interaction designer, especially in terms of dynamics. However, ask anyone who has done both which is faster, cheaper, and enables them to explore more alternatives in a given (limited) amount of time, animation or comics / storyboard approaches? Correction. Don't ask the question. The answer is obvious.

If you are still not convinced, or need another example, look at the excerpt from the airline safety card shown in Figure 106. Yes, this information could be shown in a video. But by having it on paper, the nervous passenger can study it for as long as he or she wants. And on the way to a crash landing on water, I might prefer the card to waiting for the flight attendant to play the video again!

As an aside, see *Design for Impact*, by Ericson and Pihl (2003) for a fascinating history of airline safety card design. There is a lot to be learned about visual language by looking at the 50 years of the genre covered in this volume.

This card is certainly not a sketch. It is in the visual language of illustration and, using the vocabulary introduced on page 121, is a description drawing. But don't be lulled into thinking that the techniques that it employs do not apply to ideation. For example, through the use of arrows, the incorporation of people and context, it captures the operation of the door, as well as the intended procedure for its use in an emergency.

These are all techniques that are equally valuable and relevant in the design phase. One can easily imagine sketching a number of alternatives for discussion and evaluation. The key thing is, with a sketch using this technique, one would never lose sight of the fact that it is an evacuation process that is being designed, not just an airplane door.

To the extent that they can capture motion, the still image approaches of storyboarding and comic book art enable the whole motion to be captured in a form in which it can be placed on the wall. There it can be absorbed in a glance, and compared, side-by-side, with other alternatives. Video can't do this. But then, it can do things that storyboards can't. Good. They complement each other. Storyboards are not better than video. They are just different. Each has its place.

As I have said before and will say again, everything is best for something and worst for something else. The question is knowing for what, when, and why. Experience and fluency, based on practice, is the key to gaining the literacy that embodies such knowledge.

So let's now talk about what animation can do, and how to do it.

Figure 107: A Simple Flip-Book Animation

Post-it Notes are an excellent medium for doing really fast sketched animations. This one is done by a child to tell a story. But the same technique can be used to illustrate the dynamics of an interface.

Flip-book: Adam Wood

Simple Animation

The only way to discover the limits of the possible is to go beyond them into the impossible.
— Sir Arthur C. Clarke

One of the distinctions that we have made between traditional product design and interaction design is the importance of considering the role that time and behaviour have in shaping the overall experience. In the previous section we saw how we could augment traditional sketching, by borrowing techniques from comic books and story-boarding, so as to better deal with such temporal issues.

It should come as no surprise that basic animation techniques can also be used to good advantage in interaction design. A number of simple animation techniques can be used to explore and illustrate the dynamics and character of a potential interface, transaction, or experience. My purpose here is to give a flavour of some of these, and a sense of how they can be used to advantage. I leave it to those more qualified than me to give the full tutorial.

One surprising thing is how simple animation can be. Another is how hard it can be. This is my advance warning about getting sucked in by this fascinating craft, and in the process, losing sight of why you are using it. We are sketching interaction, not making *Toy Story*. If you find yourself violating any of the attributes that we associated with sketching (fast, cheap, plentiful, etc.), then you should question if you are doing the right thing, or if you are the right person to do it.

The advance warning done, let's dive right in with an example of the simple side. Figure 107 shows the frames of a little flip-book animation done by a child. The good news is that the only things required for this were a pencil, some Post-it notes, and a story. The bad news is that the playback dynamics are only rough, and by their nature, flip-books lend themselves only to short sequences.

But that is fine. Sometimes that is all that is needed.

Flip-books can easily be created, circulated, and played back on a computer. Recently I was working on a concept that involved using George Fitzmaurice's *Chameleon* technique out in the countryside. For whatever reason, I wanted to show the handheld actually moving, rather than use the double-exposure and blurring technique that George used in Figure 7. The way that I did it was sketch the simple stick figure seen in Figure 108 on my Tablet-PC.

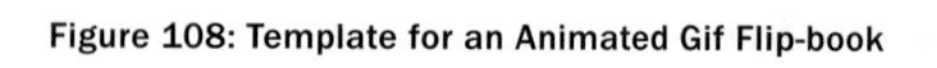

Figure 108: Template for an Animated Gif Flip-book

Figure 109: Simple Flip-Book Chameleon Animation

This image contains only what is common in all frames of my animation. Therefore, nothing is included that moves from frame-to-frame. I then loaded this image into a drawing program and made four variations, each showing the hands and eyes in a different location. The four resulting frames are shown in Figure 109.

I then used a generic image program to create an animated flip-book from the sequence. Note that by basing all the frames in the animation on the image in Figure 108, they were all registered, thereby eliminating any frame-to-frame jitter (something harder to do with paper).

I then incorporated the animation into my larger presentation and e-mailed it to my colleagues. This part took longer to describe to you than it did to create. And, despite my paucity of drawing skills, I was able to get my message from Canada to colleagues in the United Kingdom and the United States in very short order.

When you do want to employ such short animated sequences, generally the most efficient way to save, transmit, and view them is in a format called an animated gif. Besides knowing how to make one, you really only need to know three things about this type of image file:

- They are the digital equivalent of a flip-book animation. It is an image format that contains an ordered sequence of still images, rather than just one image.
- Things like PowerPoint, web browsers, and other programs treat them like ordinary still images, rather than like video files, for example. That is, the flip-book plays, but there is no need to launch some kind of media player. This can be really helpful if you want to play back more than one sequence at the same time.
- The speed of "flipping" through the book is a function of the size of the images and how many there are in the "book". The more and/or bigger, the slower the playback.

You can create animated gifs with some general imaging programs (I used *Microsoft Digital Image Suite 2006*, where they are called flipbooks), or inexpensive special programs, such as Alchemy Software's *GIF Construction Set*.

The ability to make both paper and electronic flip-books is something that I have my students practice, and something that I highly recommend you try.

The animations seen so far have both been hand-drawn—one on paper, the other on a computer. The range of materials that can be used can be far richer, and can include photographs, paper cut-outs, or some combination of these.

I really like the paper cut-out and collage techniques that we see in cartoons such as *South Park*, or the animations of *Monty Python* troupe member Terry Gilliam (McCabe, 1999). They can teach us a lot about how to do effective, simple, and inexpensive animations that are consistent with the criteria that we have adopted to characterize sketching. They are also really useful for people like me who are drawing-challenged.

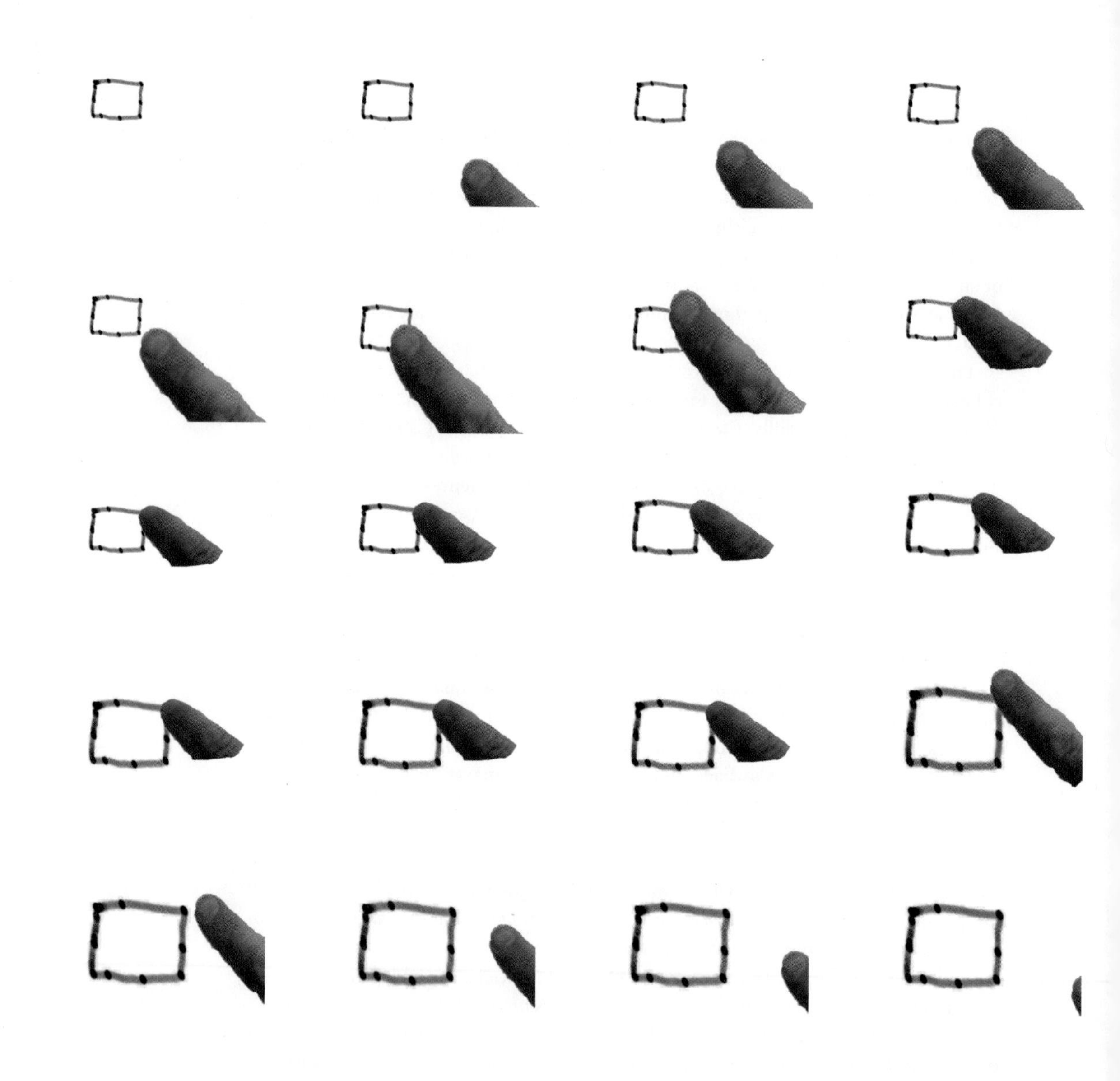

Figure 110: Simple Animation of Scaling Rectangle with One Finger
Note the use of a hand-drawn rectangle, and the contrast in medium between the drawn rectangle and the photographed finger. The animation is made up of only 20 stills, yet it conveys the concept.

For example, look at the two simple animations shown in Figures 110 and 111. I did these to illustrate two contrasting ways to manipulate the size of a graphic object, in this case a rectangle. The sequence shown in Figure 110 reflects the conventional approach used in virtually all computer graphics systems. Here, you "grab" one corner of the rectangle and "stretch" it using a mouse, stylus or—as illustrated here—your finger. The second animation, Figure 111, shows a bimanual/two-finger technique for doing the same thing. Here we can use our fingers to grab two diagonally opposed corners of the rectangle and then simultaneously position them. The approach is far closer to how we manipulate things in the real world, yet far from what we typically do while interacting with computers. It is also generally hard to implement, so few people have ever seen it on a PC, much less tried it. Hence, these two short animations provided a simple way to show the two techniques side-by-side.

Both animations are very short, 20 and 22 frames, respectively, and took less than 30 minutes to make using just a laptop computer, common inexpensive software, and a digital camera.

One reason that I included these examples is to show that you *can* show useful things with short animations. The other was to explain why I find the *South Park* and Gilliam collage type of animation interesting and relevant.

In these animations I used two types of source material. The first was a single rough hand-drawn rectangle, which was then transformed and used in all the frames. The second class of material was photographic. I shot my left and right hand's index finger in two poses: extended and bent. In the resulting animations, I used the extended pose when the finger is moving toward or away from the rectangle corner, and I used the bent pose to indicate when I was holding the corner for purposes of dragging it. (I didn't use or need a tripod here, but generally you do if you are making a flip-book with photographs.)

The rough quality of the animations is intentional. This includes the use of a hand-drawn square, the lack of care in "cutting out" (rotoscoping) the fingers from the original photographs, and the not-so-smooth motion in the animation. These are all incorporated in order to be consistent with our characterization of a sketch as being in a distinct vocabulary, and appearing to be cheap and disposable.

Nobody is going to confuse these clips with a real system. On the other hand, nobody looking at the resulting animations will fail to see the significant difference in the directness between the one- and two-fingered techniques.

Finally, the choice of mixing media—the hand-drawn rectangle versus the photographed hand—establishes a clear figure-ground relationship that helps disambiguate what is being manipulated from what is doing the manipulation.

Technically, I did the animations brute-force, using simple cut-and-paste techniques, frame by frame. This made sense, since each sequence was so short, I could have assembled the resulting frames as an animated gif flip-book, as in the previous example. Instead, I

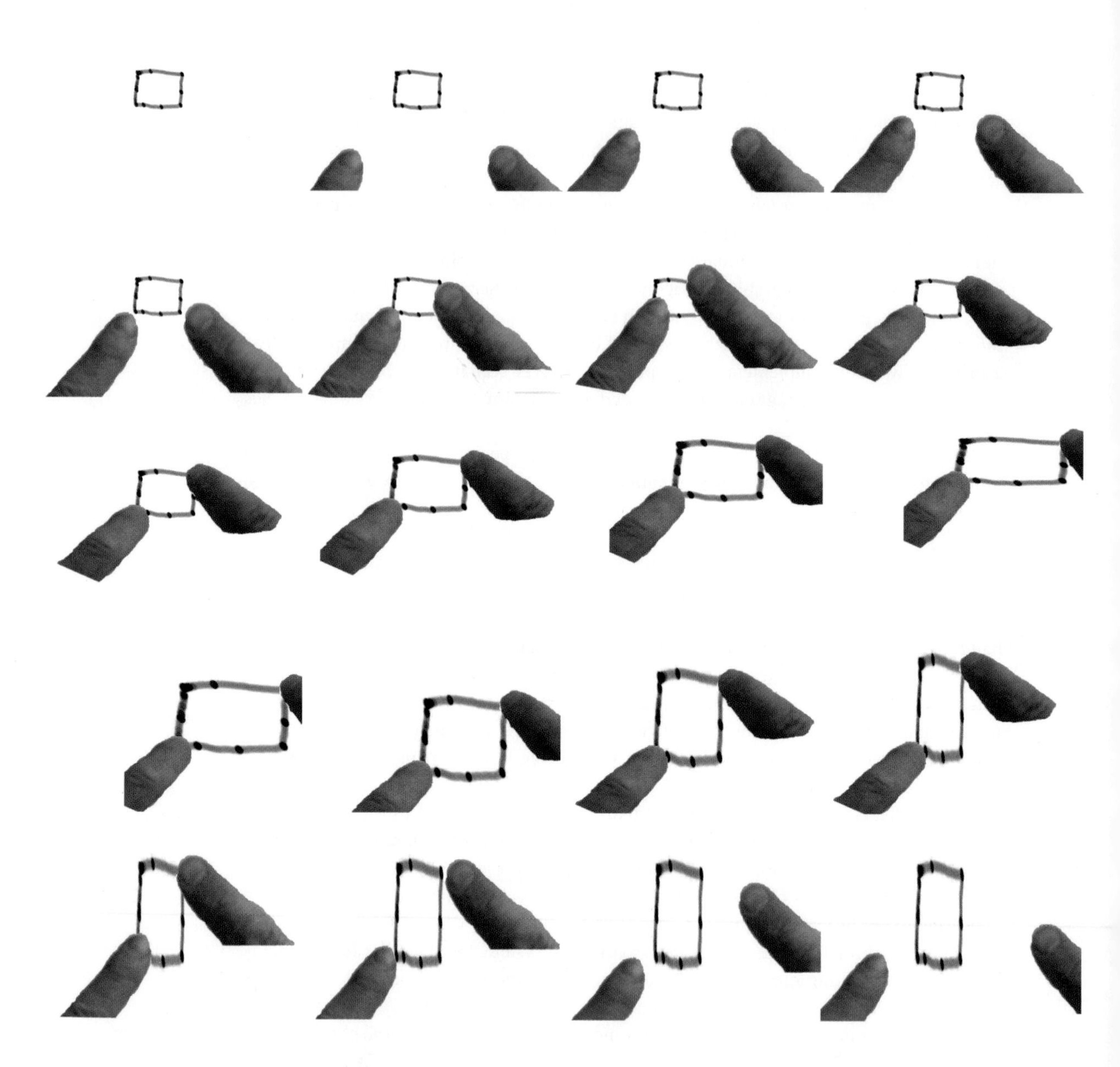

Figure 111: Using Two Fingers to Resize a Rectangle
Note that in contrast to the animation seen in Figure 110, when using two fingers the manipulation is far more like what we have in the physical world, where we can simultaneously resize and reposition the rectangle in one integrated and coordinated gesture.

loaded them as single frames into a simple video editing package (I used Microsoft *Movie Maker*, but could almost as easily have used Apple's *iMovie*). I then rendered them out as a video.

If sequences get longer, then a frame-by-frame approach becomes too expensive in terms of time and effort to afford the quick turnaround that one needs in ideation. One alternative in such cases is to use tools that let you animate objects in the scene, rather than draw frames. Let me illustrate what I mean with the simple example in Figure 112.

Here I am using a technique called *picture-driven-animation* by its inventor, Ron Baecker (1969). Instead of drawing individual frames, Baecker introduced the notion of using hand-drawn lines to define both *objects* that were seen, and *motion paths* along which those objects would move. Figure 112 shows me using this technique in *PowerPoint* to animate a sketch of using George Fitzmaurice's Chameleon technique to view a map.

In the actual animation, one sees a few different elements. In the background, only partially visible, is a representation of a virtual map of western Canada. Above that there is a drawing of a hand holding a small PDA, on whose screen a portion of the map is clearly visible. Finally, there is some hand-written text explaining things.

What is also seen in the figure, but not in the animation, is the dotted line that starts at a green triangle in the middle of the hand-held display, and then extends down to the left, ending in the red triangle. This is the motion curve.

When the animation is played, the hand and the PDA display move along the path, which just happens to follow Alexander Mackenzie's route to the mouth of the Bella Coola River in 1793 (the first overland trip to the Pacific north of Mexico). The hole through which the details of the map are seen is a white layer with a circle whose colour is set to 100% transparent.

The length of time that it takes to make the move (which equates to the number of frames), can be set and/or changed by the user. Hence, by drawing what is essentially one frame, plus a single motion curve, I get something that is comparable in complexity to what took me 22 frames in the example shown in Figure 111. More to the point, it took less than a tenth of the time. By appearance and execution, it is definitely a sketch.

Now there are other programs, such as Adobe (Macromedia) *Flash* that are very good at supporting this kind of animation, and that are far more powerful for this purpose than a program like *PowerPoint* (which was not designed as an animation package). However, that power and generality comes at a greater price in terms of either money and/or learning curve. So, although the specialized tools are useful and sometimes the only means to do something, whenever possible, I tend to use simpler, more ubiquitous tools like *PowerPoint*, which have a lower barrier of entry.

I think that a great literacy-building exercise is to have students (not to mention yourself) do things like implement the same animation using each of the techniques that I have described: as animated gifs, frame-by-frame in a video program, and as picture-driven ani-

Figure 112: Picture-Driven Animation Using PowerPoint

Here I have drawn two key elements: (1) a hand holding a small display, and (2) a dotted line showing the path along which I want the hand to move. The resulting animation will show how one can pan over a virtual map by physically moving the handheld device—an example of the Chameleon technique that we saw earlier.

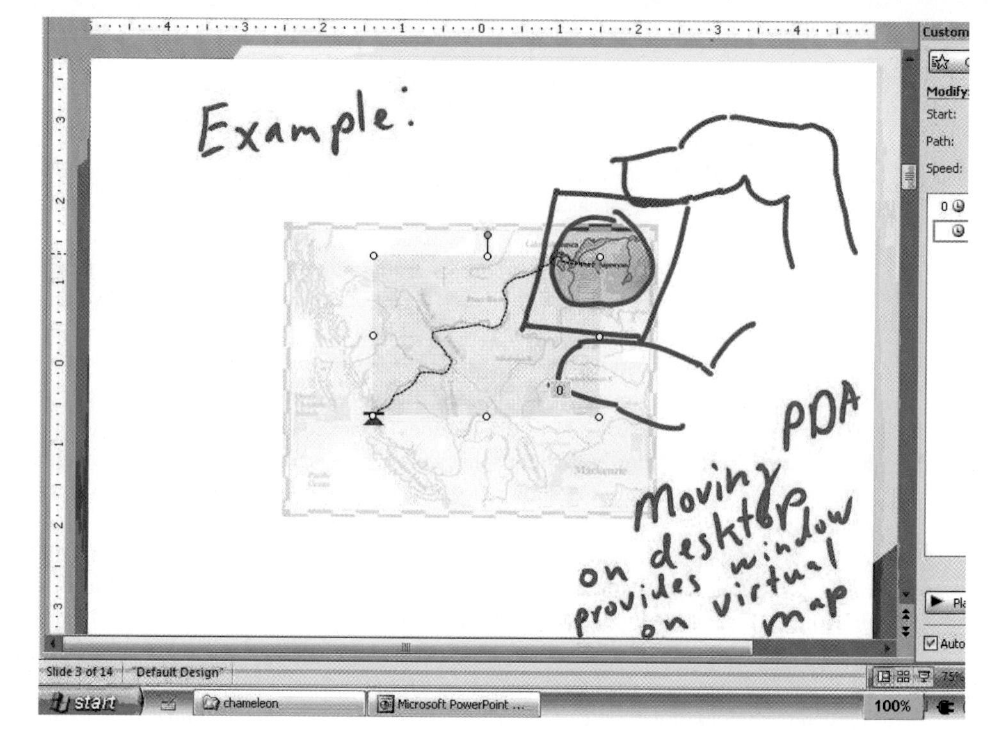

Figure 113: A Sketch of Walking

This sketch only skims the surface of the complexity of walking. Just getting the gait right in animating one style of walking is hard enough. Now imagine fast, slow, heavy, light, sad, happy, or any other type of gait. Think of the technique that it takes to get that motion right. Having considered that, ask yourself, "Why would I expect it to be any easier to get the dynamics of my interface right?"

Image: Williams (2001, p.103)

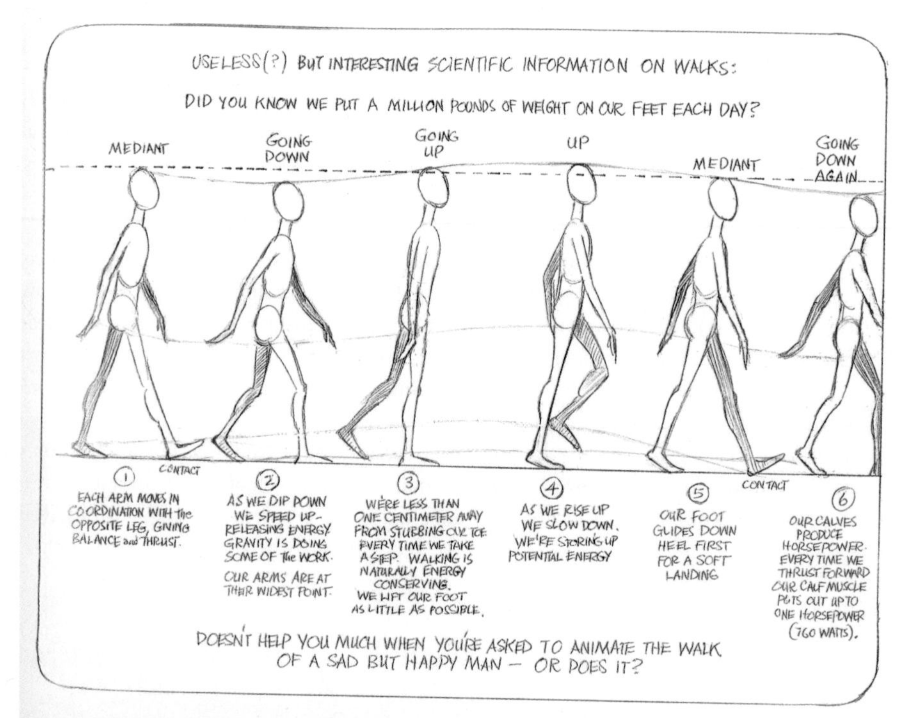

mation, for example. Only then will a deep *experiential* understanding emerge of what to use when and why.

Of course, seeing or even making an animation is not as high on the scale of experience as it is actually to manipulate the rectangles using the two techniques—something that would be possible only if they were actually implemented in code. But then, at the early stages of design, the priority is generally exploring alternatives rather than refining any single approach; so as long as animation is significantly cheaper and faster than implementation, it is a critically valuable tool.

Now it is around this issue of exploration and variation that I want to spend a few minutes before moving on. Take a look at the walking sequence shown in Figure 113. It is from quite a good primer on animation written by Richard Williams, the animation director of the feature film, *Who Framed Roger Rabbit?*

On the one hand, even if you are an interaction designer, I suspect that you will rarely be called upon to produce a credible animated walk cycle. So it is understandable if you are wondering what this example has to do with the task at hand.

The least that you are going to get out of this brief discussion is perhaps a better appreciation of the animator's skill next time you see someone walking in a cartoon. And hopefully you will be tempted to learn more by reading Williams' book, and others like it. But I want something more. Think about it. Walking is something that we do every day and are presumably pretty intimate with. Yet Williams (quite rightly) spends over a third of his book talking about how to animate a walk. It is just really hard to build the character that you want into a walk cycle and have it work. Motion is really complex and though it has huge potential to express a myriad of attitudes and emotions, it is equally complex to control. Even if we can really draw and are really good at reading the character of people's movement, it is really hard to author in an animation.

My point is simple. I think that it may take the same level of skill on the interaction designer's part to get the dynamics and character of an interface right as it is for an animator to do the same with a cartoon character. And this is true, regardless of whether we are speaking about any or all of the perceptual, cognitive, social-emotional, or economic levels. These techniques can be applied to each.

At this stage we are going to leave simple animation and move up the cinematic food-chain and look at other techniques for visual story-telling. Each builds upon the principles of the other, and even though we mainly talk about them in isolation, remember that mixing and matching is always a viable option to consider.

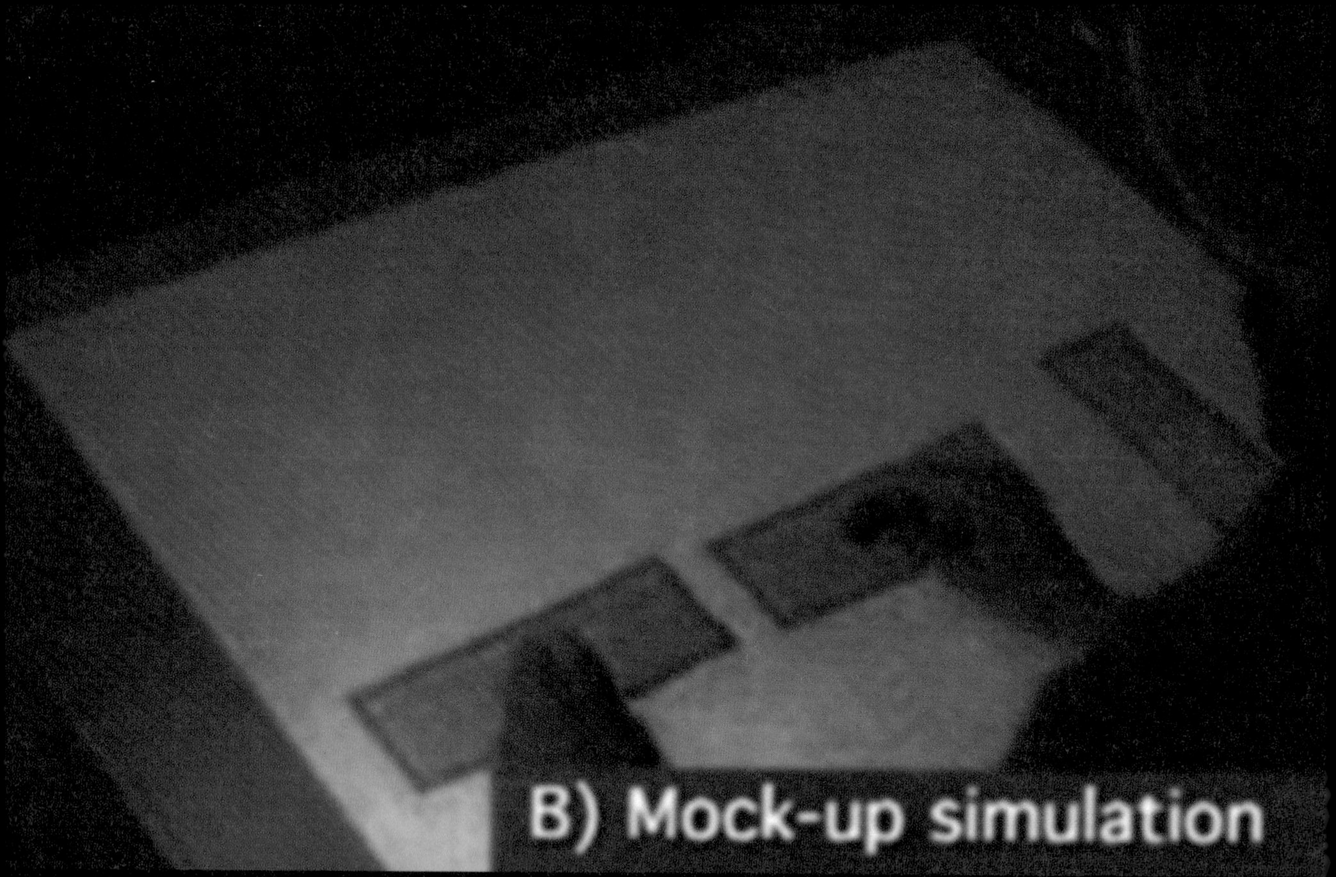

B) Mock-up simulation

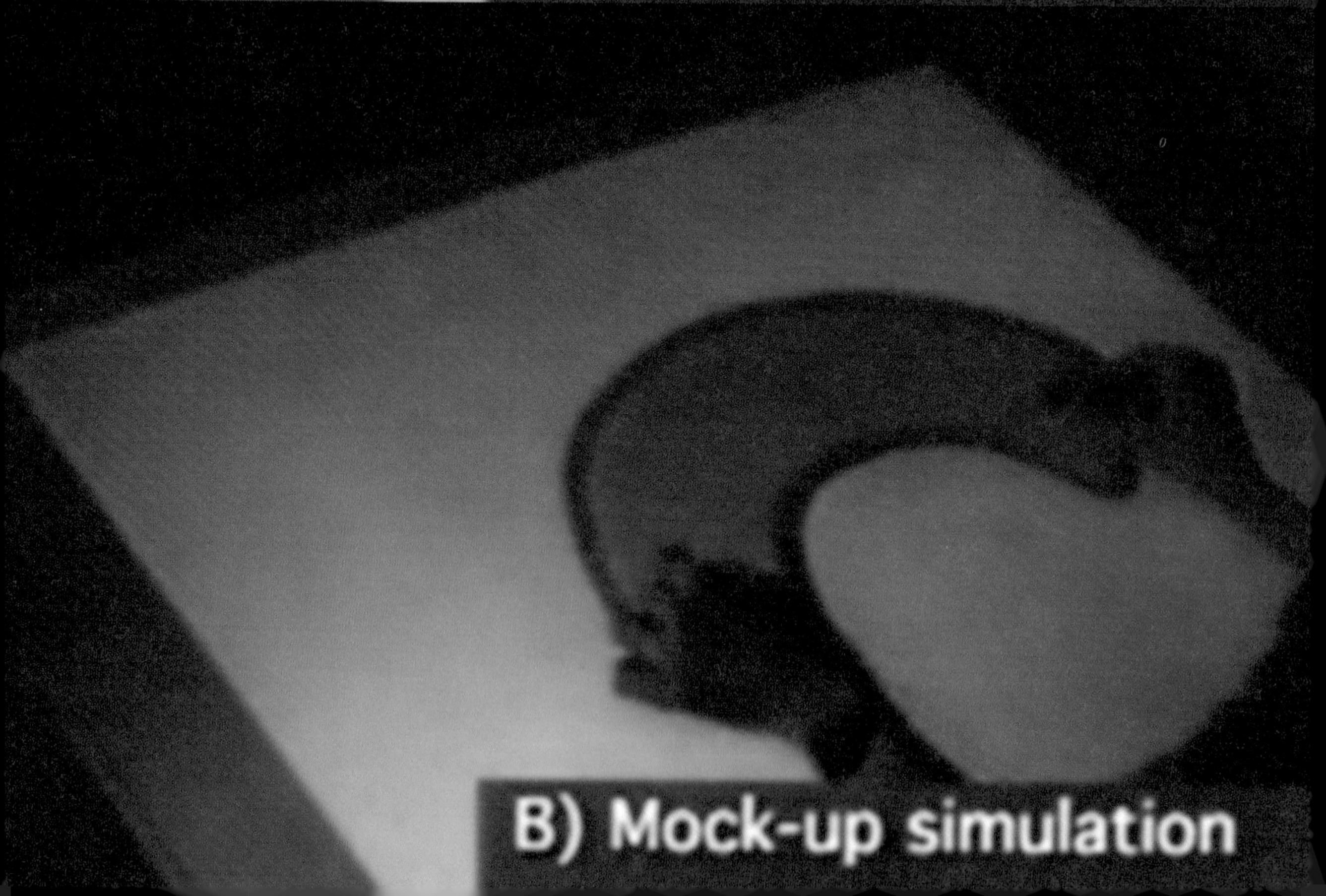

B) Mock-up simulation

Shoot the Mime

Design is a funny word. Some people think design means how it looks. But, of course, if you dig deeper, it's really how it works. To design something really well, you have to 'get it.' You have to really grok [understand] what it's all about.
— Steve Jobs

Bricks and the Graspable Interface

Having seen some examples of how we can sketch designs using simple animation techniques, I want to go one step further and introduce how those animations can be introduced into the context of the larger world and living people.

Simply stated, the idea is this.

- Using techniques like we have already discussed, make a simple sketch animation of your system's interface in action. Do not include any representation of the user or their body.
- Play back your animation on a display that is representative of the intended target system.
- Have a person mime along with the animation playback. Have them do so in such a way that their actions are exactly what they would be if they were really performing the transaction that you animated.
- Videotape them doing so.

If the mime is good, it will appear to the viewer of the video that the person is actually controlling the system. This is true even if the interface is clearly a cartoon, or sketch, since when things are less real, it is easier to suspend one's disbelief. What is also really nice about this technique is that through a little bit of theatre or set-building, you can make the display or the context on which you "perform" the interaction evoke the environment where you envision the design being used.

Let me give you an example. Suppose that you wanted to explore the concept of interacting with a new kind of desktop computer, where the desktop itself is the display, and you interact using both hands, holding little physical objects shaped like LEGO bricks.

This is something that George Fitzmaurice explored during the research for his PhD thesis (Fitzmaurice, Ishii & Buxton, 1995; Fitzmaurice 1996). His purpose was to explore the concept of "graspable" or "tangible" user interfaces—something that now, 10 years later, is a pretty hot topic.

Figure 114 shows two frames from George's video, which he made following the steps just outlined.

Before he tried to animate anything, he did a lot of pencil and paper work—the things that you would expect any designer or researcher to do early in the process. When he had an idea that he wanted to explore more deeply, he created an animation of what the screen would look like. To do so, he used a program that was then called Macromind *Director*.

He animated various scenarios, such as positioning individual rectangles, scaling them, working with multiple rectangles at the same time, and using bimanual techniques to deform them in different ways.

Figure 114: Performing for the Camera

Figure 115: The Active Desk

As an envisionment of desks of the future, the Active Desk used an overhead projector and an LCD panel to rear-project the computer display onto a transparent stylus-driven digitizing tablet covered with a translucent velum material.

Rather than a standard desktop computer, George envisioned his technique being used on quite large desktop displays. At the time, I had been working on a kind of electronic drafting table, something that we called the *Active Desk* (see Figure 115). It itself was a kind of sketch. It was a working prototype that we had mocked up by rear projecting onto a translucent graphics tablet, using an overhead projector and an LCD panel.

So George used the Active Desk as the display in his videos. It is worth emphasizing that the impact of seeing the animations on this form factor was very different than viewing them on the desktop machine on which the they had been created.

Once he had the animations and the display, George played the clips back over and over, all the while practicing his mimed control actions. When he had it at a level that would be convincing to a viewer, we shot the video.

There is something of particular importance that you can see in both the video and the frames taken from it shown in Figure 114. Notice that in doing the animations, George followed one of the cardinal rules of sketching. He used a very distinct cartoonish style in the animations—a style that clearly labeled what was being shown as a sketch, not the real thing. And even then, he still took care to explicitly label the video demo as a mock-up simulation.

By so doing, the focus of the viewer was kept on the interaction style, not hijacked by some red-herring issue about the graphics.

To wrap up, what I like about this example is how George put a nice spin on what were pretty commonly used animation techniques. Playing them back on a regular computer monitor would have had little impact, and certainly done little to stimulate conversation, much less understanding, about the techniques under investigation. However, he added in a human who appeared to be *using* the technique. Moreover, the interaction took place in a physical context that almost nobody else had ever seen before. George did all this through the creative use of available resources. The only hard part was having the idea. Realizing the video sketch was relatively simple. It was fast, inexpensive, extremely effective, and enabled him to work through a number of scenarios. And significantly, it didn't involve writing a line of code.

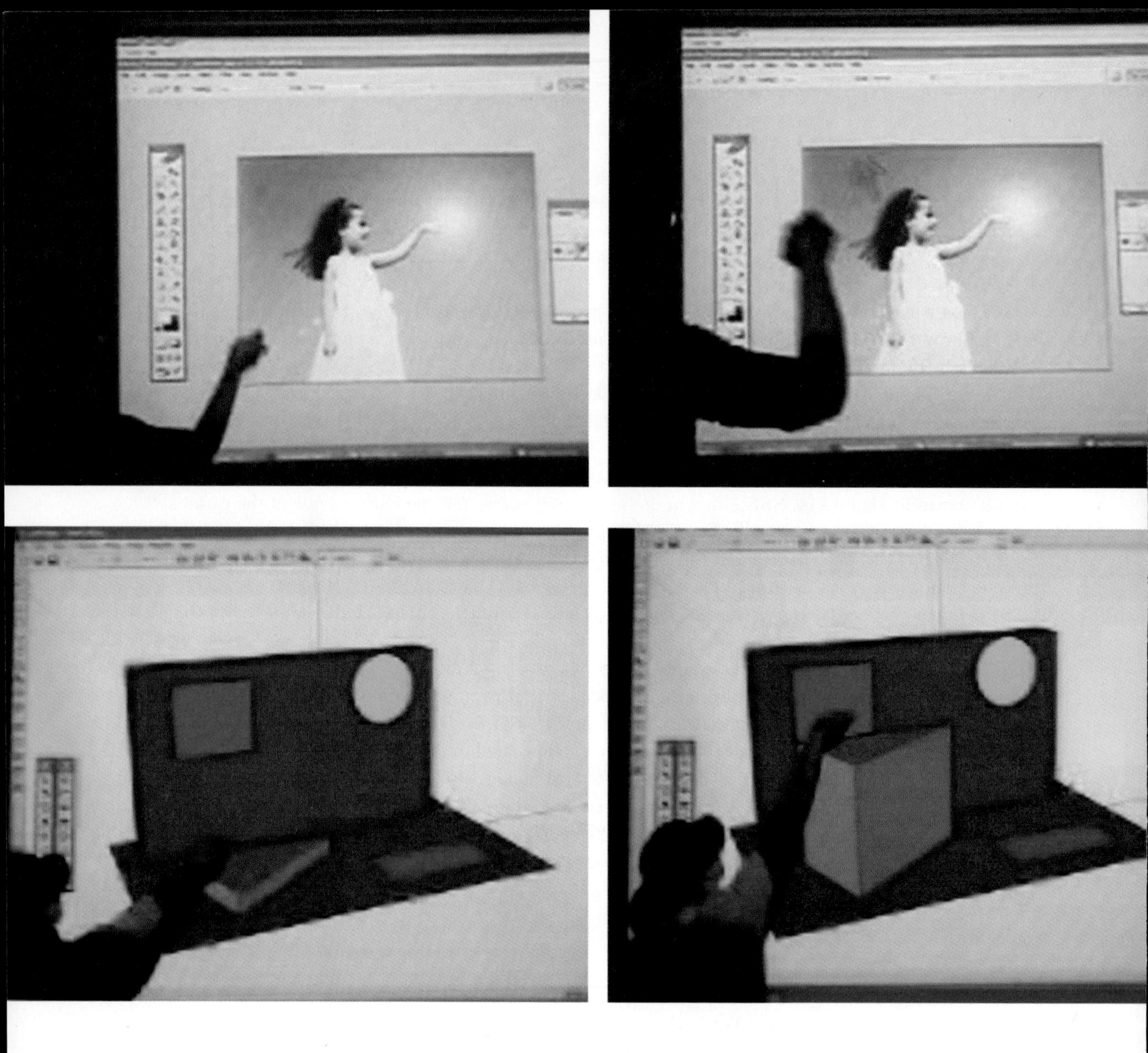

Figure 116: Wizard of Oz Remote Cursor Control

The large purple cursor's position is controlled by a mouse, and its size by the scroll wheel of the mouse. An off-camera mouse operator synchronizes with the hand motion and proximity to the screen of an on-camera person (see inset bottom left). The impression is that the cursor position is controlled remotely by hand position, and its size by distance from the screen

Figure 117: The Wizard Uses Hand Gestures to Create 3D Models

Handing It to the Wizard

In the previous example I spoke about how fast and simple it can be to make such videos. I want to give two quick examples to emphasize this point. At the same time, I want to echo our earlier discussion of the *Wizard of Oz* in order to show some examples of how it can be tied in with what we have been discussing.

Both examples come out of a third-year course that I was teaching on Experience Design at the Ontario College of Art and Design in 2004. The first is a group project involving the whole class. The other was a short follow-on assignment from one of the students.

The purpose of the in-class group exercise was to give the students their first experience in using video for interaction sketching. To do so, I had them do a simple *Wizard of Oz* system, then script, shoot, narrate, and edit the resulting video. Furthermore, they had to do the whole thing in 45 minutes. And they did. Figure 116 illustrates the final result.

The scenario that they were assigned was to make a video sketch of a large wall-mounted display whose cursor could be manipulated by pointing with one's hand, and whose cursor size would automatically grow as the operator moved further away from the screen, thereby always remaining visible.

After a brief discussion, one small group built a simple *Flash* program that implemented a dummy cursor whose position was controlled by the mouse, and whose size was controlled by the mouse scroll wheel. Meanwhile, a second group worked on the script, and a third on how to work the video camera and set up the shoot. The fourth group consisted of two people. One operated the mouse and the other was the on-camera user. While the others were working, they practiced coordinating their movements so that the mouse movement would follow the other's hand movement in front of the screen.

The resulting video is never going to win an Oscar. But it did, and does, make the point that such things can be done in very short order. The point is to just do it, and learn the craft using simple ideas that don't matter. Therefore one can focus on developing the technique so that one is fluent with it when the content *does* matter.

And, just to make the point that such simple exercises do work and can then be applied, let's look at what one of the students, Jasmine Belisle did as a follow-on assignment. Her project, illustrated in Figure 117, was to investigate how one could use hand gestures to model 3D geometry on a large screen.

Her approach was very similar to what we did in the class project. She also adopted a *Wizard of Oz* approach of having an off-camera person operate the system in such a way that the mouse action was synchronized with Jasmine's hand gestures.

There are two positive lessons from this example. The first obvious one is that she was able to do it, despite never having shot or edited video prior to that week. Second, in a way that could be discussed and critiqued by others, she was able to explore, create, and demonstrate a vocabulary of gestures that might be considered for such a system.

There is also a caveat. The value in practice, as opposed to a pedagogical exercise, is dependent upon being able to control an actual system so that it exhibits the behaviour and interactions that you want in your design. This is often not possible. In such cases one obvious solution is to revert to animation rather than *Wizard of Oz*.

The overall take away from all this is that we can combine the techniques that we have discussed, and by having fluency in all of them within the team, it is easy to switch techniques to the one most appropriate, and know when to do so.

That being the case, let's add yet another technique to our repertoire.

sketch
a move

sketch a move
cars & cups & Toys...
BIC

sketch a move
jump

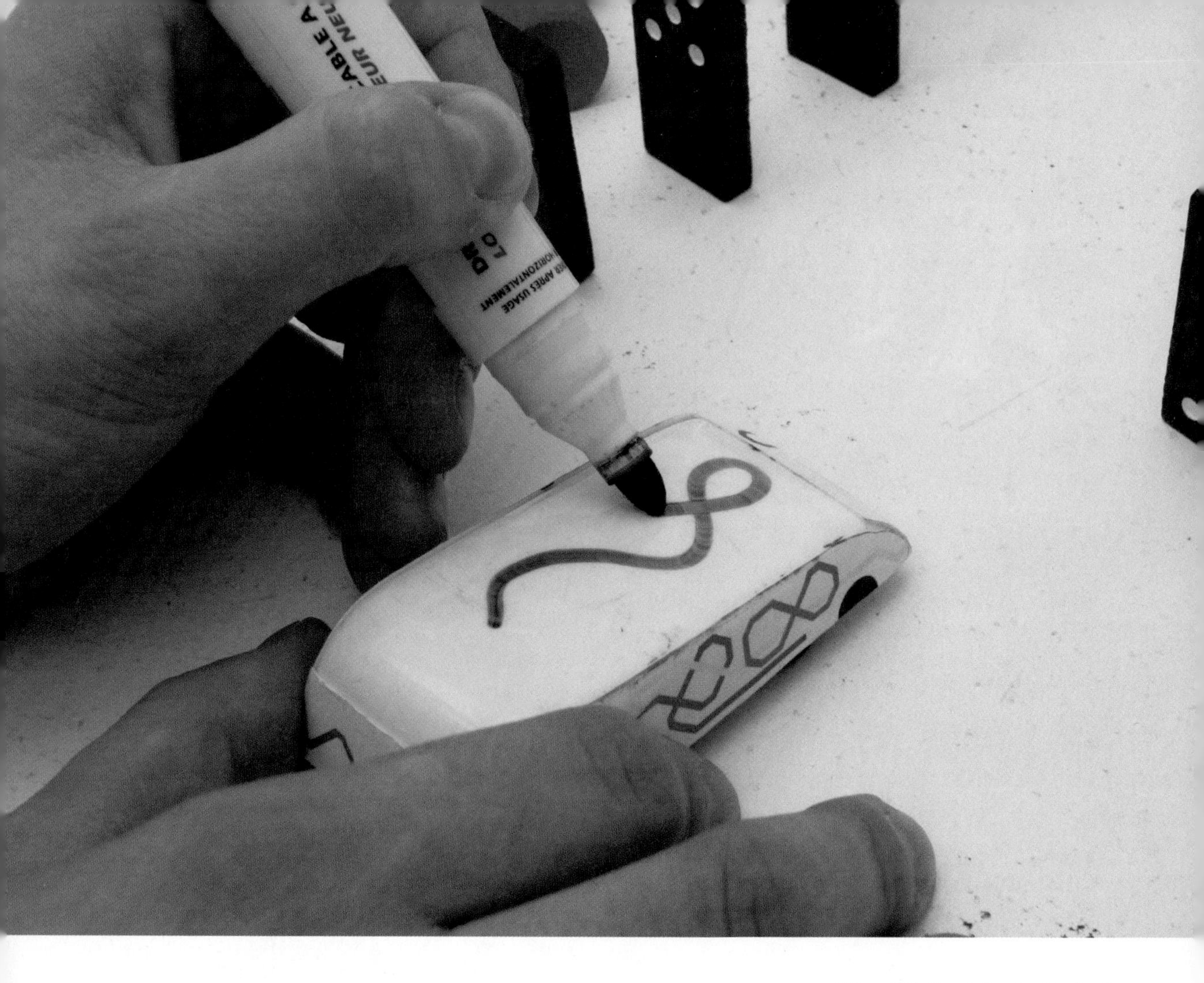

Figure 118: Drawing on the Car Roof

The path that the car would go would be determined by the pattern drawn by a child on the roof. In this case, the car would follow a serpentine path, as determined by the squiggly black line made with the marker on the top.

Photo here and in the previous three spreads: Sketch-a-Move, Interaction Design Project by Anab Jain and Louise Klinker

Figure 119: Replica Toy Car

This toy is an accurate 1:32 scale model of the real car. Contrast this with the car shown in Figure 118. One would never mistake it for the real thing; it is clearly a mock-up.

Sketch-A-Move

One of the greatest pains to human nature is the pain of a new idea.
— Walter Bagehot

The previous two examples were important because they helped augment our ability to focus attention on the person and the context of interaction, rather than just the system or product

Our next example continues down that path and is interesting in that there is no computer or high technology (other than a video camera) involved at all.

The example comes from two students at the Royal College of Art in London, Anab Jain and Louise Wictoria Klinker. It is one of the nicest examples of video-based sketching that I have ever seen. Their project is called *Sketch-a-Move*, and it was done in 2004, during the first year of their MA studies in Interaction Design.

Sketch-a-Move explores a new way of playing with toy cars. The novel spin that they put on the play was the idea that the child could determine the path the car would take by drawing on its roof (a variation on picture-driven animation that I am sure even Baecker never imagined!). This is illustrated in Figure 118 and the previous three spreads. In the figure, the person playing with the car has drawn a squiggly line, which indicates that the car should follow a serpentine path.

On the other hand, the child could have just drawn a straight line along the roof. In this case, the car would go straight-forward or backward, depending on which direction the line was drawn.

Of course, there were (and are) no toy cars that actually work this way. So Anab and Louise had to make a video to explain their concept. Most of the accompanying images are extracted from it.

I want to spend some time talking about this video, since it epitomizes so much of what I consider best practice.

First, look at Figure 118 again. In particular, look at the car that is being drawn on, rather than what is being drawn. Notice that it is just a rough caricature of a toy car, rather than the real thing. For example, contrast it to the commercially available toy car shown in Figure 119. There is no comparison. One is clearly a product, and the other is just as clearly rendered in the visual vocabulary of a sketch. What is important to realize is that this is no accident.

Sketch-a-Move was sponsored by Mattel. Anab and Louise could have used real Hotwheels toy cars in making their video. But they didn't. This was not only a conscious and explicit decision, it was the *right* decision. Why? To answer, we just need to go back to our original list of characteristics of a sketch. Doing so will remind us that:

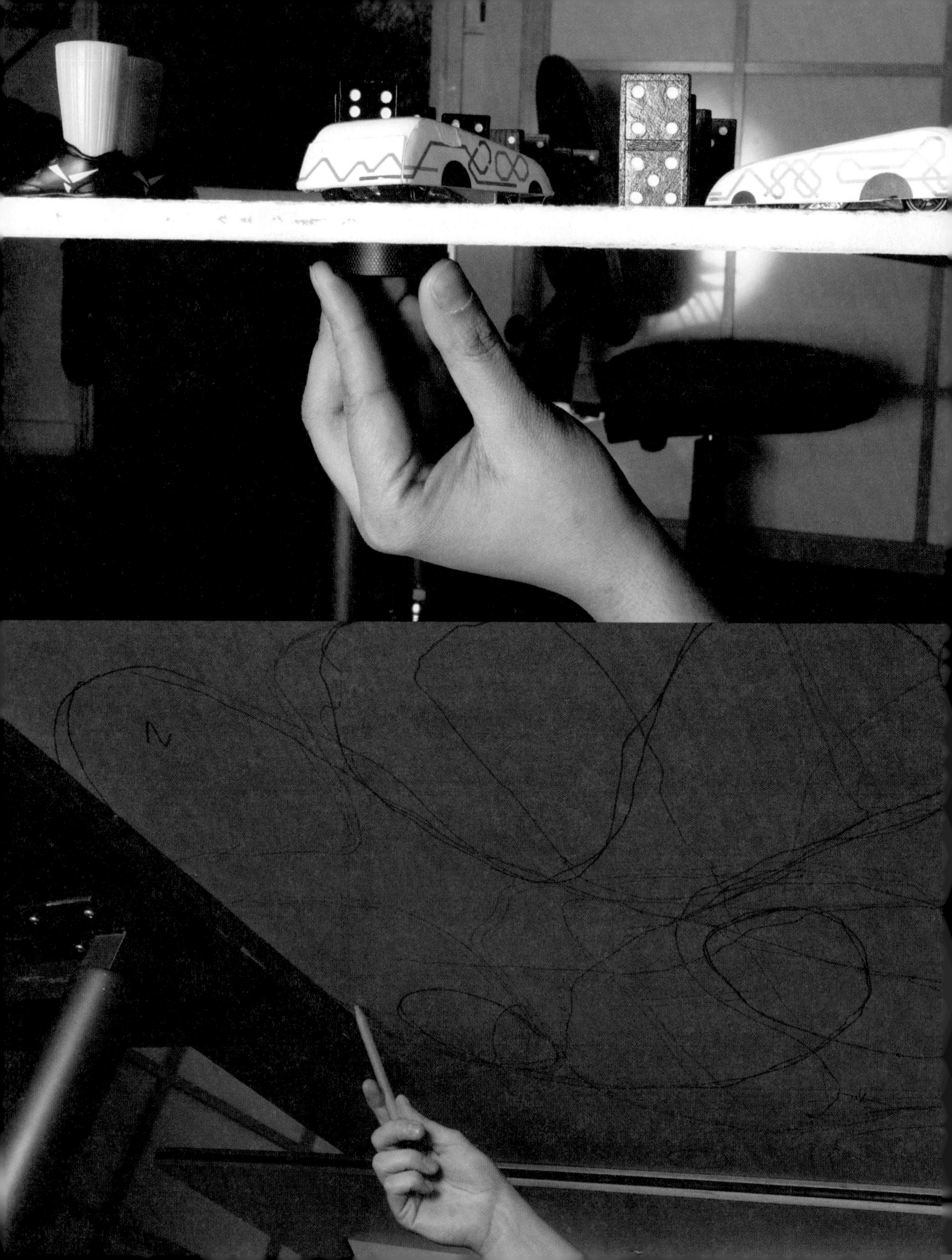

Sketches have a distinct vocabulary that differentiates them from finished renderings.
They are not rendered at a resolution higher than is required to capture their intended purpose or concept.
The resolution or style of the rendering should not suggest a degree of refinement or completion that exceeds the actual state of development, or thinking, of the concept.

These are the reasons that Anab and Louise made the decisions that they did. They were sketching, and they were (and are) extremely careful in all their discussions, presentations, and writing to make sure that nobody gets the impression that the functionality shown in the video is real. For example, when I was talking to them, they clearly knew that I was aware that the video was illusion. Nevertheless, they reminded me more than once that this was the case. It was that important to them, and this concern is what drove the sketch-like style of their representation.

Having brought up the topic, let's look at how well the video meets some of the other criteria that we have identified as characterizing sketches. At the top of our list, we said that sketches were:

Quick
Timely
Inexpensive
Disposable

As we have already discussed, these criteria are interrelated. So, as we look at the *Sketch-a-Move* video, it is worth doing so with an eye to the technique used. How hard was this to make? How long did it take? Of course, this is a lot easier to determine if you actually know the techniques used.

Before I tell you, it is worth seeing if you can figure it out on your own, just from watching the video clip itself. Before you do so, I'll tell you three things. First, it does meet the criteria of being quick and inexpensive. Second, based on many of the web discussions about this video (and there are several), most people who try to figure out how they did it get it wrong. Third, nevertheless, I am not the only one who was able to get it right, just from looking at the tape.

So, how did they do it?

First, let me tell you how they did not do it. They did not do it by stop animation. That is, they did not shoot it frame-by-frame as a series of still images that are then played back at video rate in order to make a moving image. Not in any of the shots. Not at all. With a sequence this long, doing so would have been a Herculean task taking weeks, if not months. And it would have meant that the quick, timely, and inexpensive criteria were violated.

The fact is, they shot the whole thing using regular motion video. For some of the shots this is obvious. But notice that the obvious motion shots (such as when Anab and Louise are drawing on the car), exactly match the resolution and lighting of the other shots, such as those of the cars moving. This is a big hint that the same camera and technique is used throughout.

Figure 120: Under the Table Sketching

The upper photo show the paths of the cars drawn on the bottom of the table. The cars, which have magnets inside them, are then pulled along the table top using a second magnet, shown in the lower photo, under the table. By dragging the cars along the designated paths, their motion can be controlled, while the person doing the manipulation remains out of sight of the camera.

Photos: Anab Jain and Louise Wictoria Klinker

What they did is place a magnet in the bottom of each toy car. Then they drew the path that the car was to follow in the corresponding position on the bottom of the table. When they were shooting a scene, someone went under the table and moved another magnet along that path, and in so doing, dragged the car along with it, right in front of the video camera. This is shown in Figure 120.

This way, there was no need to use any complex animation techniques to make the Styrofoam cups or dominoes fall naturally. They really were falling naturally.

By using this technique, they could change paths, and try various alternatives, very quickly. And because of this, they were able to instill a sense of fun and dynamism into the resulting video that would otherwise have required the talents of an extremely skilled animator—certainly skills that would be way beyond what one could reasonably expect from someone who was not a professional.

Having looked at how they did it, let's now look a bit deeper into the sense of play and fun that the video evokes.

One of the real risks in using video techniques is rooted in the fact that this is essentially a passive medium. You don't interact with a video clip. So, how do you gain a sense of "experiencing" the toy cars, when all you are doing is passively watching them on a screen? Well, in one sense, you don't. For example, you have no real sense of what they feel like, what it is like to draw on them, or if where the car goes is where you intended it to go, given what you drew. You can get none of this visceral type of information from a video.

But on the other hand, what I love about this video is that it is a fantastic example illustrating how much further you can go along the path of experience, than merely using the medium to document the basic concept.

Think of it this way. You (and I) would have understood the basic concept if they had simply shown a video of drawing a few different paths on the car roof, and shown the resulting movement of the car. That is, the basic concept was clear once you had seen the video clip associated with the frames associated with Figure 118, perhaps augmented by the clip that followed in the video of the circle and broken circle.

However, they didn't stop there. What the video is really about (to my eye at least) is the fun you could have if you actually had cars that worked this way. I simply don't believe that you can watch this video without having a real sense of experiencing what this would be like. The fun of the examples, and even the fun that Anab and Louise were clearly having making it is simply too contagious to miss.

You see, toys are not about toys. Toys are about play and the experience of fun that they help foster. And that is what this video really shows. That, and the power of video to go beyond simply documenting a concept to communicating something about experience in a very visceral way.

And although this type of video is not experience "in the wild," so to speak, notice that throughout the video, Anab and Louise have populated nearly all the scenes with a cast of characters and props that reflect the environment that the cars would find themselves in, if they really existed. Just look at the crowd of spectators seen in Figure 122. They didn't have to be there in order to explain the operation of the cars, or the way that they played football. Yet they are essential to effectively place the play and experience in context. It is these imaginative flourishes of detail that transform this sketch from the pedestrian to the exceptional. As we have discussed already, a huge part of design is story-telling, and each of the techniques that we discuss expands our vocabulary for doing so. For me, Anab and Louise can hold their own with the best of them.

Figure 121: It's about Play Rather Than Cars

By showing the cars in use, the video is far more about communicating the fun of play resulting from their functionality, rather than just the functionality itself.

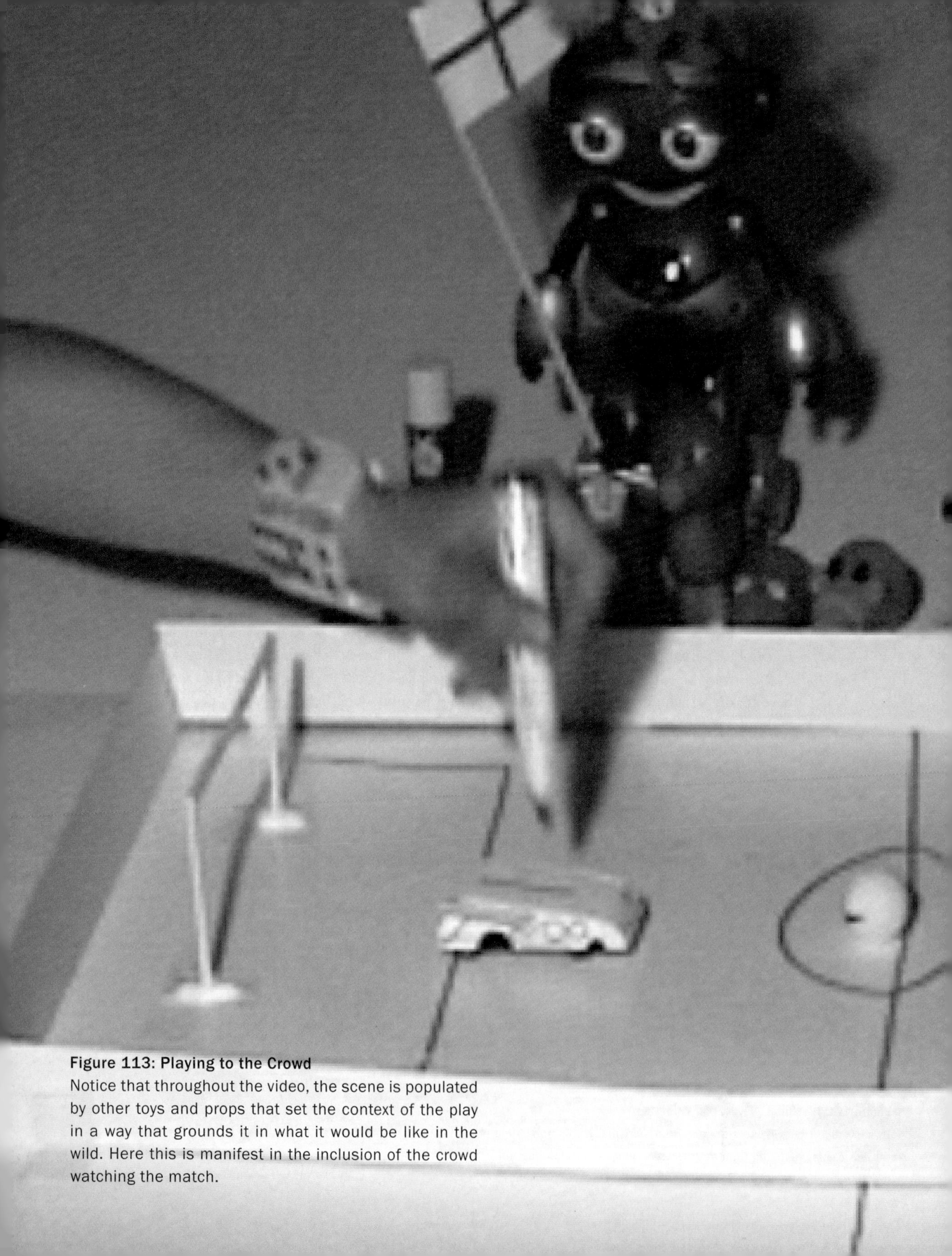

Figure 113: Playing to the Crowd
Notice that throughout the video, the scene is populated by other toys and props that set the context of the play in a way that grounds it in what it would be like in the wild. Here this is manifest in the inclusion of the crowd watching the match.

Let me wrap up my discussion of this example with a quick anecdote about how I learned about it.

In February of 2005, I was asked to give a lecture to the Interaction Design students at the Royal College of Art. My talk was about sketching; that is, it was based around the topic of this book. That evening I went home to Cambridge. A few hours later, I read my e-mail. Among all the spam and assorted other messages was a note from my friend Saul Greenberg. He was writing to point me to a web site that contained an example that he thought would be perfect for the book. It was the *Sketch-a-Move* web site. I loved the project, and when I looked at where it was done, I was completely taken aback by the fact it had been done by people who had been in my lecture that very day, without my knowing about their work. The good news was that I was going back down to the college in March to meet with students one-on-one, so I would have a chance to talk to Anab and Louise in person, which is what happened.

The reason that I am telling you this is to make an important side-point about the process of design.

Based on what I just told you, you might come to the conclusion that this was a rather interesting coincidence—an accidental timing of events that is pretty improbable. And, from one perspective you would be right. But actually, I don't think so. This was no accident.

There is an "old" saying that I made up:

If you want to be hit by a car, go play in the traffic, don't stand in a cornfield.

This is one of the most important rules of being a successful designer. Sure, there is a chance that in a cornfield a car could get out of control and find its way to hit you. But it is not very likely.

The point that I am trying to make is that I approached writing this book the same way I approach most things. I did everything I could to place myself and what I was thinking about at the busiest intersection of ideas, people, locations that I thought might be able to contribute to making it better. Sharing the various drafts of the manuscripts with people who might have good input was one approach. Talking to Saul was another, as was speaking at the Royal College of Art.

Could I have predicted that this particular example would have come from that particular place and that I would hear about it from that specific person? Of course not.

On the other hand, could I have predicted that this kind of thing would happen? On the other hand, could I have predicted that this kind of thing would happen? Without a doubt. Even more than that, the underlying take-away point is that I "designed" my behaviour precisely so as to ensure that it did happen. If you are a designer, so should you. The risk of someone stealing your ideas, or of making a fool of yourself, by sharing your ideas before they are done is nothing compared to the certainty of ending up with impoverished results if you don't.

Figure 123: Under|Control: An Interactive Sketch for a Wrist Computer

This is a detail of a wrist computer concept sketch developed by Nima Motamedi, one of my students at the Ontario College of Art and Design.

Extending Interaction: Real and Illusion

... long involvement with an unsolved problem can easily produce rigidity of outlook, a slow response to new ideas, and it is often the case that one with fewer inhibitions is better equipped to tackle it than one with greater experience.
—Eric Shipton

In this next example, I want to illustrate how one of my students at the Ontario College of Art and Design, Nima Motamedi, applied the technique seen in the *Sketch-a-Move* example, to implement an experience model of a wrist computer concept, as well as create a video of its use.

A close-up of his sketch is shown in Figure 123. It is made up of a foam-core "sandwich" with a pull-through strip as the "filling." A fore-arm wearing a watch is drawn on top of the foam-core, and a hole is cut out where the watch display would be. A sequence of watch screens was printed on the strip, as illustrated in Figure 124. Hence, when the strip is pulled through the two sheets of foam-core, the screens appear in sequence in the window of the watch crystal.

Nima's concept is fairly novel. It involves two separate surfaces in any interaction, which is why two pieces of foam-core are used. One is the display surface, which is on the top of your wrist, just as with a conventional wristwatch. The other is on the underside of your wrist, where the buckle of your watch-strap normally is located. In this design, however, the buckle incorporates a touch-sensitive surface that is used to control the device. The idea is that by having the control at the back, you can manipulate the screen cursor (or other graphical widgets) without your finger obscuring any of the information on the (necessarily small) screen.

What I love about this example is how Nima's sketch afforded the ability to explore and experience this novel mode of interaction. Independently, he came up with the same technique that we saw earlier in the *Sketch-a-Move* project. To appreciate the simplicity and elegance of the solution, and how it can be employed in a new context, look again at Figure 123. Notice that the cursor in the photo is a little cube. This is actually a little magnet.

Now look at the underside of the model, shown in Figure 125.

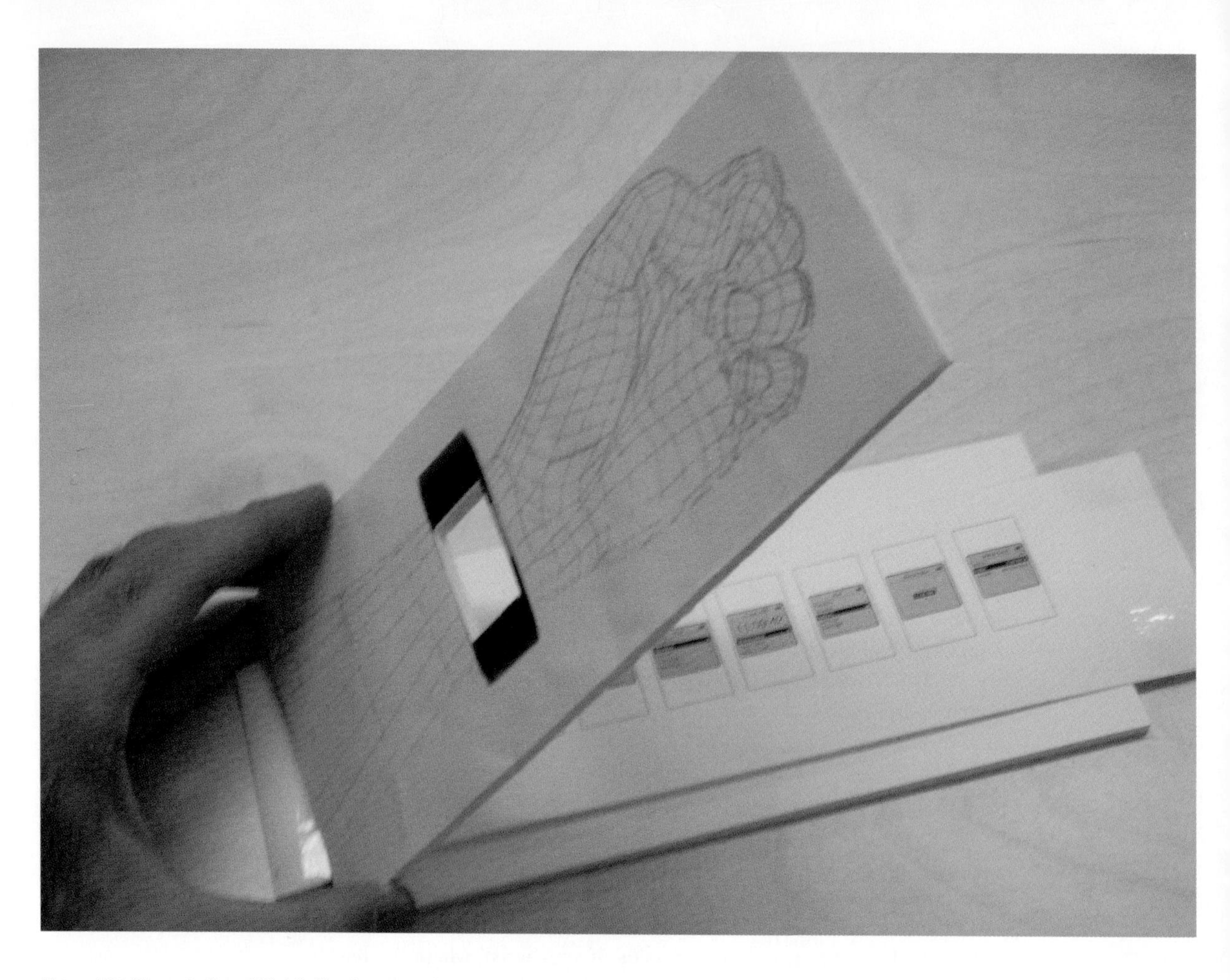

Figure 124 (Upper Left and Right): The Anatomy of the Wrist Computer Sketch
The foam core sandwich, shown disassembled in the right-hand image, has a pull-through strip with a sequence of screens drawn on it.

Figure 125 (Lower Right): Using a Map Pin as a Surrogate Finger
The magnet cursor on the top-side display (illustrated in Figure 114) is manipulated by moving a map pin around within the boundaries of a window opening on the bottom side of the foam-core model, since the magnetic force of the cursor passes through the pull-through paper strip and acts on the metal tip of the pin.

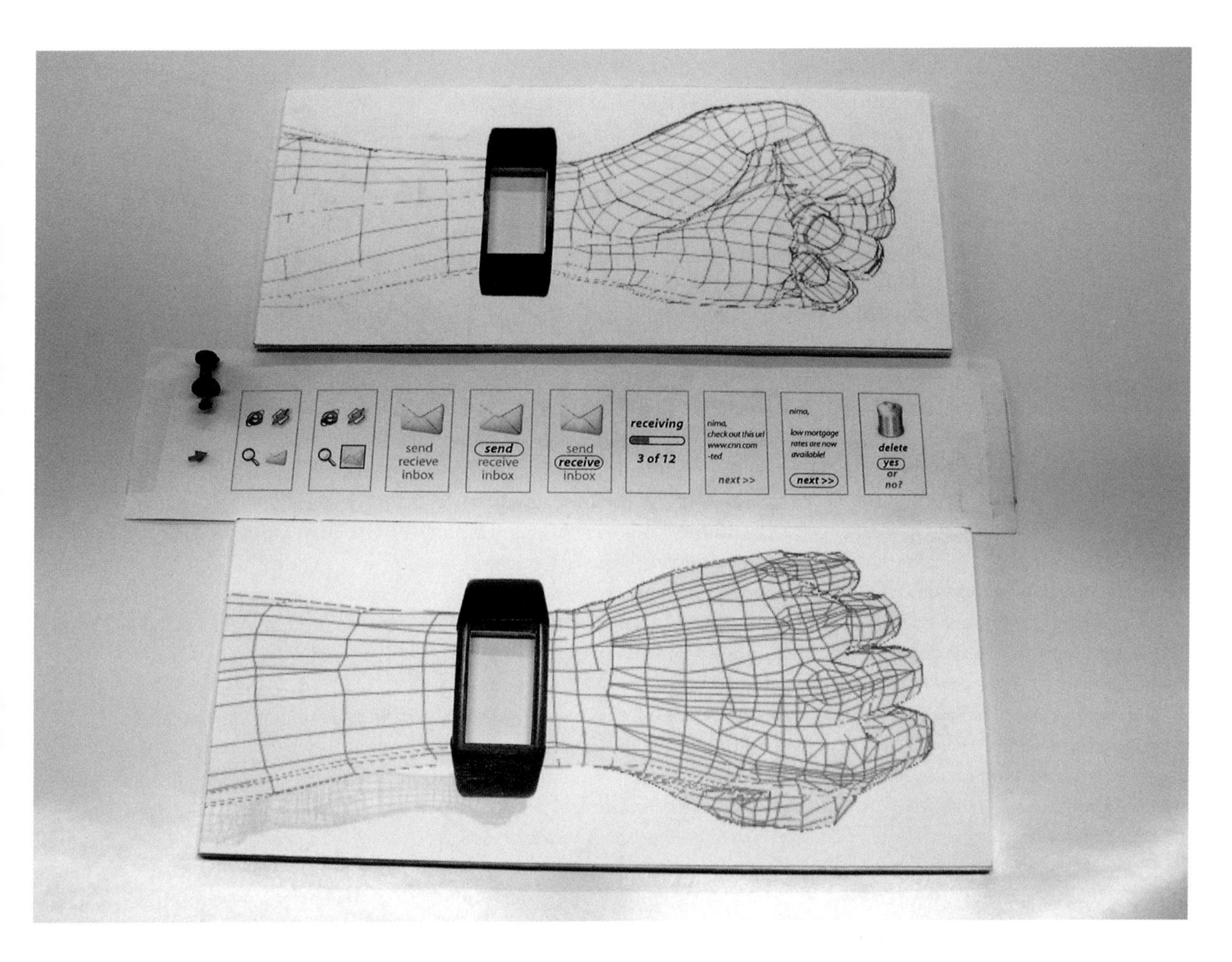
send
recieve
inbox
send
receive
inbox
send
receive
inbox
receiving
3 of 12
nima,
check out this url
www.cnn.com
-ted
next >>
nima,
low mortgage
rates are now
available!
next >>
delete
yes
or
no?

receive
inbox

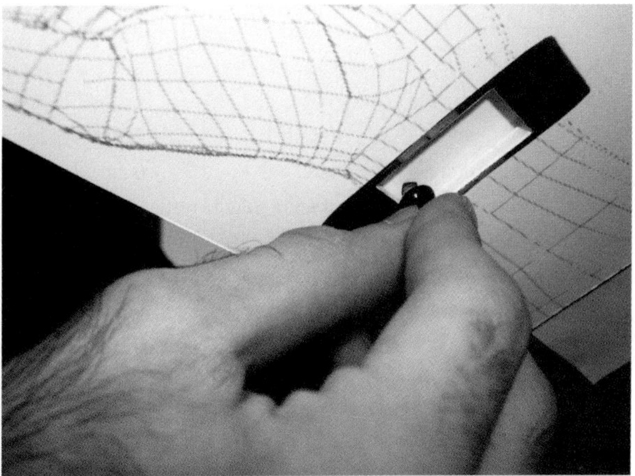

Figure 126: Raw Video with Transitions
The raw video consists of repeated segments of interacting with a screen, followed by an unwanted segment where the screen is changed.

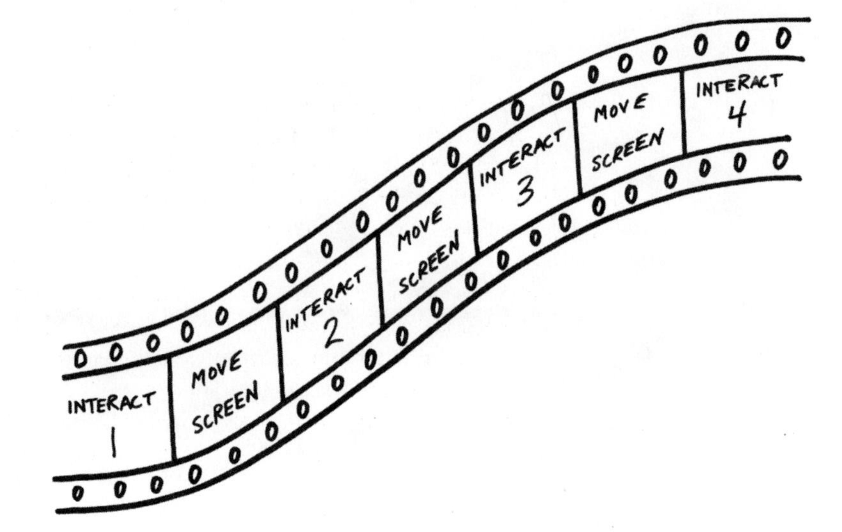

Figure 127: Pulling Out the Unwanted Screen Transitions
By reassembling the material so that just the interactions are there (see Figure 128), what one sees on the final video is the illusion that it is the actions of the magnetic cursor that are causing the screen changes.

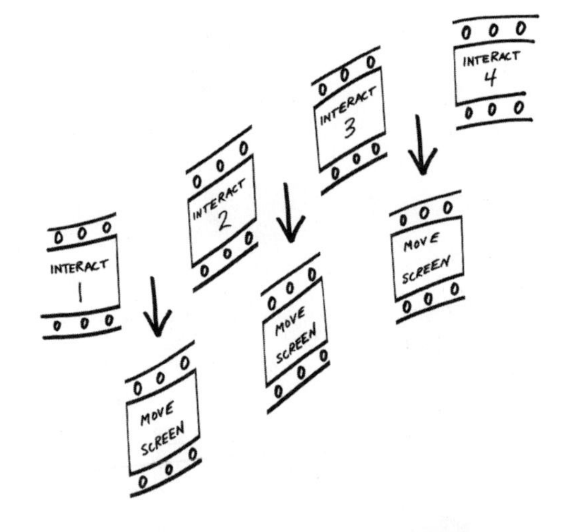

Figure 128: Note that instead of having instant transitions, Nima had the option to render them as dissolves, wipes, or any one of a number of other standard video effects.

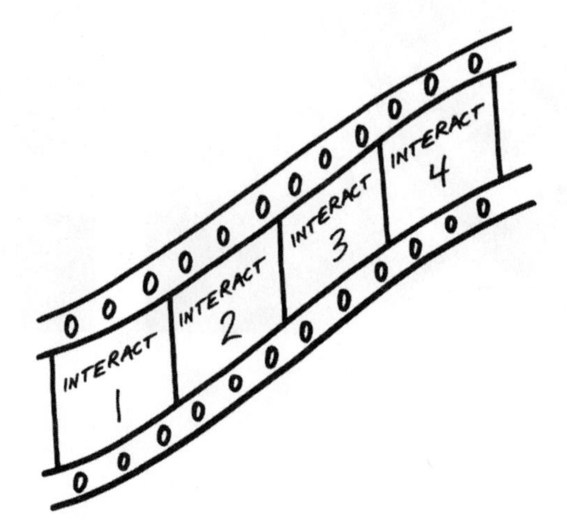

Notice that there is a window on the underside of the model, directly below the top-side display. In the model, instead of moving your finger over a touch-sensitive surface, you move a map-pin over the surface of that lower window. Since the tip of the pin is metal, the magnetic force of the cursor follows the pin's motion. Even though this is not touch sensitive, by cutting the window into the underside, its edges nevertheless provide tactile feedback as to the boundaries of the interaction. This feedback is an excellent surrogate for what one would feel with the finger in the envisioned product.

Due to his attention to these subtle details, and doing so in a simple but elegant model, Nima has provided a sketch that affords an almost instant appreciation for the viability and validity of the underlying concept. Of greatest significance, he does so by letting you achieve this through direct personal experience.

But this is only part of the story. Using the model interactively gives you a sense of the feel and usability of the cursor control, but it doesn't give a good sense of the flow from screen to screen. It is just too awkward and disruptive to pull the screen strip under the window. Yes one can get a sense of the sequence, but not the dynamics.

To accomplish this, Nima utilized another sketching technique, one that we have already seen: video. In this case, he first shot himself working through the sequence of pointing to an item on a screen, selecting it, then pulling the screen strip through to the next screen, then interacting with what was now displayed. He kept doing this from screen to screen. Schematically, the resulting video looked something like what is shown in Figure 128.

Here we see a segment showing Interaction 1, followed by a segment showing the screen changing, then Interaction 2, then another screen change, and so on. The problem with this is that we want the screen changes to appear in the video the way that they would in the real product; that is, we want them to change instantly. The way that Nima fixed this was to simply cut the transition segments out using video editing software. This is illustrated in Figure 127.

By reassembling the material so that just the interactions are there (see Figure 128), what one sees on the final video is the illusion that it is the actions of the magnetic cursor that are causing the screen changes.

Hence, by combining two techniques, Nima was able to explore his design in a shareable manner to a far richer degree than if he had just done the physical model, or the video, alone. The rule is this:

A number of simple complementary techniques is usually more effective and economical than one comprehensive approach.

Figure 129: A Problem of Context

Reading a newspaper through a hole in a piece of matte-board is used as a metaphor for the problems of reading and browsing large documents using the limited resolution of a computer display.

Source: Apperley & Spence (1980)

The Bifocal Display

The later in the process that a mistake is detected, the more expensive it is to fix.
— Fred Brooks Jr.

In 1983 I saw a video that was a revelation. It had the same effect on me as a user interface designer that seeing the first Starwars had on most of us when it was first released in 1977. The video in question was done by Mark Apperley and Bob Spence at Imperial College, London. The first part of the video was called *The Bifocal Display* (Apperley & Spence,1980; Spence & Apperley, 1982).

Now this may not be a title that would inspire most people to run off immediately to their local video store. But for me it changed everything, since it showed a new way to develop a concept and use video to illustrate it clearly, cheaply, and effectively. That is, in the parlance that I use today, it taught me how to sketch using everyday practical materials (a kind of 3D collage), and video—something that I had never contemplated before.

At the time the video was made, there were no high-resolution bitmapped displays such as we are used to seeing now. All you could get on a standard display were 24 lines of text, with only 80 characters per line. This made browsing and reading documents even more difficult than it is today. The nature of the problem was explained and put into context in the video by pinning a broadsheet newspaper page up on a corkboard, and then trying to read it through a square hole in a large piece of matte board. As with displays of the time, the hole revealed about 24 lines of text, and was one column wide (which is about 80 characters). This is illustrated in Figure 129, which consists of frames taken from the video.

Thus, by the analogy to the everyday task of reading a newspaper, the video makers were able to effectively communicate the problems of reading text on a computer, especially the inability to have an overview of the larger document, and glean some sense of context.

The next part of the video then presents a novel approach to addressing the problem, something that they call *The Bifocal Display*. The basic idea is this: don't have all the text in the reading window flat. That is, have only the section of the text immediately around the words that you are reading mapped onto a surface that is parallel to that of the window. Then, have the parts preceding and following what you are reading mapped onto a plane that recedes into the distance.

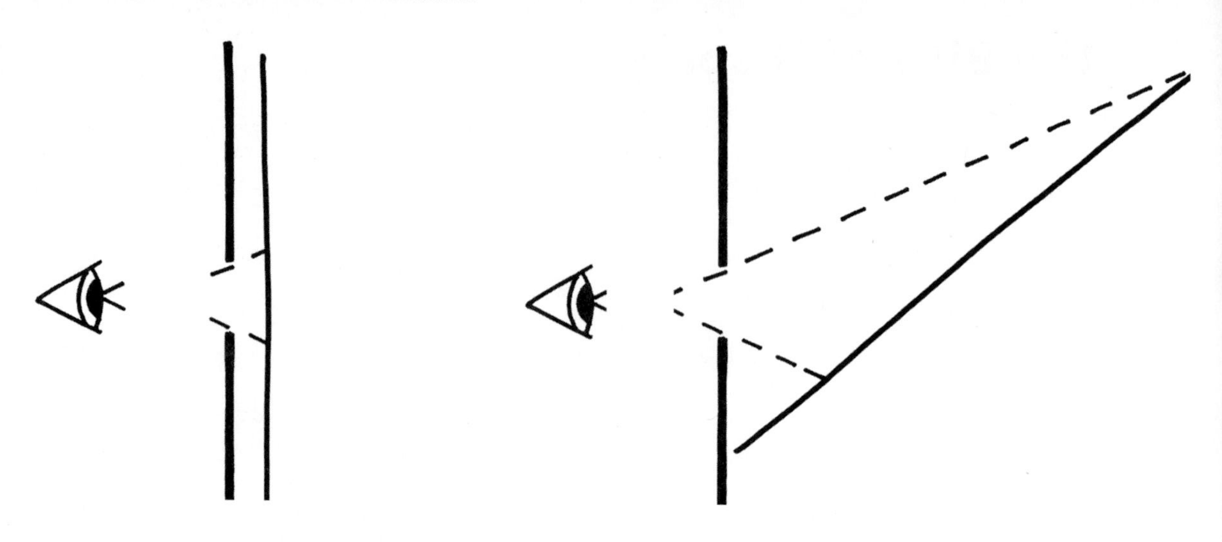

Figure 130: Extending the Range of View

The Bifocal Display exploits the observation that if a newspaper column adjacent to the other side of a hole in the matte board is parallel to the matte board, only a small part of the column will be visible. This is illustrated in the left-hand drawing. If the newspaper column is tilted away from the hole, far more of the column is visible, but it will be harder to read. This is illustrated in the right-hand drawing.

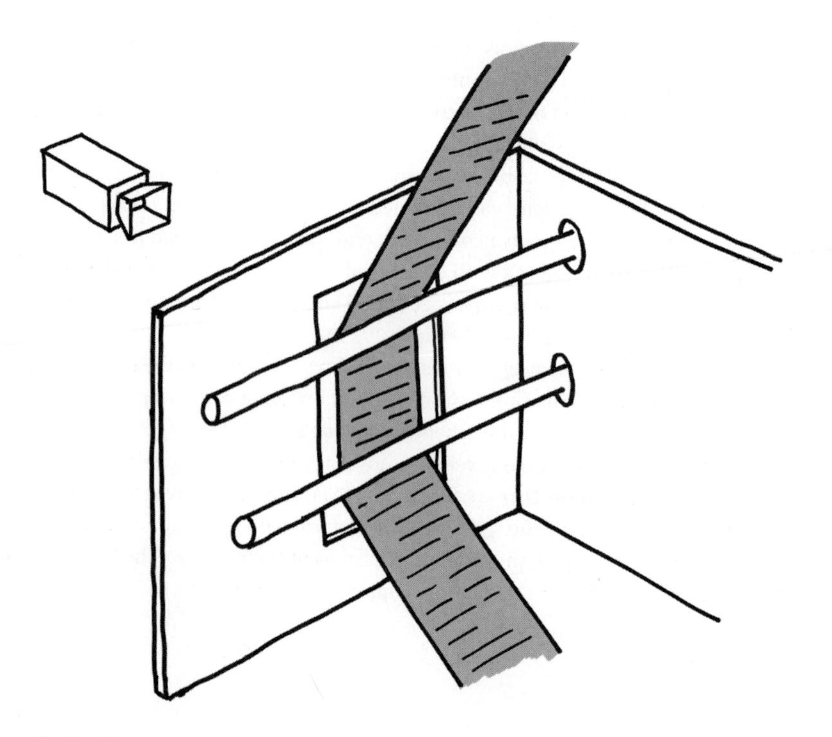

Figure 131: Schematic of a Bifocal Display Video Sketching

This is perhaps better understood using Figure 130. Look at the drawing on the left first. Imagine that the matte board is the same one that we saw in Figure 129. If we hold a newspaper column right up against the back of the hole in the board, we will be able to easily read, but not too much of the overall column is visible.

Now imagine that the newspaper column is tilted away from the matte board at about a 45° angle, as is shown in the right-hand drawing in the figure. In this case, one will be able to see far more of the column, but as it recedes away into the distance, it will be harder and harder to read. Coming back to a reference that I made earlier in this section, it will have the same visual effect as the credits in the original *Starwars* movies.

What *The Bifocal Display* does is combine the properties of both the left- and right-hand parts of Figure 130. The main part of the newspaper text is parallel to the viewing plane, as in the drawing on the left. However the parts of the column preceding and following this main section recede into the past and the future, respectively, in a manner consistent with that shown in the drawing on the right.

How this was all put together is shown schematically in Figure 131. The newspaper column is stretched over two horizontal rods mounted behind a square hole (the "window") in some matte board. The part between the rods is parallel to the window, and the parts above and below recede in a manner similar to that illustrated on the right in Figure 130. A video camera is set up on the far side of the matte board, looking in at the newspaper column, through the window.

What is actually seen by the camera is shown in the upper image in Figure 132. The lower image in the figure shows the set-up with the front matte board removed, thereby exposing the horizontal rods.

What is interesting is how rich the affordances of this approach to sketching the concept are. For example, you could explore a number of important aspects of the concept by manipulating things such as:

- The percentage of the display that should be flat for reading versus the percentage that recedes back to provide context
- The gradient at which the warped part of the column should recede
- The impact on text versus images, or other types of material
- The effectiveness of the approach when used in scrolling horizontally versus vertically
- How content could be altered in order to enhance the effect

All of these can be explored simply by manipulating the properties of the basic components illustrated in Figure 131. This is one of the examples that I think every student of experience design (including those already working and teaching) should actually reproduce and experiment with, rather than simply viewing the tapes and reading commentaries on them, such as mine. I think that it is that rich and important.

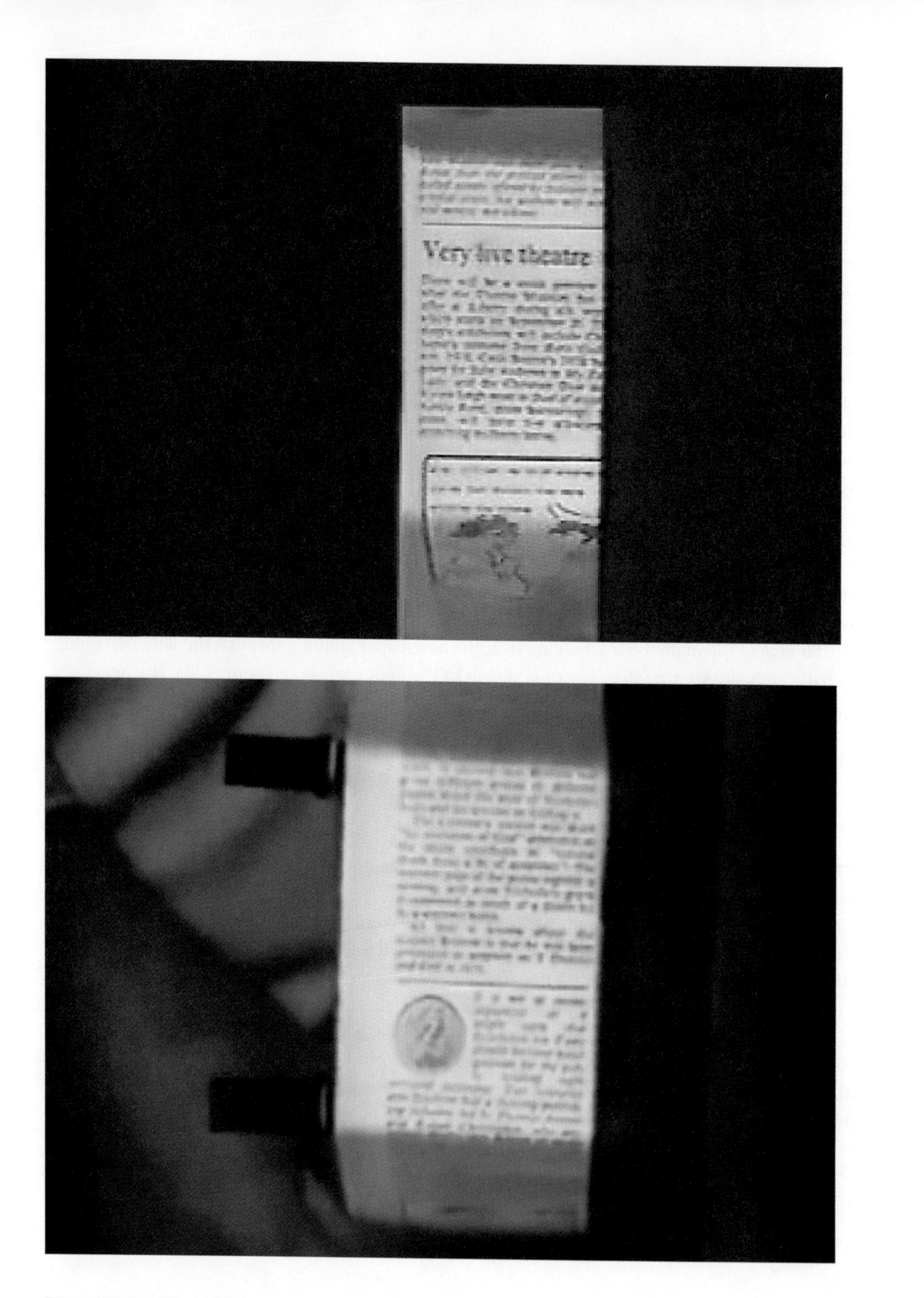

Figure 132: The Bifocal Display

The solution offered was a kind of "fish-eye lens," which let you see the part of the text that you were reading in detail. What came before and after the part that you were reading "rolled off" into the distance (top frame). The lower frame exposes the rods over which the column is being rolled to get the distortion.

Source: Apperley & Spence (1980)

Figure 133: Level of Detail Control

When scrolling, documents are coded by colour and an easily identifiable feature (upper frame). When scrolling stops, the details of the document in the reading position are revealed (lower frame).

Source: Apperley & Spence (1980)

Figure 134: The Perspective Wall

An information visualization concept in the tradition of the Bifocal Display, as illustrated in Figure 133.
Source: Card, Robertson & Mackinlay (1991)

Let's see how Apperley and Spence explored some of these variations in their original video. Figure 133 shows two more frames from the video. The first thing to notice is that in this example they are scrolling horizontally rather than vertically. However the main question that they are exploring is, "Can content can be manipulated to enhance the power of the technique?" As a result of their experimentation, they suggest that this may well be the case if one suppresses detail and highlights key salient features of the material that is receding.

In the frames illustrated in Figure 133 they have used two techniques. First, they show that if different types of pages have different background colours, that these can easily be seen, even if the content of the receding page cannot be read. Second, they suggest that the level of detail of a page be reduced as it recedes, and that certain key features be made more salient. In the example, the orange page labeled TELEX would have its full text displayed when it was in the central viewing plane, as shown in the lower frame. However, when it is receding to the left or the right, the page contents would be reduced, for example, to a simple large T, as is illustrated in the top frame of the figure.

One small point to note in this example is how they have used simple Post-its to implement this level of detail simulation, thereby providing a good example of how "paper prototyping" can be used with video to good effect (more on this later).

Just to wrap up this discussion of the bifocal display, I want to leap ahead a number of years, and juxtapose this work with that which came later, yet explored the same space. The first is some work in information visualization undertaken at Xerox PARC by Card, Robertson, and Mackinlay in 1991. The other is some work undertaken by Sheelagh Carpendale and her colleagues in what she calls *Pliable Display Technology* (Carpendale, Cowperthwaite & Fracchia, 1995; Carpendale 1999).

Because of the emergence of ever-more powerful computer graphics at ever-lower prices, the 1980s and 1990s saw an expanding interest in what was known as *Scientific Visualization.* This combined the power of computer graphics and simulation to render complex scientific data in a visual form that would hopefully make it more comprehensible and lead to new insights and capabilities. Two good overviews from that time can be found in McCormick, DeFanti & Brown (1987) and DeFanti, Brown & McCormick (1989).

At the same time, Card, Robertson, and Mackinlay were asking themselves, "If these visualization techniques work for scientific data, why not explore their use in helping us better understand other types of complex data, such as we might find in the normal workplace? They, probably like you and me, were perfectly aware that computers had contributed to a great deal of complexity in the workplace, and that it was only fitting that they should likewise be used to help manage it. Hence, they began their program of research into *information visualization.*

Among the techniques they explored was something they called *The Perspective Wall,* shown in Figure 134. In looking at the figure you will see immediately why I am discussing it here. This is clearly a direct descendent of *The Bifocal Display.* However, this version was interactive, it

WITH AN OPEN HAND

JERRY BAUER

How much Money do you need?

DISCLOSURE

Show us the cliffs that we can't climb

Simon Jones

SECURITY POLICY

Don't shut the firewall too tightly

Fighting For Their Lives

THE COMPANY

MARK R. TIGGES

WITH AN OPEN HAND

JERRY BAUER

How much Money do you need?

DISCLOSURE

Show us the cliffs that we can't climb

Simon Jones

SECURITY POLICY

Don't shut the firewall too tightly

ling the political process.

Alliance leadership candidates freely admit it costs at least $1-million just to throw one's hat into the ring. With this hurdle, it is no surprise that high-end donors wield tremendous power over who gets to even enter the race.

The privacy argument is especially repugnant when one considers the amount of public subsidies that parties enjoy during elections, in the form of access to free broadcast time on the public's airwaves, tax credits for donors, and expense reimbursements for candidates and parties.

ghting For

THE COMPANY

Figure 135: An Example of the Bifocal Concept

With the page on the left, one has a good overview, but it is hard to read the actual text because it is so small. If you magnified the whole page, then you would be able to see the text that is visible, but you would lose the overview of the page. The page on the right shows Carpendale's Pliable Display Technology used to implement a form of the bifocal display. It echoes back to what we have already seen in Figure 132.

Source: IDELIX Software Inc.

used state of the art 3D hardware and software techniques, and enabled Card, Robertson, and Mackinlay to explore aspects of the technique that went well beyond those that could have been investigated using the implementation by Apperley and Spence.

Even though this later work represents a real contribution to our understanding of the concept, I think that it is really worthwhile to compare the two approaches: the sketch versus the implementation. Both have their strengths and weaknesses. But I think that it is pretty clear that it is orders of magnitude cheaper and faster to answer some questions using the sketch, than it is with the implementation of Card, Robertson, and Macinlay. And, just as clearly, there are things that the sketch could not handle that the implementation could. The point that I have made before, and that I am going to make again, is that Card, Robertson, and Mackinlay's implementation was, relatively speaking, *really* expensive in time, money, and effort. Hence, before making that investment, you had better make best efforts to convince yourself that you are investing in the right direction. My argument is that the techniques demonstrated by Apperley and Spence are key tools that we should have in our arsenal that we can draw on to do that particular homework.

To close our discussion, I briefly want to introduce one more example, that of the *Pliable Display* Technology, developed by Sheelagh Carpendale and her colleagues. Her work greatly expands upon that of *The Bifocal Display*. In particular, it looks at a broad range of 3D distortions (or lenses) that can be used to achieve the similar end of enabling one to see in detail what one is interested in, without losing a view of the surrounding context, or periphery. In many ways, her work can be viewed as a super-set of *The Bifocal Display*.

This, of course, means that it can also be used to implement some of the things that we have already discussed, as illustrated in Figure 135. Here we see the future and past of the column receding back to the page, as in *The Bifocal Display*. We also see that the *The Bifocal Display* has been incorporated into the display of the entire page. So, in some way, it is an amalgam of some of what we saw in Figures 129 and 133, combined. Finally, note also that with Carpendale's technology, the area of the column that is being read has been magnified. This, however, comes with the incumbent problem that the magnified column obscures the columns to either side, despite maintaining context above and below. I will leave it to you to learn more about her work in order to see how she solves this problem.

Finally, let me conclude with the following observation. I believe that a qualified interaction designer should be able to replicate the basic video of *The Bifocal Display* in 30 minutes or less. That includes building the physical framework illustrated in Figure 131, shooting the video, transferring the video to a computer, and editing the final result. If they can't do so, they likely don't have the right facilities, the right skills, or they are prototyping and not sketching. Not only knowing about this work, but being able to duplicate it is, to me, basic literacy for the profession.

Try it. Here is the recipe:

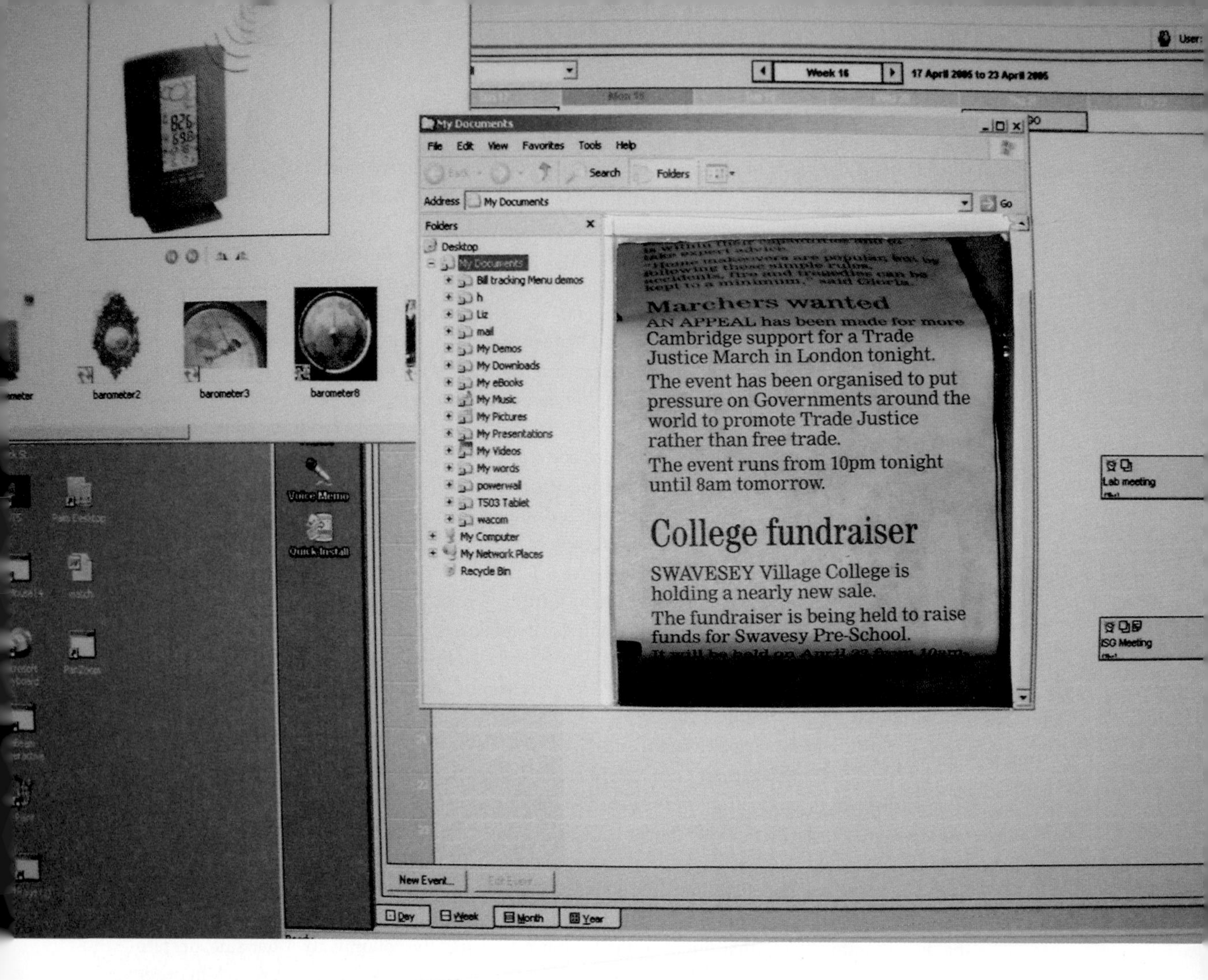

Figure 136: Reproducing the Bifocal Display Sketch

This was done in about 30 minutes using existing materials and tools that were found within the immediate area of our work group in Cambridge.

Step 1: Cut out newspaper column

Step 2: Make paper guides

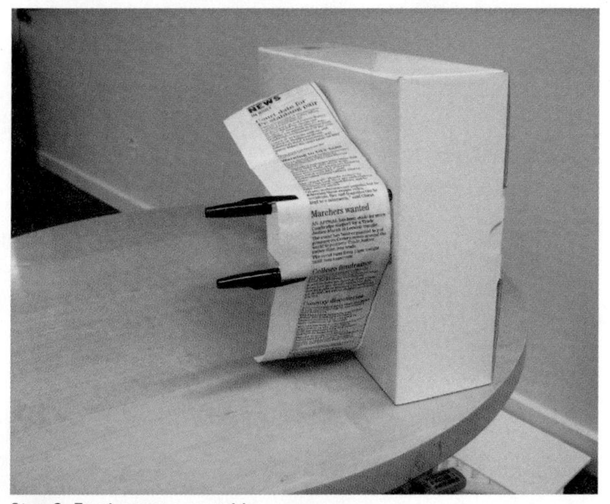

Step 3: Feed paper over guides

Step 4: Create “Window”

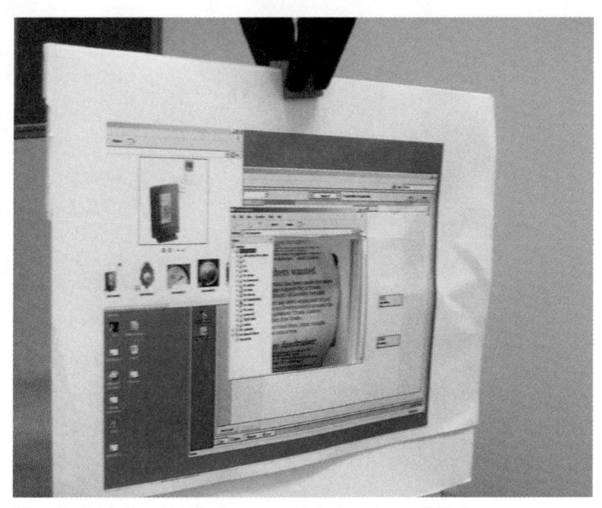
Step 5: Add “Desktop” of you operating system of Choice

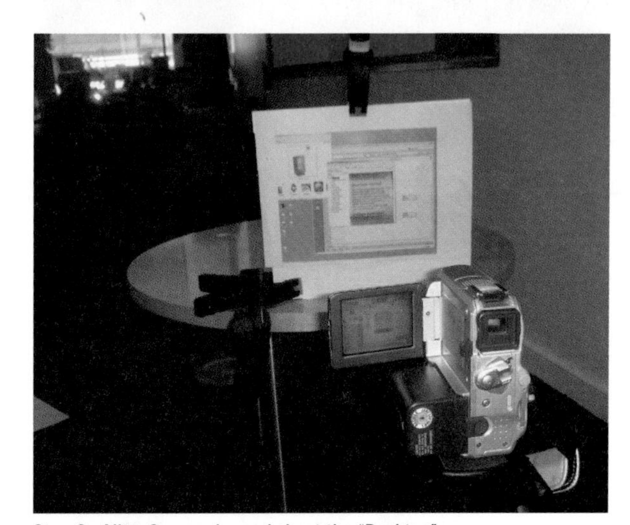
Step 6: Align Camcorder and shoot the “Desktop”

“This is the Xerox 895X Matter Duplicator.”

‘One simply takes an original object...

Video Envisionment

Play is the highest form of research.
—Albert Einstein

By now we have encountered a number of examples where video has played a role in mocking up, sketching, and capturing various design ideas. There is a subspecies of this genre that I want to spend some time discussing, for a number of reasons. It is what can best be called Envisionment Videos. As their name suggests, these are clips that are created to communicate some holistic view showing an envisioned system in context. Unlike many of the examples that we have seen, they go beyond just presenting a demo. Rather, they usually are built around a narrative that tries to capture a way of working with technology, as opposed to the design specifics of the device itself. Actually, *Sketch-a-Move* could well be considered an example. Overall, they are very much part of the story-telling tradition of design.

The Xerox 895X Matter Duplicator

Let me start by giving you a really simple example that was made in a workshop that I did in Cambridge England in 1988. It is a simple and playful tape called *The Matter Duplicator* by Randy Smith. The video gives a "live" demonstration of the Xerox 895X Matter Duplicator, frames of which are shown in Figure 137.

Randy demonstrates how one makes a copy of a screwdriver simply by placing an original in the input hopper of the 895X copier (which bears an uncanny resemblance to a regular photocopy machine), and pushing the green *copy* button on the console. Sure enough, after he has done this, he is able to retrieve an exact copy of the original from the output tray.

Of course, the tape's style and content ensure that nobody will ever take it seriously as a demonstration. I include it here for three reasons. First, I just think that it is funny, and thought that you might enjoy it, despite the poor technical quality. Second, it is a nice and short introduction to the genre. Third, it is a simple but effective example of the type of thing that one can do in a short workshop to learn how to make such things—from both sides of the camera. This last point is not to be taken lightly. It is doing such simple "finger exercises" that develops the technique that lets one pull off what we saw Bob Spence do in *The Bifocal Display* video.

Figure 137: The Matter Duplicator

This short video gives a demonstration of the Xerox Matter Duplicator, a product that lives only in the warped mind of my friend and one-time colleague Randy Smith. Here he demonstrates that a screwdriver placed in the input hopper can be copied as simply as you copy paper documents today.

...inserts it into the input compartment...

...and applies a digit firmly and vertically on the green square...

...when the process is complete...

...lo and behold, the duplicate, and the undamaged original, for you to screw around with, in the future.

Office of the Professional

Having just mentioned Bob Spence, let me use another video of his as my next example. *Office of the Professional* is the first example of envisionment videos that I am aware of. It was made at Imperial College, London, in 1980 by Mark Apperley and Bob. Figure 138 is a frame from one of the scenes in the video.

The tape consists of a number of short scenarios built around the types of things that a professor does in the normal course of events. These include interacting with colleagues, examining experimental data, looking at student records, making travel arrangements, and filing information. What the video brings to the equation is a cinematic manifestation of how future technology might be integrated into our environment and lives so as to facilitate the performance of such tasks.

It is interesting to view such a video more than 15 years after it was made. For example, there are curious anachronisms, such as the circa 1980 telephones, references to sending a TELEX, and an Apple][computer, all in the midst of scenarios involving now current interaction techniques such as speech recognition, pen-based input, touch screens and tablets, remote pointing onto the wall, and advanced graphical visualization.

But such technological inconsistencies are, and were, not the point. The video made no pretence of predicting the future (although it was remarkable for its insights, it is worth remembering that it was done more than 10 years before Mark Weiser published his classic 1991 paper on ubiquitous computing and two years before the launch of the Apple Macintosh).

Significantly, the real purposes for making the tape are clearly articulated in the commentary that serves as a kind of coda at the end. Key among these were:

To show the bifocal display in context: This in itself is an example of why I have distinguished envisionment videos from, shall we say, demonstration videos. These may not be the best terms, but contrast the form and content of *The Bifocal Display* video and *Office of the Professional*. The former went deep into a particular problem and worked through a possible solution. The latter paints a much broader context, physical and social, into which the bifocal display can be placed. We learn far less about the display's characteristics in it. But on the other hand, we gain a deeper perspective of why we might want to, or how it might fit in with the larger social and physical ecology in which it might find itself. Bob and Mark understood this, even then, when they were pioneering the use of video. A lot of us could still benefit by learning this lesson.

Demonstrate the use of intuitive skills rather than taught skills: This is also really telling in that it is centred on the human user, not the technology. In many ways this presages an attitude not popularized until the publication of books such as Norman & Draper's *User Centered System Design* (1986) and Norman's *Psychology of Everyday Things* (1988). The notion here was to communicate, by example, how future systems could be designed so as to exploit existing everyday skills, rather than demand the acquisition of fundamentally new skills, and incurring the attendant costs.

To capture a particular view of the future: Yet keeping in mind, as Bob Spence says on the video, that what might seem futuristic today "has the habit of coming next Wednesday." This harkens back to something that we discussed much earlier in the book. Even when designing for the immediate future, we must do our best to anticipate the future into which our design has to evolve. I characterize this as walking with one's head in the clouds and feet deeply entrenched in the mud.

Provide food for thought for those doing research and designing tomorrow's systems: The video is clearly envisionment, and has the production values of a concept rendering, or sketch. Nobody would confuse this with product, or a real system. This speaks to one of the prime attributes that we have associated with sketching, namely being a critical component in a conversation, by both what is shown, and what is not. This video is not telling us about the future. Rather, it is inviting us to have a conversation about it, and giving us a good place to start.

If it is not clear already, I really like this video, and I have a lot of respect for those who made it. It is a good example of the importance of knowing our history and traditions. Even today there is about as much to learn from this video as there was when it was made.

Figure 138 (Following Spreads): Office of the Professional

The first envisionment video that I ever saw.

Source: Apperly & Spence (1980)

Worker: "...I need one more reference on manufacturing efficiency..."

Boss: "Right, I know just the one you need, and the transaction's here..."
(begin retrieving data by pointing to bulletin board on the sidewall)

(the target file on the bulletin board is lit up by a laser pointer)

Boss: "Let's see, I think it is in 1978...no it must be in 1977...Yes..."
(The requested file appears on the monitor. Boss leaves through files on the screen to find the pertinent information. Once file is found, he presses a button on the desk...)

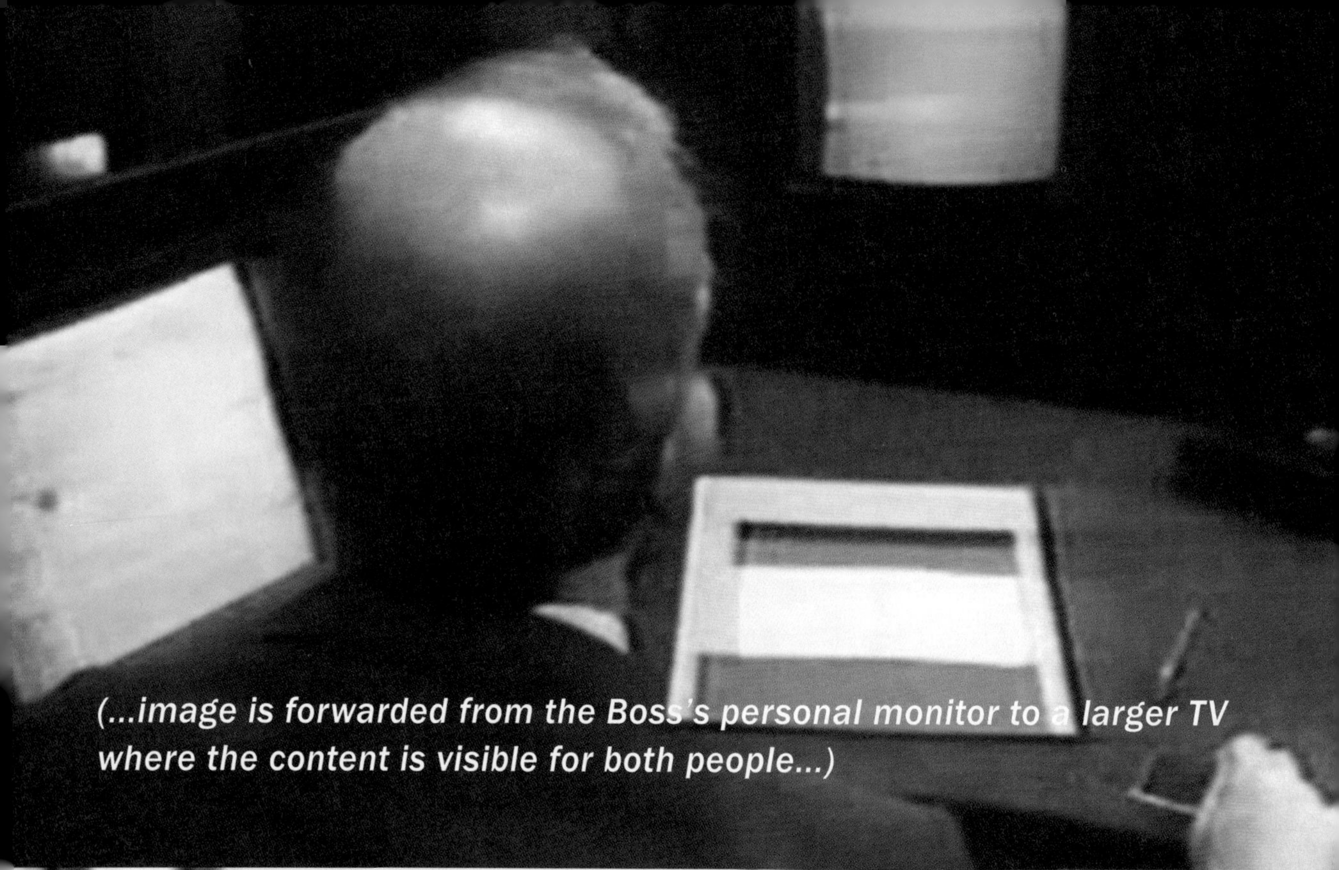

(...image is forwarded from the Boss's personal monitor to a larger TV where the content is visible for both people...)

Boss: "Yes, that's it. Right..."
(The Worker see the data, nods in affirmation.)

Boss: I'll leave a copy of that in your pigeon hole, COPY."

(Boss points to wall where the worker's picture is located and copies the file over to the worker's file through voice activation.)

Worker: "Oh, thanks boss, now I'll be able to get on with it."

The Knowledge Navigator

The next example that I want to talk about is a marked contrast to *Office of the Professional.* It is a video called *The Knowledge Navigator.* It was made in 1987 by Hugh Dubberley and Doris Mitch for a presentation by the then president of Apple Computer, John Sculley.

It is the contrast between it and *Office of the Professional* that led me to include it. But first, by way of background, take a look at the accompanying sidebar, where Hugh Dubberley gives a first person account of the making of the video (See Figure 139.)

The Knowledge Navigator is certainly an envisionment video, but it is equally certainly not a sketch. The first hint, even without seeing it, is its budget and production schedule, as outlined by Hugh. Yes, they were pressed for time for what they were doing. But they were also not doing something that was quick, timely, plentiful, or disposable. This was a video that was created in the visual vocabulary and with the production values of the TV commercial or, as Hugh describes it, a science fiction film. And, as such, they did a good job.

On the other hand, this, and many of the similarly produced films that it spawned, also gave rise to some real issues, some of which are hinted at in Hugh's account. Remember what I said earlier about good stories being like a kind of viral marketing—they get retold over and over? Well, so it was with *The Knowledge Navigator.* None of us had ever seen anything like it before, and everyone wanted a copy to show. But when this happened Apple lost control. Out of the context of Sculley's presentation the visual vocabulary and production values were too persuasive: people being people inevitably started to believe that this was a system that Apple was actively working on. Not way down the road as a vision, but right now. "Why else would they spend all that money on this and tell us this story if that were not the case?" they might ask. And if you actually persuaded someone that it was not something in the works, that almost made things worse. Raising expectations that high and then not delivering meant that there was all the further for them to fall.

I included this example as a reminder of what we said much earlier about the properties of a sketch:

Clear Vocabulary: It must be distinguishable from other types of renderings.

Minimal Detail: It must include only what is required to render the intended purpose or concept. Superfluous detail must not distract attention to the sketch's purpose.

Appropriate Degree of Refinement: A sketch should not suggest a level of refinement beyond that of the project being depicted.

Even if *The Knowledge Navigator* had been quick, timely, inexpensive and disposable, it still would not work as a sketch. It was about telling, not asking, and what it said, it said with conviction. And it was too vulnerable to misinterpretation—the wrong kind of ambiguity. *The Knowledge Navigator* is an extreme case, but hopefully that makes the point all the easier to see. The thing to take away from this example, besides an important part of our history, is a vigilance in deciding upon the visual vocabulary and types of production values that you use in your own work, as well as a more critical eye in terms of how you read the work of others.

Seamless Media Design—Clearboard III

I want to conclude this thread of our discussion with one final example. I include it because it caused me problems. In many ways, it seems to break the very principles that I just spoke about.

The example is an envisionment of a future incarnation of a shared electronic drawing board. It presents a concept called *Clearboard*, first developed at NTT Labs in Japan by Ishii & Kobayashi (1992). What Ishii and Kobayashi did was build a drawing surface that two people could draw upon at the same time, one on each side of the surface, such that they could see each other through their common drawing. A frame from the envisionment video illustrates this in Figure 140.

So what is the problem? Well, in watching the video, it seems to me that it may well have cost significantly more to make than *The Knowledge Navigator*. The production values say anything but, "This is cheap, fast, plentiful, and disposable." Just the opposite. So why is it when you watch it, it does not conjure up memories of TV commercials past? Why is it that I never once heard anyone suggest in person or in writing that this was a representation of something that NTT was making as a product?

These are not idle questions. If we are going to use the language of video and film in our work, then it is really important that we think about such things and develop our literacy to the point where we are alert and vigilant enough to both ask the questions and be able to sustain a reasonable discussion that helps us gain some insights about them.

In this, I am not suggesting that this video is a sketch. Far from it. It is an envisionment, it is not part of a TV show, and not a TV commercial or product announcement. Beyond that, it is may not matter too much about what it is. It is just important that we know what it is *not*.

For me, the reason that this video does its job so well is rooted in the role of stylization that is so much a part of ancient Greek and Noh theatre. I leave it to you to watch *Office of the Professional*, *The Knowledge Navigator*, and *Seamless Media Design*, to come to your own insights about the relationship between content, intent, visual style, and production values. I just want to prompt the ensuing conversation.

Figure 139 (following spread): Sketches of Frames from the Knowledge Navigator
Both the video and numerous still frames from it can readily be found on the Internet.

Figure 140 (second spread): Clearboard III
Credit: NTT

Knowledge Navigator

Desertification in the Sub Sahara

NEW YORK ⇆ TOKYO

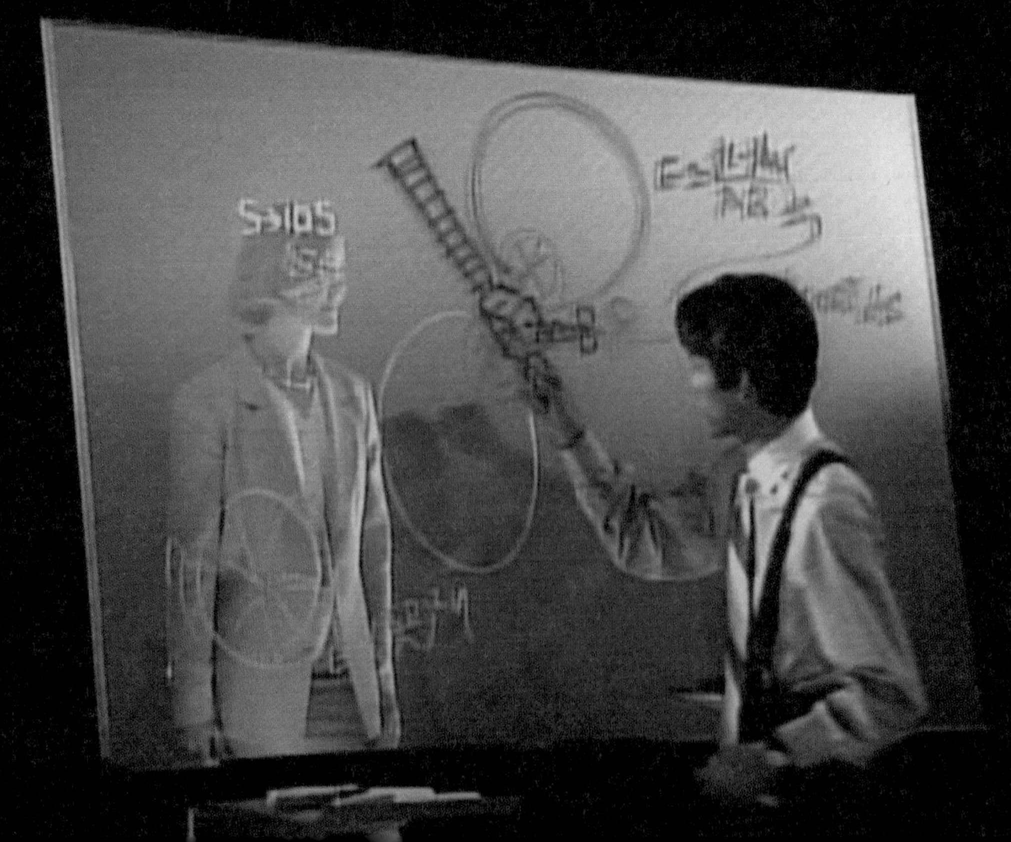

The Making of the Knowledge Navigator

by Hugh Dubberley

We made the Knowledge Navigator video for a keynote speech that John Sculley gave at Educom (the premier college computer tradeshow and an important event in a large market for Apple). Bud Colligan who was then running higher-education marketing at Apple asked us to meet with John about the speech. John explained he would show a couple examples of student projects using commercially available software simulation packages and a couple university research projects Apple was funding. He wanted three steps:

1. what students were doing now,
2. research that would soon move out of the labs, and
3. a picture of the future of computing.

He asked us to suggest some ideas. We suggested a couple of approaches including a short "science-fiction video." John chose the video. Working with Mike Liebhold (a researcher in Apple's Advanced Technologies Group) and Bud, we came up with a list of key technologies to illustrate in the video, e.g., networked collaboration and shared simulations, intelligent agents, integrated multi-media and hypertext. John then highlighted these technologies in his speech.

We had about 6 weeks to write, shoot, and edit the video—and a budget of about $60,000 for production. We began with as much research as we could do in a few days. We talked with Aaron Marcus and Paul Saffo. Stewart Brand's book on the "Media Lab" was also a source—as well as earlier visits to the Architecture Machine Group. We also read William Gibson's "Neuromancer" and Vernor Vinge's "True Names." At Apple, Alan Kay, who was then an Apple Fellow, provided advice. Most of the technical and conceptual input came from Mike Liebhold. We collaborated with Gavin Ivester in Apple's Product Design group who designed the "device" and had a wooden model built in little more than a week. Doris Mitch who worked in my group wrote the script. Randy Field directed the video, and the Kenwood Group handled production.

The project had three management approval steps:

1. the concept of the science fiction video,
2. the key technology list, and
3. the script.

It moved quickly from script to shooting without a full storyboard—largely because we didn't have time to make one. The only roughs were a few Polaroid snapshots of the location, two sketches showing camera position and movement, and a few sketches of the screen. We showed up on location very early and shot for more than 12 hours. (Completing the shoot within one day was necessary to stay within budget.) The computer screens were developed over a few days on a video paint box. (This was before Photoshop.)

The video form suggested the talking agent as a way to advance the "story" and explain what the professor was doing. Without the talking agent, the professor would be silent and pointing mysteriously at a screen. We thought people would immediately understand that the piece was science fiction because the computer agent converses with the professor—something that only happened in Star Trek or Star Wars.

What is surprising is that this piece took on a life of its own. It spawned half dozen or more sequels within Apple, and several other companies made similar pieces. ...These pieces were marketing materials. They supported the sale of computers by suggesting that a company making them has a plan for the future. They were not inventing new interface ideas. (The production cycles didn't allow time for that.) Instead, they were about visualizing existing ideas—and pulling many of them together into a reasonably coherent environment and scenario of use. A short while into the process of making these videos, Alan Kay said, "The main question here is not is this technology probable but is this the way we want to use technology?" One effect of the video was engendering a discussion (both inside Apple and outside) about what computers should be like.

On another level, the videos became a sort of management tool. They suggested that Apple had a vision of the future, and they promoted a popular internal myth that the company was "inventing the future."

The Mug Metaphor
Interface
Lucia Terrenghi
FLUIDUM

The Mug Metaphor
Interface
Lucia Terrenghi
FLUIDUM

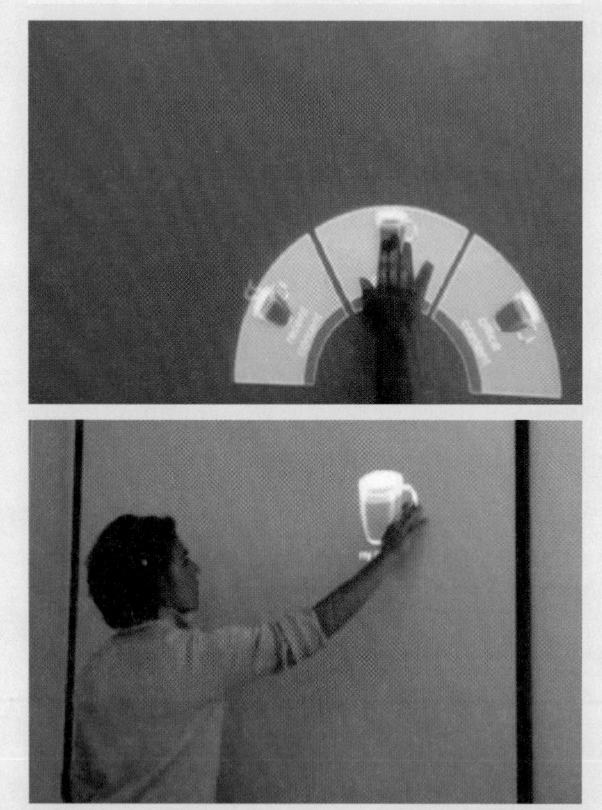

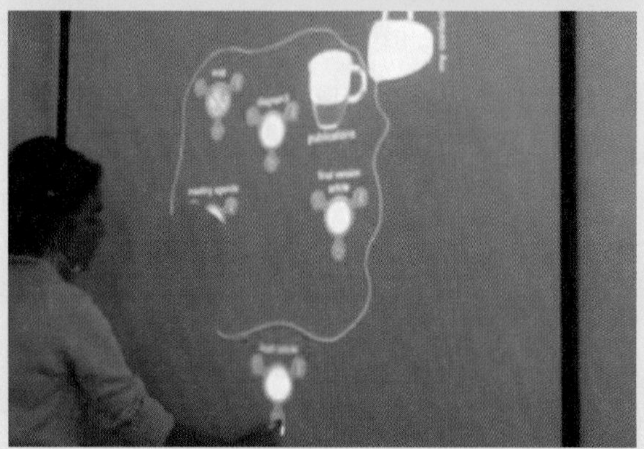

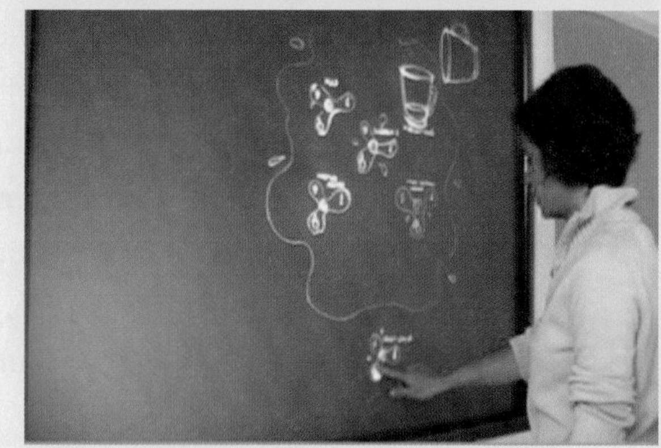

Rendering Style and Process: A Brief Case Study

In 2005 I was in Rome giving a talk. While there I met a young Italian designer, Lucia Terrenghi. She was a PhD student studying at the University of Munich with Andreas Butz. She showed me a video, several frames of which are shown in the left-most column. It struck me as interesting, but I also felt that it suffered because of the style in which it was rendered. I suggested that it would be interesting if she redid it using a sketch style of rendering. We talked a bit, said goodbye, and I thought little more of it. Then, about 2 weeks later I got an email with the original video as well as the remake, matching frames of which are shown in the column to the right of the previous one. I was as impressed as I was interested. And as it provides a good opportunity for you to compare the impact of rending the same material different ways, I have included this example, along with its story, as told by Lucia and Andreas.

Lucia: My PhD thesis deals with using the hands to directly manipulate information on large interactive surfaces. On a train I came up with the idea of using physical objects and their affordances as a metaphor. For example, a mug could be a ubiquitous container of information from which information, like a fluid, could be stored, and poured in or out of. While on the train I sketched the concept on a piece of paper and then on the whiteboard during a team meeting the next day. We agreed that the idea had potential and opened up several other possibilities. Consequently I sketched a more detailed storyboard and asked Sebastian (a student working with us) to make an interactive animation using Flash. Sebastian interpreted my sketches and gave a realistic visual appearance to the items. When Andreas saw the flash animation though, he was not completely satisfied with the result.

Andreas: When I saw the Flash demo, I liked it for the ideas that were made more explicit, and for the way in which it was able to communicate the basic concepts and functionality. Nevertheless, something didn't feel right. The student who had implemented the demo had interpreted Lucia's concepts in his own way and I was somewhat disturbed by the varying level of detail among different visual elements. While the mug was rendered almost photographic with reflections and shading, the menu backgrounds were flat and in a solid color. The kitchen sink in the left hand menu was an actual photo, but the laptop in the same menu was drawn in rectangular shapes. We discussed this and I told Lucia that somehow I had liked her original sketches on the whiteboard better. This was because they were somehow more inspiring and open.

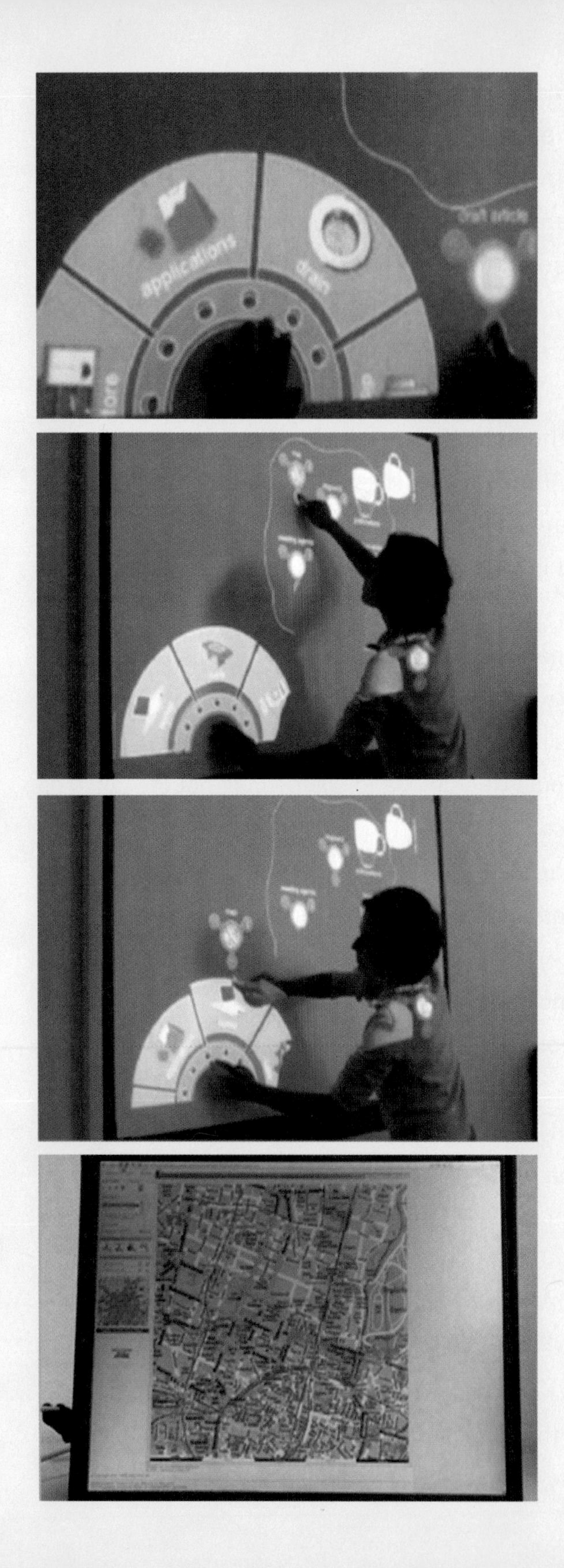
applications

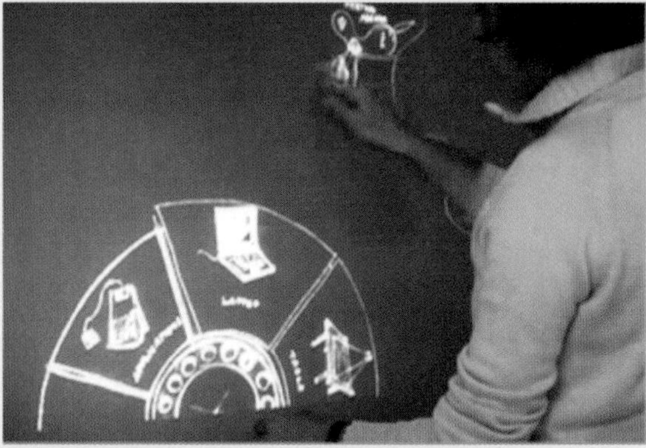

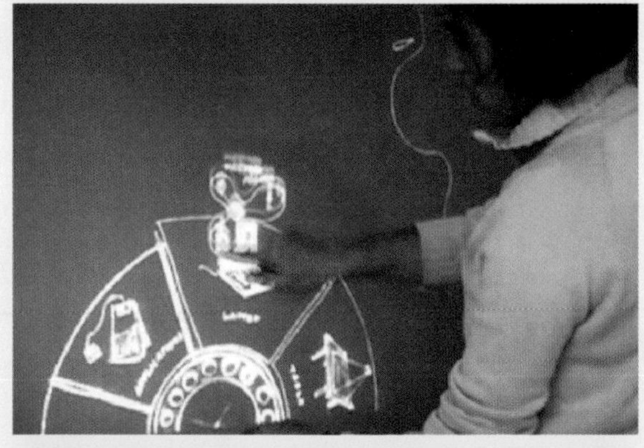

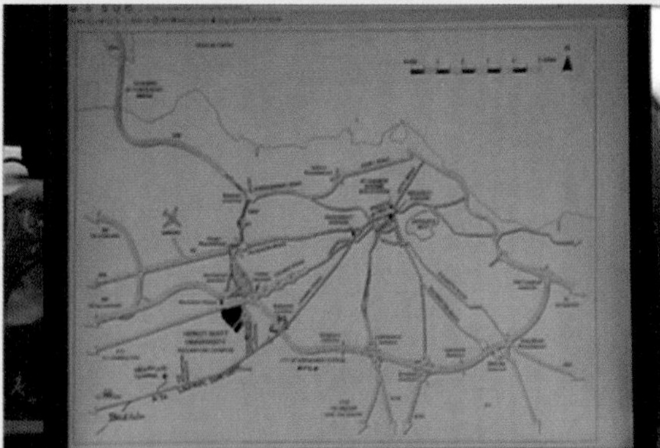

I suggested redrawing all visual elements in a coherent style to get rid of the disturbing inconsistencies. However, we wanted to show the work at a workshop, and there was little time. So instead of changing the rendering, I suggested that the time would be better spent making a movie where Lucia simulated the interaction as if it was on a large wall-mounted display. It struck me that it would be more effective than simply showing it with a mouse using a conventional PC. So the time was spent doing this using the existing version of the Flash file.

Lucia: I presented the same movie during the Doctoral Colloquium at INTERACT'05 Conference in Rome. I also showed it to Bill, whom I met there. He immediately suggested remaking exactly the same movie, but using a hand-sketched representation instead of the existing realistic one. I recalled Andreas' comment about the sketches on the whiteboard and told Bill this was an interesting "coincidence" – that after seeing the original Flash animation, my supervisor had had a kind of "nostalgia" for the earlier whiteboard sketch. This probably reinforced Bill's (and my) intuition that there was something about the style of representation that was worth exploring. Hence, we wanted to make the second movie so that we could compare them. Back in Munich I sketched the elements of the interface on paper, scanned them, and then replaced the symbols in the existing Flash file. We then re-shot the movie with exactly the same storyboard as the previous one.

Significantly, many people to whom we showed both versions assumed that the sketch version was the older, and that the more realistic version a remake. While this was not true for the actual movies, it was true when thinking of the original sketches on the whiteboard.

Figure 141: Paper Sketches of Oscilloscope Front Panel

Two frames from a videotape made during usability testing of oscilloscope front panel concepts. The usability of the numerical keypad is being tested. In each version, I have circled the location of the keypad.

Image: Tektronix Corp.

Figure 142: Interacting with Paper Front Panel

This is a frame from the same video. In this image, and the second one in Figure 141, notice how the hand gesture used by the test user is exactly what would be used with the physical control represented. This is a subtle but important point that illustrates the power of the affordances of the controls, even on paper.

Image: Tektronix Corp.

Interacting with Paper

... success every time implies that one's objectives are not challenging enough.
—Mick Fowler

We now move back to the world where sketches are interactive and they can be experienced first-hand. Curiously enough, to start things off, we are going to go back to where we started: with traditional sketches on paper.

Example: Scoping Out Tektronix

Sometime in the 1980s I was invited to do some consulting for Tektronix, a company in Portland Oregon. While I was there, I had the opportunity to speak with some of the people who designed their test instruments (core products of the company at the time). Earlier, I had published a paper on iterative design (Buxton & Sniderman 1980). So in typical fashion, a little too self-important, I asked them, "So, do you do iterative design as part of your process for developing the front panels for your oscilloscopes?" Knowing me well enough by then, they promptly answered, "Yes, of course. We do exactly one iteration per device, and then we ship it."

Okay, I got what I deserved. But after explaining the facts of life to me, such as the costs (in time, labour, and dollars) of building prototypes of various panel designs, they showed me what they *did* do. It was a revelation.

This was, I believe, my first exposure to what some people call "paper prototyping," but which I simply refer to as *paper interfaces* (more on why, later). What they did was have one of the designers make a quick drawing of the proposed front panel, and then bring in expert users to test them. The tests were videotaped for later analysis and comparison. What the sessions consisted of was a member of the design team explaining a particular scenario to the user, and then having the user perform the appropriate task using the sketch as a proxy for the planned product, all the while talking aloud explaining their actions and intentions, as well as asking any questions. Figures 141 and 142 are frames showing two different panel designs from one such video.

What is interesting in viewing the tapes is that the scenarios are imaginary. There is nothing displayed or moving on the paper oscilloscope's screen, and during the tests, nothing changes in the drawings. As long as the appropriate scenarios and users are selected, this turns out not to be a problem (but it may, of course, restrict what tests or users could be employed).

In the case of the two screens shown in Figure 141, what is being tested is the location and layout of the numerical keypad. Due to cost and space constraints, the mechanical engineering team had proposed locating the keypad underneath the display, as illustrated

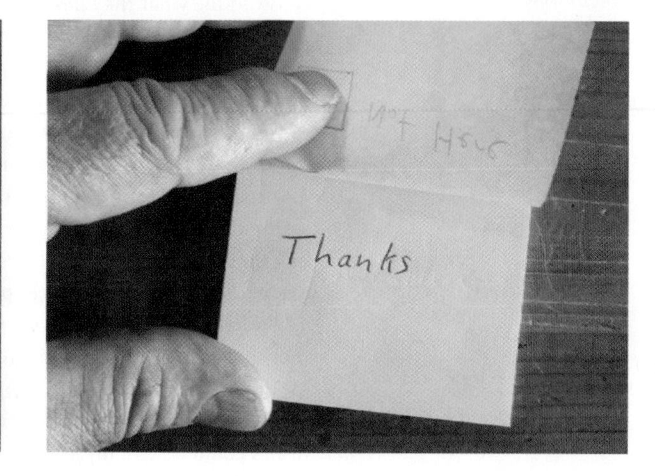

Figure 143: A Simple Finger Exercise

One can create and experience an interactive paper interface in two minutes with nothing more than Post-it notes and a pen. Push a button to go to a particular page. Push the wrong button and return to the first page.

in the top image. Using that sketch, expert users were asked to perform various tasks that involved the keyboard. However, as can be heard on the video, they frequently asked questions like, "Which keypad?" or "Now what keypad do you mean?" or "Where is the keypad?" This prompted the designers to change the position and layout of the keypad, as per the second image. When this revised design was subjected to the same tests, the questions about the keypad simply disappeared.

In retrospect, the problems with the initial version of the keypad may seem obvious. But of course, hindsight is 20-20. Their root is in our previous experience with telephones and pocket calculators. These have acclimatized us to expect numerical keypads to be arranged in 3x4 arrays, as in the second, as opposed to the 8x2 arrangement seen in the first.

By using paper, the design flaw was caught early, and an alternative design was quick to generate and test. Because of the speed of the technique, and the compelling nature of the results, the revised design was able to be incorporated into the product before the mechanical tooling needed to be ordered.

Finally, there is one other observation that I want to make. Look at Figure 142 as well as the second image in Figure 141. In each I chose a frame from the video that includes the user's hand. What struck me when I first saw these clips was how powerful the affordances of the drawn controls are. There can be no confusion on the part of the operator that this is just a piece of paper. Yet the hand posture over the rotary and push-button controls is exactly what one would encounter with the actual physical device. To me, this speaks volumes as to the power of paper to evoke real-world behaviours.

Interactive Paper Interfaces

With the Tektronix oscilloscope example, we saw an example of interaction with a static sketch. Now we introduce paper interfaces that are *dynamic*. That is, what you see in the sketch changes depending on one's interactions with the drawn interface. Like the Tektronix example, the technique requires someone who assumes the role of the target *user*, and someone from the design team, the *facilitator*, to guide that user's experience.

The basis of the technique is very simple. If the user interacts with something on the sketch, that should cause a change in what is seen. For example, if the user pushes a button, the facilitator makes the appropriate change happen. This might be accomplished by something as simple as replacing one sketch with another. A simple two-minute exercise, such as that shown in Figure 143, lets you get a feel for this. Make a simple interface with a pen and a Post-it note pad, and be your own facilitator. Then try it on someone else while you facilitate.

The fact that my example is trivial is a good reason to replicate it, rather than not to. Because it is virtually contentless, you can focus on the technique. And, because it is so simple, the contrast in your (cognitive) understanding before and your (experiential) understanding after will be all the more marked.

	Facilitator	User
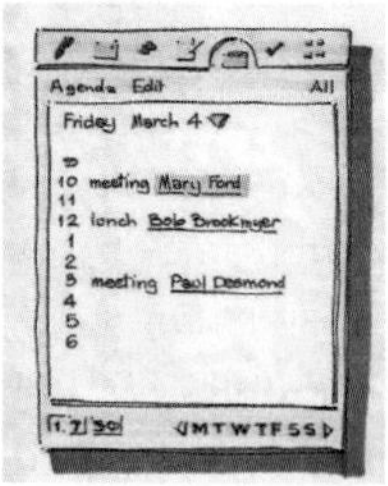	(Start: Sketch 1.a in front of user.) The sketch in front of you shows the screen of your PDA. I want you to send a message to your 10:00 am appointment. For this exercise, to do anything, just touch what you think is appropriate on the screen, and tell me what you are doing or thinking as you go along.	
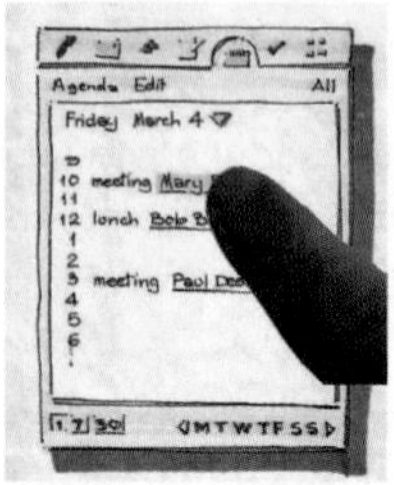		Okay. I assume that you want me to send a message to Mary Ford, since she is my 10:00 am appointment. So I will touch her name.
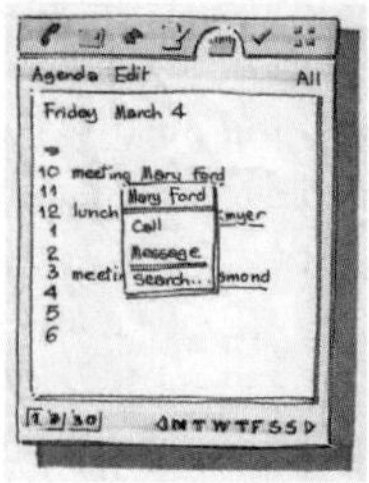	(Facilitator replaces sketch 1.a with 2.b)	Now I see a menu that lets me either call her or message her.
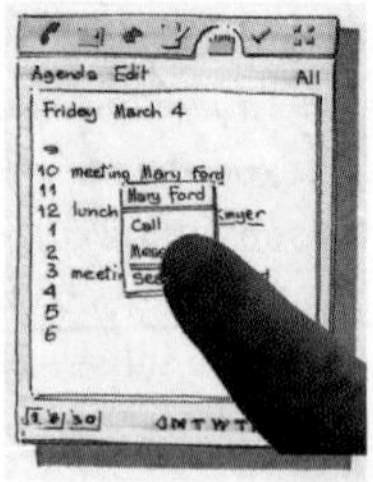		So, what I will now do is touch "message" on the menu.
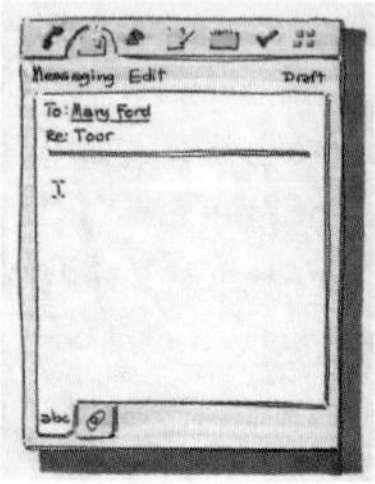	(Facilitator replaces sketch 2.b with 4.a)	Okay. I now see a screen that lets me send a message to Mary Ford. What now?

Table 2: Example of Interacting with Dynamic Sketched Paper Interface

The user is guided through a transaction with the agenda originally seen in Figure 20. Each time the user's action would result in a change in a real system, the facilitator gives the user an appropriate replacement sketch with which to continue.

Back to Our Agenda

Now let's apply this kind of technique to a less trivial example. To do so, we will repurpose the drawings of agenda screens that we saw in our discussion of storyboards, as illustrated in Table 2.

In this scenario, the facilitator starts by showing users an initial screen image, 1.a (with the screen map at the bottom cut off). The user then is asked to send a message to his 10:00 A.M. appointment. Each time the user takes an action the facilitator makes sure that the current screen image is replaced by the new appropriate one.

Read the cells in the table left to right, top to bottom. In the left column of each row, you see what the subject sees, as well as what he is pointing at and selecting with his finger. In the middle column you read what the facilitator says, and in the right column, the remarks by the user. The session is hypothetical. I made it up just for the example, but it is representative of what you will encounter in the field, nevertheless. Try it for yourself and you will see.

In the example, notice how a very simple change in representation (adding some text and a finger, and cropping out the screen-flow map) enabled the images of Figure 99 to communicate different aspects of the interface design. Literacy in such techniques and representations is one of the most important tools in the repertoire of the experience designer—knowing what to use when, with whom, and how.

The example shown in Table 2 is still verging on the trivial. Again with sketching that can be a good thing. Unlike grade school, doing anything beyond the minimum required to achieve your objective is a waste, not something that gets you a gold star. Fast and simple is what lets us afford to explore multiple concepts. But at some point we will need to go deeper. One approach to doing so is to take our sketches electric. That is, use a program like *Flash*, *PowerPoint* or *Director*, for example, to make a computer-based interactive sketch. In this case, Ron Bird built an interactive program in Director to capture the concept. A screen snap of this is seen in Figure 144.

In this implementation, Ron did a couple of clever things. First, he did not try and implement the interface in a way that the user could use it. It would simply have taken far too much work to bring either the design or the implementation to a point where one could let users loose on it. All that he wanted to do at this stage is capture some of the basic concepts. Instead, the interaction involves the user selecting aspects of the user interface that he or she wants to see. The options are listed in the text column on the right side of the figure. When the screen snap was taken, the user had selected the second item, and the cursor can be seen there.

The trick to doing this was to distinguish the real cursor (what the user selected the right-hand list item with), from the representation of the cursor driving the sketched interface (represented by the sketch of the stylus).

What I also like about this example is how Ron preserved the sketch-like quality of the graphical rendering that we saw earlier in his paper sketches. This is no coincidence. He did the images on paper and then scanned them in to the computer. The effect reinforces the fact that these are not finished ideas.

Later, as the testing converges on a design that they have more commitment to, it can make sense to progress to the next step and make a more interactive model that can be

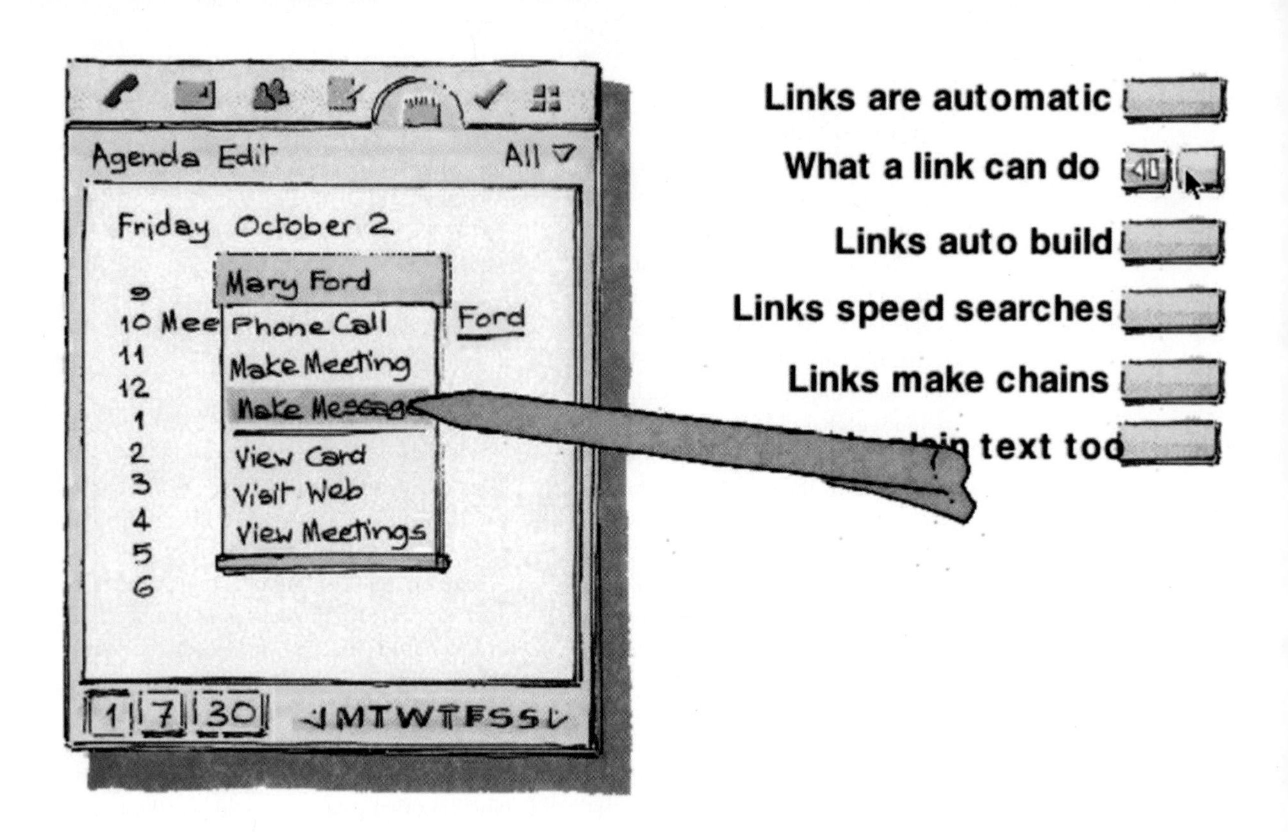

Figure 144: A Computer-based Sketch of the Agenda Interface

This is a screen snap of an interactive computer-based sketch of the agenda seen in the previous example. In this case, the user can explore aspects of the interface listed on the right (the user-controlled cursor can be seen having selected "What a link can do." This invokes an animation, where the user's interactions are represented by the graphical stylus. It is important to note that Ron has preserved the hand-drawn sketch-like character of the interface. It is clearly not done. This is just a probe.
Image: Ron Bird

operated directly by the user. But doing so brings us closer to the world of prototyping than sketching. It has its place, but it is not our primary focus now.

So instead, we are going to back-track and explore an alternative approach—one that lets us get interactive much sooner, much faster, and much cheaper—by sticking with paper.

The approach that we are going to take builds on the paper techniques used in the agenda interface just described. Using various materials that can be found at any business supply store, we can make interfaces that are far more interactive than the paper ones we have seen thus far.

To illustrate this, we are going to work through an example. But if you want to dive a bit deeper into these, a good, short introduction can be found in Rettig (1994). Try it first. Then if you really need or want to go into way more detail, see the book by Snyder (2003) which is excellent. Just don't be intimidated by any of this. Kindergarten provided you with most of the technical skills that you will need.

So here is what we are going to do next:

- Get a better taste of paper-based techniques, just so you have a bit better sense of the overall flavour.
- Talk about what is wrong or missing in most discussions about so-called paper prototypes.
- Explore how these techniques fit into the larger picture of interaction design, especially in terms of mind-set and methodology.

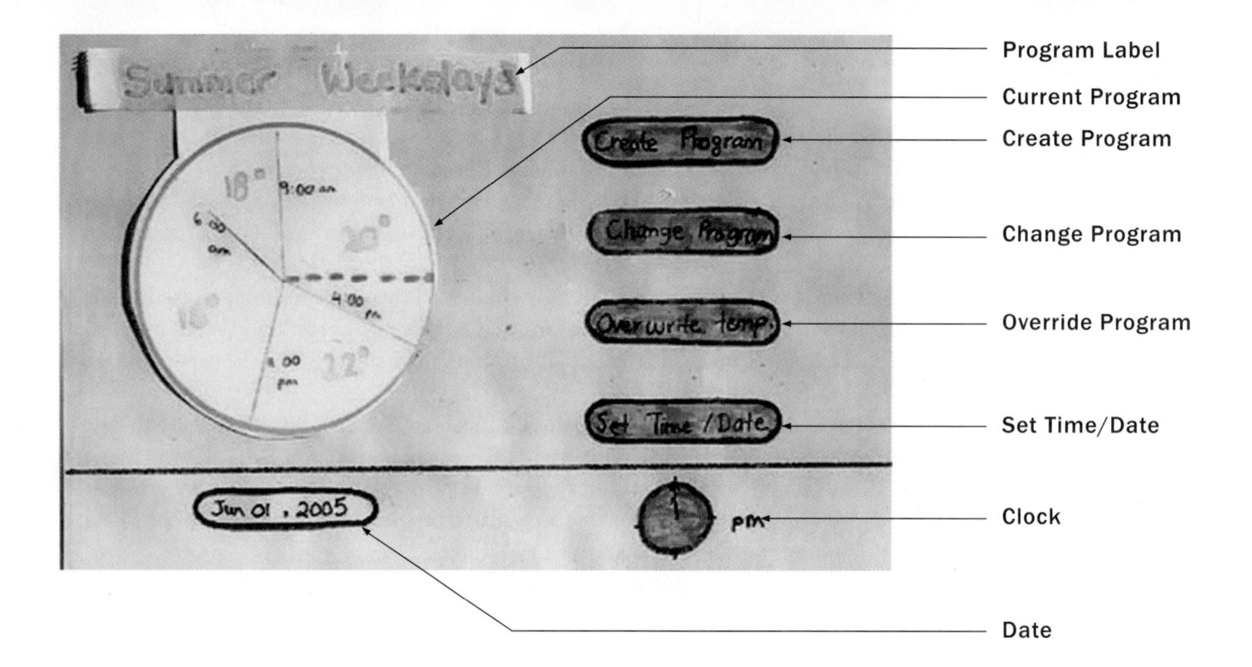

Figure 145: Paper Interface to a Programmable Climate Control System
The basic interface is made up of buttons and circular dials, and displays. The concept is that the user would interact directly on the screen by means of a touch screen.

Keeping Cool—A Home Climate Controller

The example that I am going to use was done by a graduate student at the University of Toronto, Maryam Tohidi, working with Ron Baecker, Abi Sellen, and myself (Tohidi, Buxton, Baecker & Sellen, 2006a).

Maryam's brief was to do an interactive paper sketch of a home climate control system. The system had to enable the user to do things like program the temperature for different times of day and for different days, such as weekends and weekdays, summer versus winter, and so on. One version of the basic home display is shown in Figure 145, along with labels for its main components.

The exercise is intended to simulate a system where one touches things on the display to exercise control. The design language is based around a circular dial. It can represent different things, such as the duration of the day, or the four seasons of the year. In Figure 145, the dial represents the 24 hours of the day. The "slices of the pie" represent distinct intervals of the day, each with its own temperature, as indicated by the text on the slice.

Some of the steps in creating a new program are shown in Figure 146. This illustrates some of the techniques that the facilitator can use to make the paper display dynamic, in the sense that it can respond to the user's interactions with it.

Another technique is shown in Figure 147. In this case, instead of moving tape, or using replaceable overlays, the facilitator is erasing what is written on the interface, and then writing in new information with a dry marker. The interface is paper underneath. It just happens to be covered with acetate, on which it is easy to write and erase, using markers. Yet another trick from grade school.

I have focused on the mechanics of manipulating the paper interface, not on how you might use it. In that regard, there are many options, including the following:

- If the intent is to quickly explore a concept, or show it to your colleagues, the designer might play the role of both user and facilitator, and use the paper interface to walk/talk through the design. In such cases, the designer may well revise the interface on the spot, based on comments, and then immediately start again with the new version. Such a walk-throughs might be captured on videotape for future reference, or to communicate the ideas to those not physically present.

- For informal testing, or quick probes, the designer might play the role of facilitator and work through the interface with someone representative of the intended end user or customer. Often recording such sessions on video is useful, since it is hard to capture the user's comments, or notice all of the subtleties of their interaction with the interface, if you are trying to "operate" the paper at the same time. As in the previous case, the design may be revised on-the-fly, based on comments that it generates.

- Then, there is their use in usability testing. Here the objective is more to uncover errors and determine usability than to come up with new design concepts. The purpose is to get the agreed upon design right, rather than determine, from among alternatives, what is the right design. The interface generally is tested with several users, and for results to be valid, the interface cannot be changed from user to user.

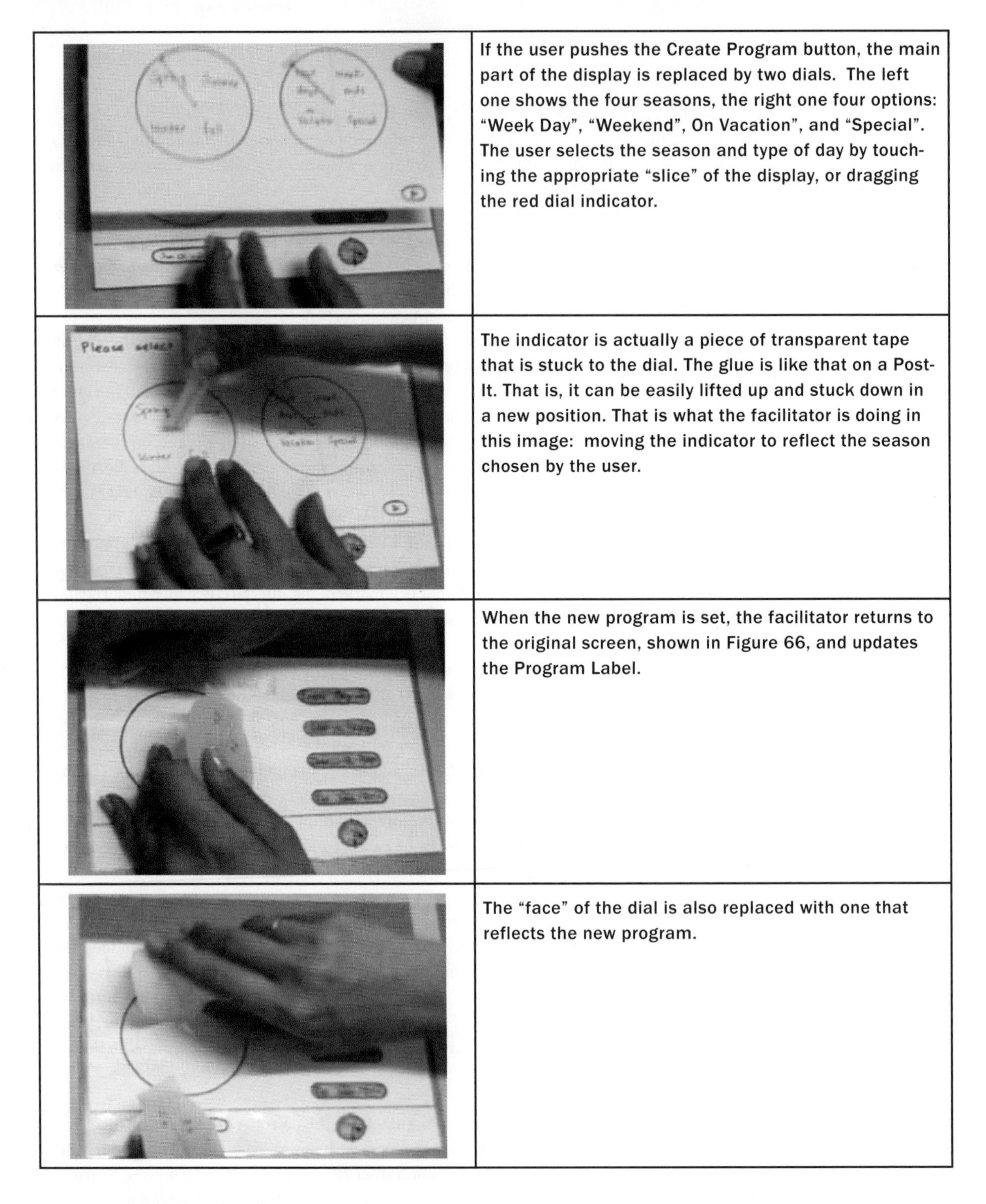

If the user pushes the Create Program button, the main part of the display is replaced by two dials. The left one shows the four seasons, the right one four options: "Week Day", "Weekend", On Vacation", and "Special". The user selects the season and type of day by touching the appropriate "slice" of the display, or dragging the red dial indicator.

The indicator is actually a piece of transparent tape that is stuck to the dial. The glue is like that on a Post-It. That is, it can be easily lifted up and stuck down in a new position. That is what the facilitator is doing in this image: moving the indicator to reflect the season chosen by the user.

When the new program is set, the facilitator returns to the original screen, shown in Figure 66, and updates the Program Label.

The "face" of the dial is also replaced with one that reflects the new program.

Figure 146: Creating a New Program

The third case typically is used later in the design cycle when there is already a fair degree of commitment to a particular design. The size of the investment at this point is higher, consequently, the process is more formal than in the previous two cases. Hence, besides the users, there is a larger team of people required.

As commonly practiced (e.g., Rettig 1994), these include the *facilitator*, who talks with the user and works them through the exercise. There is a separate person, the *computer*, whose sole job is to manipulate the paper interface in response to the user's actions and according the conventions of the design and the test procedure. There is a *videographer* (or two), responsible for capturing the session—a two camera shoot is generally best, since you can then get both a close-up of the interactions with the interface, and a wide shot showing the context and things like general body language. Then there is one or more observers who take notes during the session. Finally, there is often a greeter, who manages logistics, and essentially stage-manages the whole thing.

Hmmmm. That sounds like a lot of people and a pretty heavy-weight process. Despite the sketch-like qualities of the paper interface, it doesn't sound much like sketching when we test it according to attributes such as quick, cheap, timely, disposable, and lots of them.

In fact it isn't. I want to push on this point a little harder.

Appearances Can Be Deceiving

Hopefully you remember my earlier assertion:

Sketches are not prototypes.

As I have mentioned, the use of paper-based interfaces generally is referred to as *paper prototyping*. This is a term that I have been very careful to avoid during our discussion. Likewise, I have avoided using the terms high-fidelity prototype and low-fidelity prototype (Rudd, Stern & Isensee 1996). The reason in both cases is rooted in the earlier assertion. A prototype is a prototype, regardless of the technology that is used to implement it, and regardless of its fidelity relative to the actual product. Prototypes are not sketches and the term low-fidelity prototype is definitely not a synonym for my use of the term sketch.

So here is my cautionary note:

Just because something looks like a sketch doesn't mean that it is a sketch.

Here we can reap the benefit of having been explicit about sketch attributes. Rendering style was just one of them, and on its own, is certainly not sufficient to make something qualify. As the heading of this section says, appearance can be deceiving. This shouldn't come as a surprise. After all, that is the whole premise of the *Wizard of Oz* technique. Sometimes the deception due to appearance can work in our favour, and sometimes not.

In the case of paper interfaces, I think that the appearance can too easily lull us into the impression that we are sketching and designing, when in fact, we are using inexpensive prototypes to do usability engineering. This is certainly what Snyder's book is about, rather than design. This is not a bad thing. Both are important. We should just be conscious of the difference and its significance.

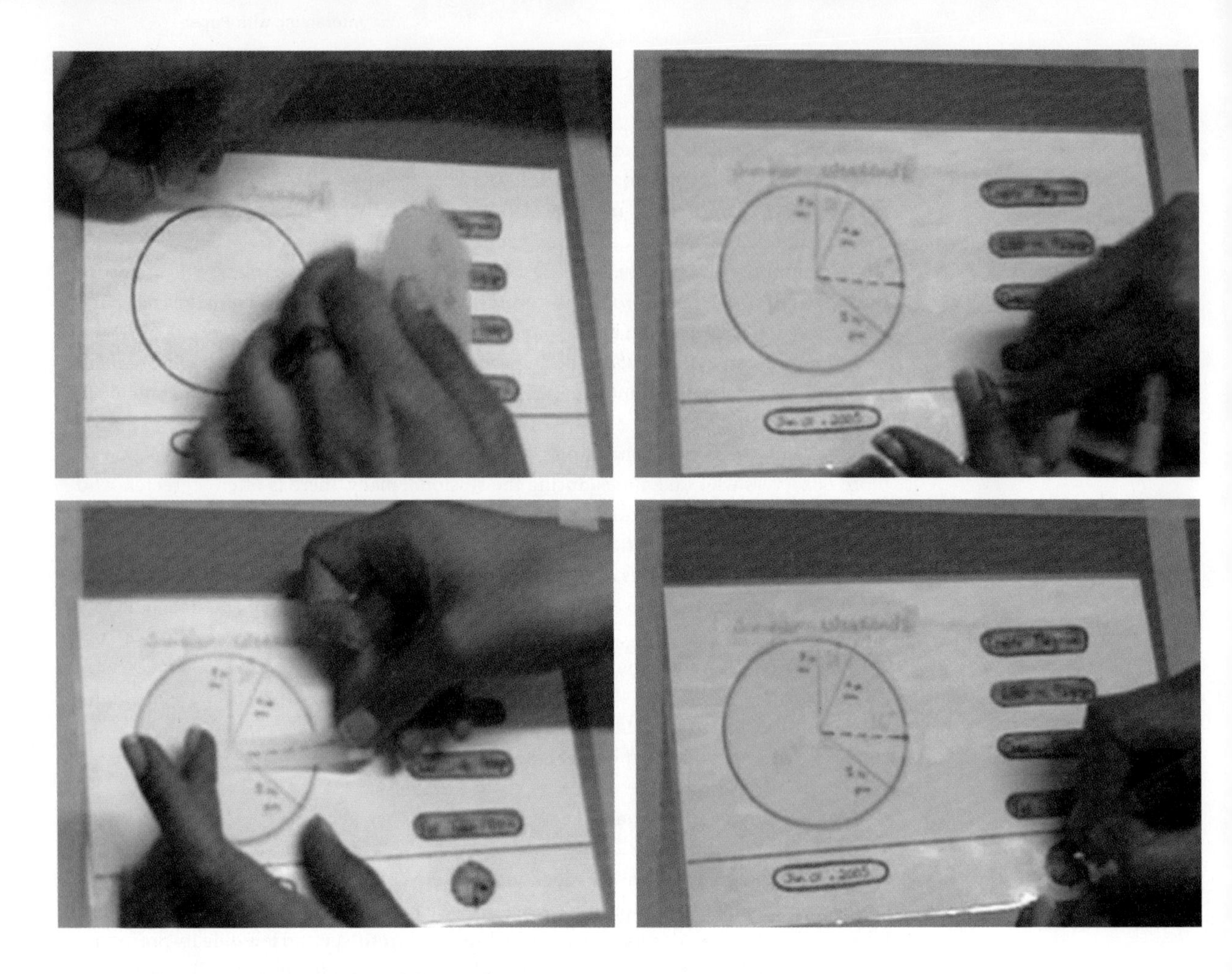

Figure 147: Changing the Display by Erasure and Writing

By covering the paper with plastic, one can write on it with a dry marker, and have what is written easily erased with a cloth when the information needs to be changed. Sometimes this is easier than having a stack of premade objects to stick down.

Sketching is not just *what* you use, but *how*, *when*, *where*, and *why* you use it. *That* is what determines if you are sketching versus prototyping. In the previous section, for example, I outlined three possible scenarios for using the paper climate controller. I would suggest that the first two are good candidates for qualifying as sketching, and the last one is definitely not. It is far more consistent with prototyping.

The importance of all this really came home to me during the early stages of researching this book. I was reading an article entitled, *Does the Fidelity of Software Prototypes Affect the Perception of Usability?* (Wiklund, Thurrott & Dumas 1992). In the authors' words:

> The objective of this study was to investigate whether the aesthetic refinement of a software prototype is related to subjects' ratings of the usability of the prototype.

This was a pretty good study, and one that is relevant to our work. After all, if I am going to advocate using rough sketches to test and explore design concepts, it would be nice to know if there is any scientific evidence that the results have any relevance to the real thing. So let me explain a bit of what they did. I'm going to do so for three reasons: first, because it is an interesting study; second, because of what I said earlier about learning about each other's traditions; third, because it feeds into our discussion of sketching and ideation versus prototyping, usability, and engineering.

Wiklund et al. took an existing software package—one that combined the functions of a dictionary, calculator, and thesaurus—and retroactively made four prototypes of it. Each of the four prototypes varied in the fidelity with which it visually reflected the actual product. Two of the four are illustrated in Figure 148. The one with the lowest fidelity was done using line-art, the next one using half-tone, the next in gray scale, and the highest fidelity one was done in colour.

They were all interactive, and each was tested and rated for three things: ease of learning and use, forgiveness of errors, and aesthetics. The key finding was:

> … the aesthetic quality of prototypes <u>within the range we varied</u> did not bias users for or against the prototype's perceived usability. (Wiklund, Thurrott & Dumas 1992)

So all of this was really interesting to me, and was relevant to my work. But it was *nothing* compared to what Wiklund and his coauthors said, almost in passing, in the discussion of their results. Here is what they wrote:

> In studies such as this one, we have found subjects reluctant to be critical of designs when they are asked to assign a rating to the design. In our usability tests, we see the same phenomenon even when we encourage subjects to be critical. We speculate that the test subjects feel that giving a low rating to a product gives the impression that they are "negative" people, that the ratings reflect negatively on their ability to use computer-based technology, that some of the blame for a product's poor performance falls on them, or that they don't want to hurt the feelings of the person conducting the test. (Wiklund, Thurrott & Dumas 1992)

It means what it means.

For those who are not accustomed to interpreting the findings of psychology experiments, I have underlined the qualification that Wiklund, Thurrott & Dumas carefully and properly included in their words on the previous page. It is always dangerous to extrapolate to the general case from a specific experiment's results. For example, despite the variations in fidelity, none of the prototypes in this study were rendered in a sketch-like style. They were all "drafted", so the experiment does not tell us if the results would have been the same if they had been drawn freehand. Yet, they are relevant, since they suggest that we may be pointing in the right direction. Consequently, read these findings as, "This looks like a good spot to go prospecting," not as "Wow, here is where the gold is, let's open a mine."

I say this as if what I just warned you about is a trap that only amateurs fall into. I wish that that was so. In the course of doing the research for this book, I have come across countless examples of authors—all of whom should know better—making claims that this paper shows this or that about "low fidelity prototypes", when what they mean by this term bears little or no relationship to what Wiklund et al. actually studied.

The experiment means what it means, and that is all. The frequency with which authors stretch the interpretation of other's results to suit their own purpose is the reason that I always read such things with a skeptical mind. When it is important to me, I go back and read the original reference before acting on the interpretation of someone else. That is also why I don't cite papers second-hand, and why I possess in my personal library, and have gone through individually, virtually every entry in this book's bibliography. It would have saved a lot of time and money to trust other people's interpretations. But I can't and won't. The savings are not sufficient compensation for the scholarship that would be lost.

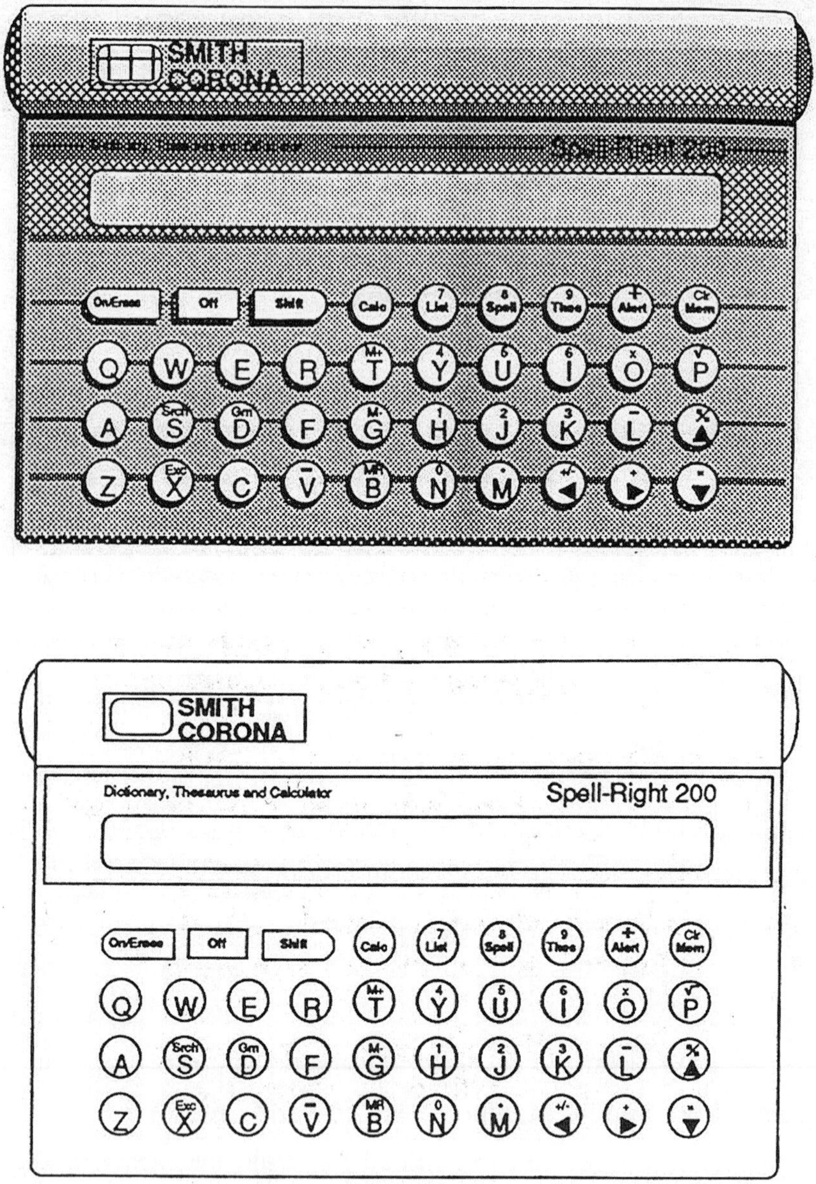

Figure 148: Line Art and Half-Tone Prototypes

These are two of the low-fidelity interfaces for the digital calendar used in the study by Wiklund, Thurrott & Dumas (1992). Contrast their formal drafted graphical style with the freehand renderings of Ron Bird's agenda sketches in Figures 99 and 144.

This really struck home with me. First, it rang a very strong note of familiarity. Second, it brought up an important issue that I probably would have not otherwise mentioned.

My immediate reaction on reading this was that Wiklund et al. encountered the kind of behaviour that they reported *because the subjects only saw one version of the dictionary.*

Just think about how the subjects' behaviour might change if they were exposed to three or four different versions of the prototype, rather than to just one. That could put them in a very different position. Their comments and evaluations would be grounded on a broader base of relevant experience. When they see just one design, judgments are absolute, but on a scale relative to what index? Seeing multiple alternatives can help set the scale so that it is relative to what is possible, rather than just to what the user has seen in the past.

Now we are no longer talking about Wiklund *et al*'s study, but a fundamental difference between design and usability engineering, at least as it is currently practiced.

From my perspective, usability engineering follows a process that can be characterized by the drawing that I have made in Figure 149. One is on a single, already established trajectory (represented by the red arrow). An iterative process (represented by the black conic spiral) is followed that eventually converges on the completed product. The trajectory is established by the basic design, which is already done, and the iterations are there in order to test and refine its implementation, and fine-tune the design.

The key thing to note is that the process is one of incremental improvement where the result of each build-test-evaluate cycle is an improved version of the previous cycle. There are not major changes in the design trajectory, or any backtracking, unless some fundamental flaw is found.

In contrast, the purpose of design is to *establish* the trajectory that we saw in Figure 149. Hence it both *precedes* usability engineering, and is *complementary* to it. Furthermore, it is distinct, because of the very different nature of the process, as I have tried to capture in Figure 150. Rather than a converging spiral, the branching structure shown in the figure is how I think of it.

In this case, the purpose of the red line is to emphasize how many branches were being explored at that given time in the process. Hence, what I am describing is completely consistent with the definition that I gave earlier:

Design is choice.

The various branches shown in Figure 150 are intended to represent the various alternatives that were explored in the process of arriving at the end design, which is represented by the branch at the extreme right. This is where the red arrow in Figure 149 begins.

If I was asked to try and say what these two figures are intended to represent in the fewest number of words, I think that I would say this:

The role of design is to find the best design.
The role of usability engineering is to help make that design the best.

Figure 149: Prototyping as Iterative Incremental Refinement

In engineering, prototyping is like a spiral closing in along a single trajectory. Each prototype is a refinement of the previous one, and takes you one step closer to the final product. Iterative prototyping is a form of incremental refinement and validation, rather than a technique of exploration.

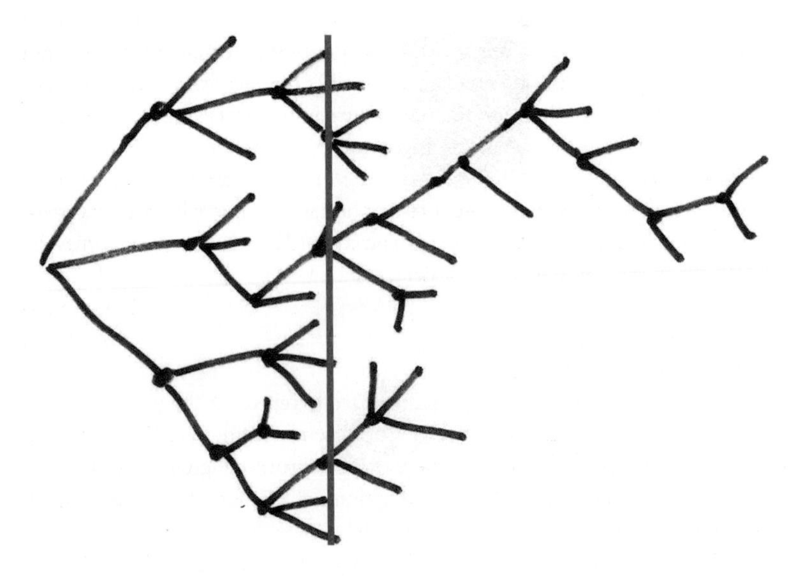

Figure 150: Design as Branching Exploration and Comparison

Design is about exploring and comparing the relative merits of alternatives. There is not just one path, and at any given time and for any given question, there may be numerous different alternatives being considered, only one of which will eventually find itself in the product.

Another way of saying this is:

> **The role of design is to get the right design.**
> **The role of usability engineering is to get the design right.**

In many ways, these are just variations of the distinction that we talked about earlier between problem setting and problem solving (Schön 1983)

So now let me go back to the study by Wiklund et al. and tell you what thoughts were triggered by the passage from their discussion.

The first was the following hypothesis:

> If we asked users to rate three distinctly different design solutions to performing a particular task, and then showed the design with the lowest rating alone to a comparable population of users, the users who saw just the poorest design would rank it significantly higher than those that saw the other two designs as well.

So why do I care if this is true? More to the point, what difference might it make to you?

- The comments by Wicklund et al. suggest what a lot of us already suspected: that we may not be able to trust the ratings that subjects give us about our designs. If the hypothesis is true, then this is at least true for bad designs—precisely the ones where their being ranked too high could have the most negative impact.
- Furthermore, if it is true, then the study would also show a way to run the studies that avoids the problem; namely, don't just show one design—rather, show three (or more). That way users are not placed in a position of being negative, since they can balance any negative comments about one design with positive comments about another.
- The real significance of paper interfaces and other low-cost techniques in usability testing would be shown to be that they make such parallel testing of alternative solutions possible. However, this is not the common practice, nor what is taught in the textbooks. The hypothesis being true could help change this.
- The hypothesis being validated would highlight the need to maintain a design mentality; that is, pursue various alternatives, rather than zero in on a particular solution, throughout the process, including usability testing.

Now I've told you what the hypothesis is, and the significance if it is true. The next thing to do is figure out if it is true. The good news is that the answer is determinable. The way of doing so is to follow the example of Wicklund *et al.* and run an appropriate study. Such is the progress of science.

We ran such a study in the summer of 2005 (Tohidi, Buxton, Baecker & Sellen 2006a). In fact, it was in the course of that study that Maryam built the paper interface shown in Figure 145. But what you have seen so far is only one of three versions that she built. The other two are seen in Figure 151. As you can see, each is in a distinct design language, or style, but all three are rendered at about the same level of resolution, and have the same functionality.

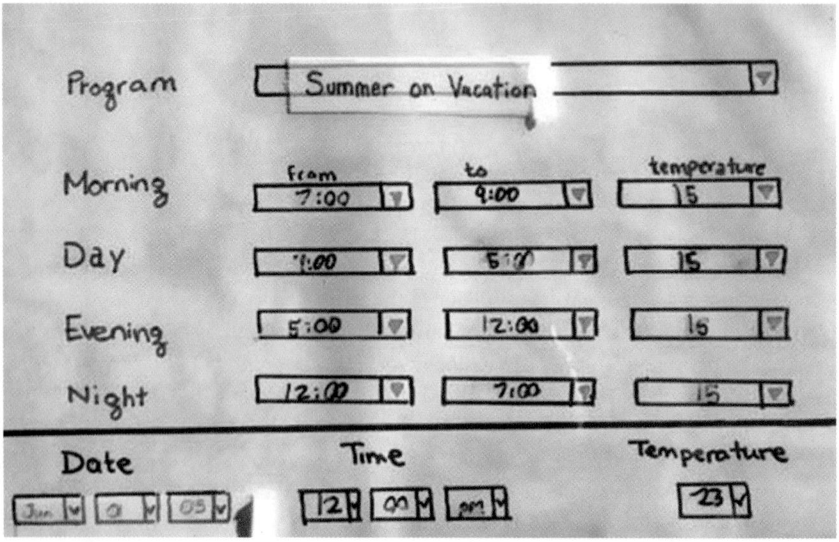

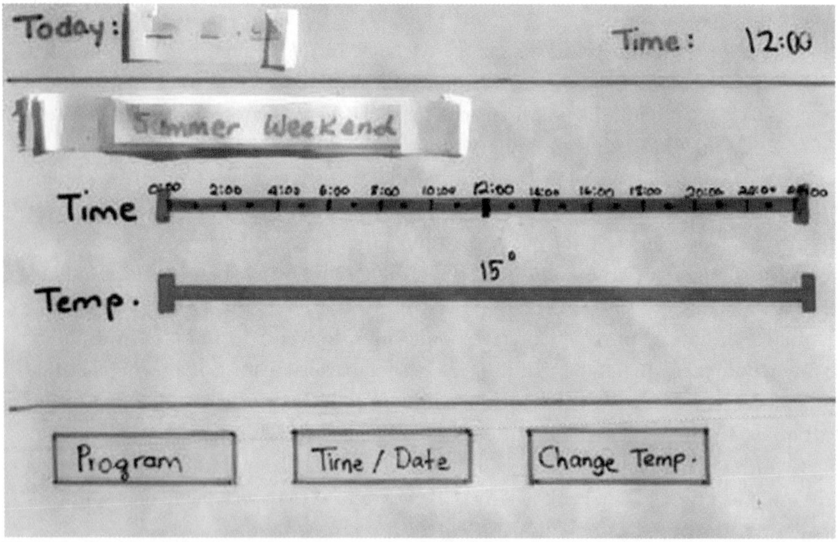

Figure 151: Two Alternative Programmable Climate Control Interfaces

These interfaces are functionally equivalent to the one shown in Figure 145, and the associated figures that follow. However, each utilizes a different design language, or style. The one in Figure 145 uses a dial-based interface. The one on top in this figure uses a tabular form-based approach, whereas the one at the bottom uses a time-line based approach.

Briefly, subjects performed the same set of task on all three interfaces, one after the other. The order in which the three were presented was changed in order to counter any bias that might influence the results. After having used all three, they filled out a questionnaire, for them to rate each.

In parallel with this, we also ran another set of comparable subjects who saw only one version of the interface. They ran the same set of tasks as those who saw all three, and were asked to fill out the same post-task questionnaire and evaluate the interface.

The quick summary of the results, at least as they pertain to our current discussion, is that the hypothesis was shown to be true. If you compare the ratings given to the lowest rated interface of the three, as judged by subjects who had seen all three, that rating was significantly lower than the rating given to that same interface by subjects who saw only it.

I have already given you four reasons why we should care about this result. But I would like to close this discussion with two overriding, related conclusions:

This study helps emphasize why we should not commit to a design too soon, and why there is value in continually exploring various options to any question.

The most important value in using these quick and inexpensive techniques is not that they save money in making a prototype (although that is not to be sneezed at). It is that they make it affordable to make and compare alternative design solutions to problems throughout the design process.

This second point is not just true for the paper-based techniques but virtually all the techniques covered in this section of the book.

Finally, there are inherent potential problems in some of what we have been speaking about, such as presenting users with so many options that they are confused or overwhelmed. What I am trying to do is build up a basic literacy around sketches and prototypes. Understanding what makes each distinct is an essential step toward making the right choices in an environment otherwise dominated by the prototype approach.

In other words, our purpose is to provide a counter-balance to statements like:

> For most applications, an evolutionary, whole-system, continuous prototype is a desirable choice for the user interaction developer. (Hix and Hartson 1993, p. 256)

Figure: 152 Are You Talking to Me?
Testing and observing are one set of tools for gaining insights about needs and concepts in the wild. Asking the right questions and having the right kinds of conversations with potential users and stakeholders is another. Here is one place where the ethnographer and social scientist can play a really important role in the design process.
Photo: Richard Banks

Are You Talking to Me?

Tell me, and I will forget. Show me, and I may remember. Involve me, and I will understand.
— Confucius

I want to close this part of the book with a post-script to the climate control study that we just discussed. You see, I didn't tell you the whole story. There was another hypothesis to the study—one that I got wrong. Very wrong.

At the start of the study, I strongly believed that the broader experience resulting from exposure to alternative designs would translate into more creative and constructive suggestions from the users than we would see from those exposed to only one design. However, after studying the videos of the experimental sessions and reviewing the end-of-session questionnaires and interviews this turned out not to be the case.

Those who saw multiple designs gave us more critical comments. A few even rejected one of the designs (something extremely rare in studies that only evaluate one design, and therefore very significant). But constructive suggestions? There was no difference between users who saw one design and those who saw three.

So, as we were writing it up, I started writing something along the lines of, "Well, why should they have given us more constructive suggestions? After all, they saw us as professional interface designers, and there was nothing in their background that would equip them to do so. We had been naive to think that things would have been otherwise."

I thought that my hypothesis had been wrong, and that through the study, I had gained a new insight. But what actually had happened is that I had just made another error, and I had learned nothing. This became all too clear as we did the analysis of some of the other data that we collected during the study—data that we were using for another paper (Tohidi, Buxton, Baecker & Sellen 2006b). Luckily, we discovered the problem before we had submitted our first article, so we were able to fix the error of our ways.

So what were these new data, and what were their relevance to the hypothesis about users being more constructive when exposed to multiple designs?

What Maryam did at the end of each user's session was ask what seemed to be a simple favour. It went something like this: "Since we finished a bit early, would you mind taking a couple minutes and making a simple sketch of your ideal home climate controller? It can be as rough as you like. It doesn't matter if you can't draw well. Just do a quick sketch on this piece of paper."

And sketch they did. Sometimes they hesitated. Sometimes they were shy. Some were gifted draftspeople, and others were pencil-challenged like me. But every one of them did a drawing, and they were a revelation.

Later, during analysis, Maryam and I spread them all out on a very large boardroom table. All 48 of them. We then sorted them by condition. Thus, we had four long rows: one for each of the three groups who saw only one design (Circular, Tabular, and Linear) and one for those who saw all three.

For the first three groups (the single condition ones), we then sorted the sketches according to how close they were to the original. We started with the ones that were most similar, and ended with the ones that were the most different.

The sketches from those who saw all three designs were just clustered by similarity. You can see the sorted sketches in Figures 153 through 156. Compare them to the three designs that the users saw, as shown in Figures 145 and 151.

What is clear in these sketches, even to a lay-person, is that the users did have original ideas about alternative designs. What we had not done in the first study, however, was let them communicate them to us in an appropriate language.

I can hear the collective yawn of the participatory design community at this stage as they tell me, "I could have told you so." Okay, *mea culpa*. But at least this episode helps bring closure to this section of the book with yet another example emphasizing that sketching is a language that supports a particular form of dialogue—a dialogue that can help all of us bring our ideas one step closer to fruition.

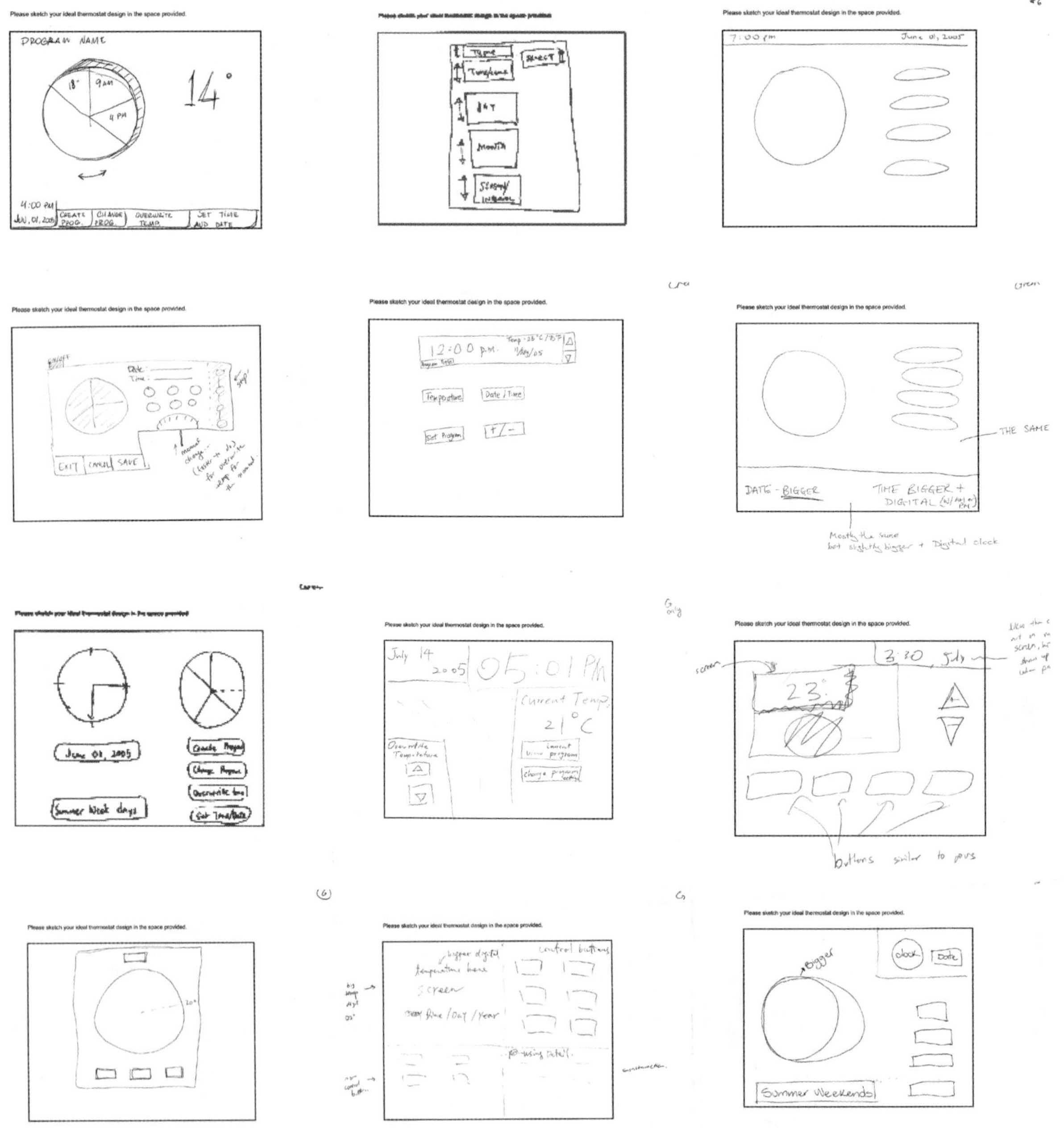

Figure 153: User Sketches from Circular Conditions

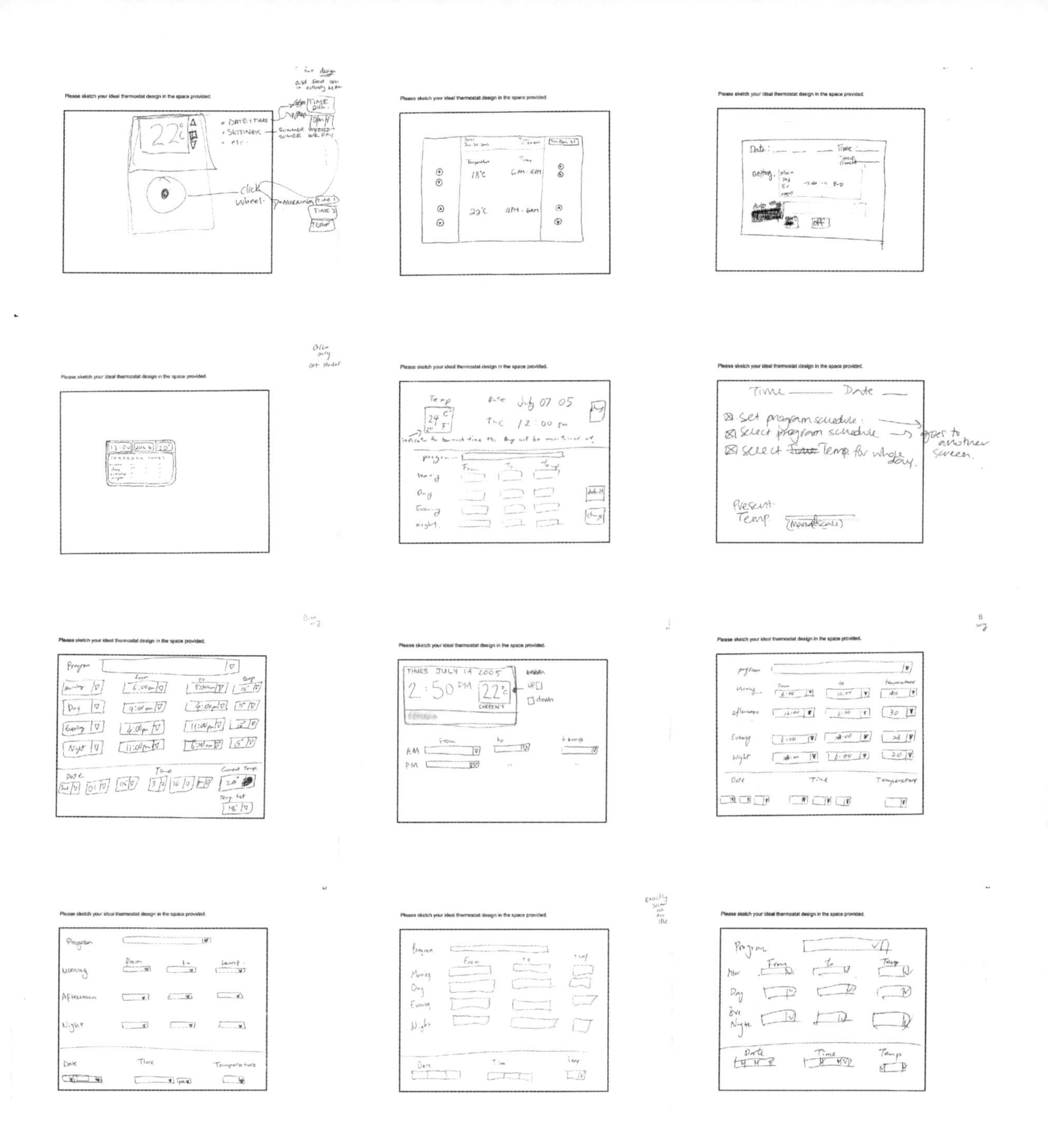

Figure 154: User Sketches from Tabular Conditions

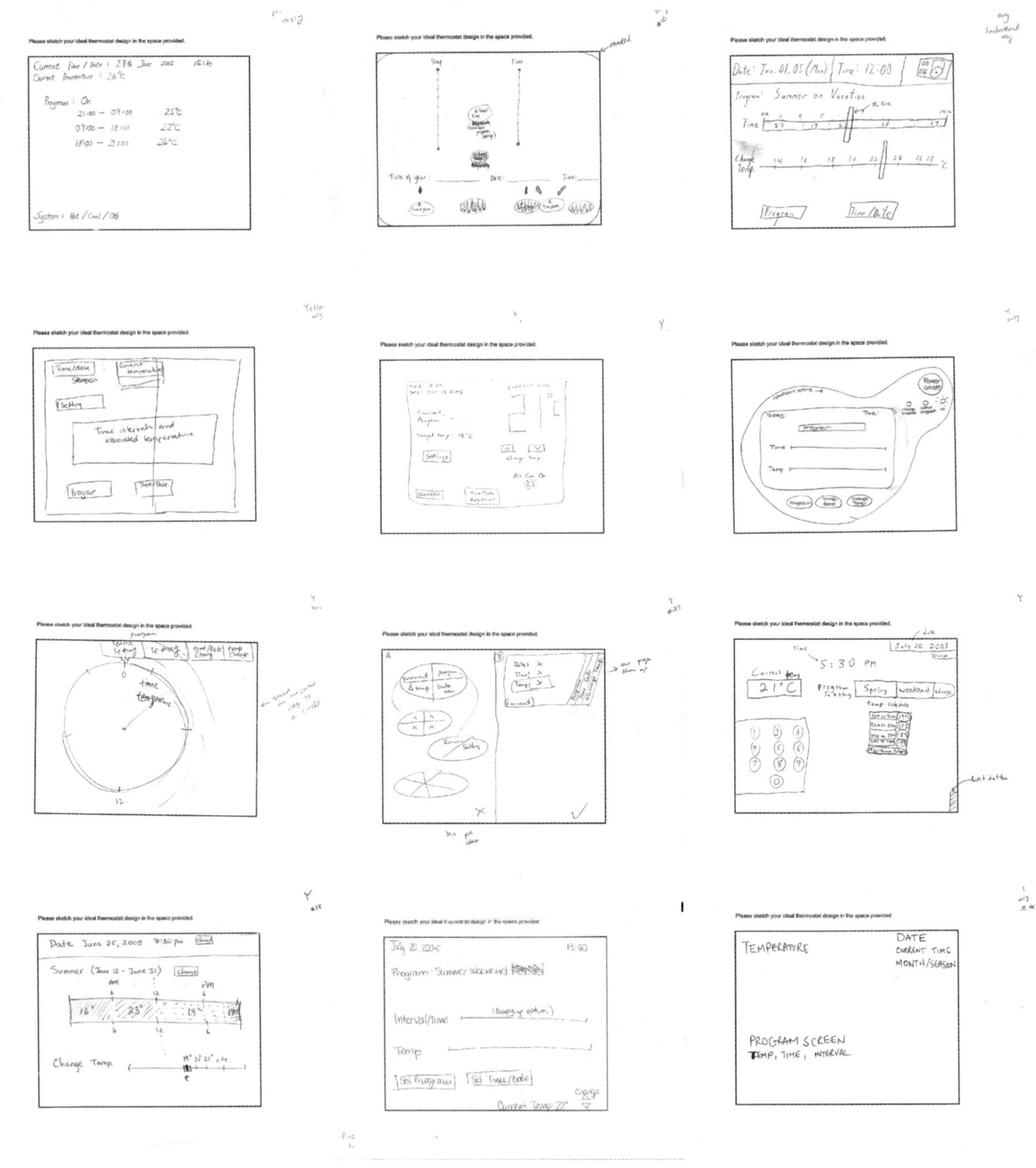

Figure 155: User Sketches from Linear Conditions

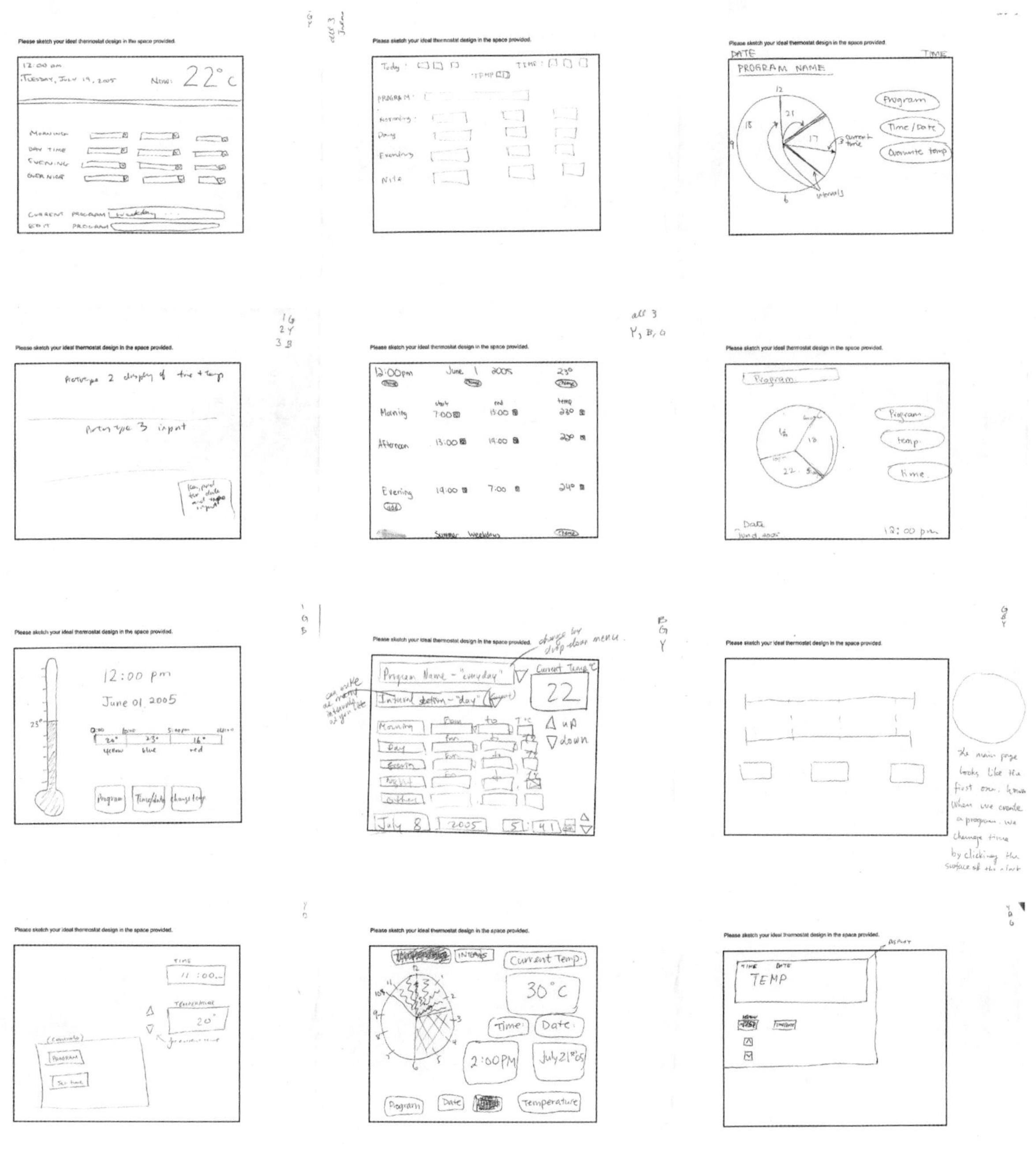

Figure 156: User Sketches from Multiple Conditions

Recapitulation and Coda

1927 – 1930: A Period of Transition in Design—Not Unlike Today

The period that we are in right now in terms of interaction design is not unlike that in which American industrial design found itself in the latter part of the 1920s. The discipline was establishing itself as a distinct and legitimate profession. Yet, for the very reason that it was not yet established and there was no defined curriculum for training practitioners. Those who established what were to become some of the most influential practices and consultancies came from a myriad of backgrounds, including illustration, architecture, engineering, fashion, painting and theatre design.

Like the interaction design practitioner of today, these early industrial designers transferred skills from established disciplines, and adapted them to the demands of product design at the time.

Here are a few examples.

It was at this time that the first car that was designed as well as engineered was sold. It was the 1927 Cadillac La Salle and it was designed by ***Harley Earl***. Earl studied engineering at Stanford but had been drawn to designing custom automobiles for famous clients in Los Angeles. There he attracted the attention of General Motors. This investment in design by General Motors was in stark contrast to their main competitor, Ford, who were still very much functioning in the tradition characterized by Henry Ford's famous statement, "You can have it in any colour as long as it is black." While Ford produced their fifteen millionth Model T that year, they also discontinued it, and shut down their production line for six months. It took them that long to come up with a competitive product, the Model A. Harley's La Salle had changed the automotive industry forever. Design, and the techniques that he introduced (such as building clay models), became the norm in the industry.

Walter Dorwin Teague was a successful graphic artist who opened an industrial design practice in 1926—a firm that is still in business today. His first project was a Kodak camera, the Vanity Kodak, that would fit in your vest pocket. Not unlike today's iPod Mini, one of the things that distinguished this camera was that it came in five colours, blue, brown, green, grey and red, with matching coloured bellows and satin-lined case.

Henry Dreyfuss is likely the most important person in terms of introducing ergonomics into industrial design. Yet, he began his career as a designer of stage sets for theatre. He opened his own office in 1929 in New York city and that same year won a competition from Bell Laboratories to design the phone of the future. If you pick up the handset of almost any payphone in North America, you are experiencing the results of his design that resulted from this competition. His firm is still in business.

Raymond Loewy emigrated to New York from France in 1919. He began his career in New York as a window display designer as well as a fashion illustrator for magazines such as Harper's Bazaar and Vogue. Like Dreyfuss, he opened his own firm in 1929. His first commission came that same year. It was the redesign of the Gestetner mimeograph machine. One of the most universally iconic of his designs is the classic Coca-Cola bottle.

It is in such people, their practice, and their development, that I see the model for the evolution and maturation of interaction design as a distinct profession.

Some Final Thoughts

Never doubt that a small group of thoughtful, committed people can change the world. Indeed, it is the only thing that ever has.
—Margaret Mead

I give a lot of talks about the material in this book—to students, professors, designers, and executives. I do these talks because I believe in what I write about and I want to help bring about change. So here is my nightmare. I live in fear that some professor will jump up after my talk and say, "I'm convinced! Show me the text books and the curriculum, and tell me who to hire to flesh out the faculty." In another version it is an executive who, in the spirit of someone who has just found religion, exclaims, "Wonderful! Help me reorganize my company. Help me recruit a Chief Design Officer, and a team of managers and designers who can execute on this vision." And finally, there is the plea from a young student who wants to follow this kind of path, and who pleads for advice about where best to pursue studies that equip him or her to bring these skills to the market.

The nightmare is not that they have bought into the vision that I am framing. Rather, it is my inability to do an adequate job in filling the needs or the requests that I have helped cultivate. There are few universities or colleges that have comprehensive programs that teach design in the way that I think or speak about it, and therefore, most of those people who have the types of skills that I look for have acquired them through nonstandard means. That is to say, we are in a period of transition like industrial design in 1928-29 (see sidebar), where the nature and need for these skills is starting to be appreciated, but collectively, we all have some tooling up to do in terms of supporting their development in any structured way.

Such is the world as I see it today. This is what we have to work with. To play this hand as best we can, we must apply the same type of imagination, energy, and innovation to reshaping the current situation as we apply to the design of products.

I want to conclude with some thoughts about this that complement the material that we have already covered.

Process versus Product

Not so long ago, I worked for a company where the executive response to almost all questions was something like, "Let's put a process in place to figure this out." The problem with this response, from my perspective, was that too frequently it was a means to avoid making a decision. Once set in motion, the process ruled the day, without any sense or understanding of either its appropriateness to the question at hand, or the way in which it was executed. Even worse, "the process" was all too often a way to abdicate responsibility. Despite the outcome being a foregone conclusion, responsibility could always be deflected from the executive to the process. Hence, when things went wrong, the VP of product development, for example, could (and often did) respond by saying, "This wasn't my decision. I had an open mind. My best people went through a process and this was the best result that they could come up with."

This is all by way of saying that I am very gun-shy of the "P" word. But on the other hand, process is a bit like technology. It takes informed design to get it right, so bad design is not a testament against design, but against bad designers. I mention this because, in essence, this whole book is about process.

Lester Thurlow is the dean emeritus of the Sloan School of Management at MIT. In 1992 he published a book called *Head to Head*. In it, he compares the investment in research in Japan versus the United States in the years following the Second World War. He characterizes Japan's strategy as predominantly an investment in process, and that of the United States as mainly in product. Furthermore, he argued strongly that the return on investment on process is significantly higher. His claim is that it doesn't matter if you invented DRAM, the VCR, or LCD panels if you lose the benefits to a country that can manufacture and distribute them better than you. The United States paid the high cost of the R&D for these products, yet reaped relatively few of the long-term benefits. Japan (and later Korea) took much less risk, yet reaped most of the benefits.

So, despite my fear of too heavily relying on process, and despite my passion for product innovation, my take-away from Thurlow was this:

Innovation in process trumps innovation in product.

In order to create successful products, it is as important (if not more) to invest in the design of the design process, as in the design of the product itself.

Have you spent millions of dollars trying to develop new interactive products in- house, and still failed—like a company that I worked for? Then, in light of the preceding, consider where you have invested your resources. Perhaps it might have been in the wrong place: product instead of process.

This relates to an article that I read in the January 2004 issue of *Fast Company* magazine (Hawn 2004). The article compared the business of Apple with that of Dell. The article questioned whether Apple's large investment in design and innovation could be justified. The argument made was that Dell, which makes bland products with little in the way of styling or superficial design, completely dominates Apple in the PC marketplace.

What the article missed is that Dell does innovate. It is just that their primary investment is in innovation in process—of manufacturing, sales, distribution, and service—rather than product (Holzner 2005).

So does the success of Dell suggest that much of this book has been wrong-minded? Does Thurlow imply that there is no need for product design, that there is nothing to learn from Apple, and it is sufficient to emulate Dell and just work on process innovation?

Obviously, I think that the answer to both questions is an emphatic "No." I would not have included the Apple Case Study if I believed the latter, and would not have written this book if I believed the former. But that does not mean that there is nothing to learn from Dell. Innovation needs to happen on both fronts, product and process. As we discussed in our earlier discussion of the mountain bike, the responsibility, and need, for creativity and innovation is a holistic one. Ultimately, design must pervade all aspects of the company.

Let me give you an example. If you were Apple and you wanted to beat Dell, how would you do it? The strategy—as opposed to the execution—is simple. Just match Dell's process of efficient manufacturing and delivery with products that are as well designed as the iMac. Which prompts me to revise my previous statement:

Innovation in process may trump innovation in product. But innovation in both trumps either.

"All" that you need is a combination of the creative spirit of a Steve Jobs and the brilliant execution of a Michael Dell.

But can such a combination work? It already has!

Besides the PC market, Apple and Dell also compete in the portable music player market. But the competition is pretty one-sided. According to the *New York Times* (August 27, 2005), Apple's iPod, which we have already discussed, has about 80% market share, and iTunes has 75% of digital music sales. Look at Figure 157, which shows the Dell product, and contrast how many of these you have seen compared to iPods.

This is the other side of the Dell/Apple story, one that is pretty informative in terms of the nature of innovation and design, and also one that is too often missed or overshadowed by the PC story.

Will Apple be able to sustain its lead? Nobody can say. However, the cycle of innovation will continue to roll on, just as it does with mountain bikes. So what does Dell need to do to respond? Invest in design—something that they have already started to do (Umbach, Musgrave & Gluskoter 2006). For one example, see Figure 158 which shows the XPS M2010, which Dell released in 2006. Also see Dell's March 2006 acquisition of Alienware—a boutique manufacturer of high performance computers known for their distinctive styling. So, the battle continues. The lesson is that without paying attention to the types of issues that we have been focusing on, one's chances of being competitive in the long term are very small.

Demo or Die: A Short-Sighted Mantra

Sketches, prototypes, models, simulations, and demos have a huge role to play in the design process, product development, and innovation. You cannot have gotten this far in the book without having some sense of how strongly I believe this. This is reflected in my day-to-day work over the years, as well as in what I have advocated in my writing.

But there is a trap in all of this. One characterization of it is captured by misappropriating the title of one of Neil Postman's books:

Amusing Ourselves to Death (Postman 1985)

A large part of the value of this type of activity stems from the fact that we get to play with what we have built. From this, we gain experience that can lead to new insights. We also gain a

Figure 157: Quick! Name This Product.

Answer the two skill-testing questions: Who makes this product? What is its name? You probably don't know. The fact that I removed any brand identification probably didn't help matters at all. But I wanted to make a point. This is the Dell DJ 30 Digital Juke Box, Dell's top-of-the-line MP3 Player. I don't know anyone who has one. When people compare Dell to Apple, they seem to talk about only the PC business, and ignore the MP3 player business, where Dell is getting their proverbial butt kicked. Looking at both gives a much more useful perspective.

Photo: Courtesy of Dell Inc.

Figure 158: Dell Home Systems XPS M2010

Dell's response to its problems in maintaining its position in the market has been to invest more in product design. This is one example.

Photo: Courtesy of Dell Inc.

communications tool that facilitates others in obtaining a visceral, as well as intellectual, understanding of a concept. Sketches and prototypes provide shared points of reference against which we can compare or contrast other ideas.

This can all be positive, and the benefits have been written about by many, such as Schrage (2000).

However, in all of this there is a risk. The trap that the associated demos can set lies precisely in one of their main strengths: they are seductive. The consequent risk is that making demos becomes confused with design or research (see Brand 1987, for example).

Demos are often great and inspiring. They can be flashy and easy to write about. Demos are a hugely valuable component of research, design, and innovation. But they are only one part of a means to an end, and certainly not the end in itself. If the reward is for the demo, that is, if the demo is the focus, then it gets elevated to a stature that overshadows the underlying concepts and insights that its existence is ostensibly to elucidate. Seeing this happening triggers one of my more impassioned cries:

"Show-and-tell" is not a legitimate form of research.

Attention to demos must be balanced with the need to understand their underlying motivation, rationale, principles, and scholarship. Their implementation must be accompanied by the consideration of a number of questions.

If a demo or prototype is good or bad, what is the reason? Based on the experience with demo, should we build demo+1? If so, what should it be? Why? What do we expect to accomplish? Why? How do we determine if it succeeds? What are our criteria? Why?

In short, without the accompanying analysis, reflection, projection, hypothesis formulation, and testing, what one gets is a shotgun-type splattering of isolated demos that are far more suited to fundraising than they are to constructing a body of knowledge, or experience on which we can build anything solid.

No matter how great the demo or the underlying idea, the chance of the student or researcher repeating that success is directly proportional to:

- The depth of their understanding of the historical, cultural, and technological context from which it evolves
- An ability to clearly articulate how their work relates to that which preceded it
- An ability to clearly articulate their underlying conceptual model
- An ability to clearly articulate their design rationale for all key decisions, in terms of what they did, and did not do
- An ability to clearly state what they learned, and what the implications are for future work

In short, unless accompanied by the skills of scholarship, the authors of even the most innovative demos are predisposed to become one-shot-wonders who are ill-equipped to go on to repeat their initial success in other problem domains.

To constitute research, and to maximize understanding and learning, we need to go beyond demos alone. One of the biggest challenges for the educator is keeping the energy and excitement of creation of a compelling demo, which is as tangible as it is valuable, while balancing it with the accompanying skills that will help form a complete designer.

Back to Sketching 101

In many ways, I think that I am getting grumpy or impatient in my old age. Be that as it may, I really do hate admitting it, and I hate becoming a conservative curmudgeon as well. But I have to say it: I am constantly amazed at the lack of traditional scholarship and sense of history that people bring to addressing these issues. One way that I have expressed this for years, probably quite unfairly, is this:

> **The closer that one gets to Route 128 in Boston, or to Silicon Valley, the more it appears that people seem to think that innovation is all about invention, and seem to ignore the critical role in this played by research and scholarship.**

How can we expect people to be mature interaction designers if they have no deep understanding of the history of the field and its techniques? Why is the (relatively short) history of the discipline not as well known and taught as the history of any other creative discipline, such as music, architecture, or painting?

Part of the answer is because nobody has written a history, and furthermore, the videos that document the dynamics of that history, which are so important, are essentially unavailable. One reason that so many of my examples are historical is to make some small contribution to disseminating knowledge about our traditions and the work of the people who helped shape it.

> **If you are going to break something, including a tradition, the more you understand it, the better job you can do.**

But I think that this is only part of the problem. We not only don't show adequate respect to the importance of the history of the discipline, we also compound the teaching of technique with the problems of design (McCullough 1996).

In contrast, when I was in music school, for example, my classes on technique (such as harmony, counterpoint, and orchestration) were completely separate from my composition classes, which focused just on the art of music. In the former, we were marked solely on our command of the rules. There were no bonus marks for writing a wonderful fugue or choral. It was just about our mastery of technique. On the other hand, in my composition classes (which were often with the same teacher), I could break all the rules of harmony or counterpoint that I wanted. I could make up my own rules. All that mattered was the quality of the music that I wrote. The assumption was that the time spent developing technique provided the foundation for the creation of the art. One built on the skills of the other, but by teaching them separately, the student could always focus on the task at hand.

The same is true in classical art and design education. There are classes such as printmaking, life drawing, and water colour, whose purpose is to lay a solid foundation in technique. This underlies the complimentary set of classes that focus on the content of the work—the art rather than the technique.

Drawing Antiquities in art class at the Department of Fine and Applied Art of Drexel University in 1904.
Photo: Drexel University Archives and Special Collections.

The drawing studio at the Department of Fine and Applied Art of Drexel University in 1892.
Photo: Drexel University Archives and Special Collections.

Drawing antiquities in drawing class at the Art Institute of Chicago, 2004.
Photo: Rachelle Bowden.

Copying the Classics

In traditional art education, drawing antiquities laid the foundation for building up the student's skills. One only progressed to life drawing once adequate proficiency had been reached. There were two underlying pedagogical principles in this. The first was to focus on technique, not composition or art. The student is not overwhelmed by trying to deal with issues such as thinking about an appropriate subject, composition, creativity, etc. These classes were about learning how to master the pencil, a prerequisite to high art. Second, by using plaster casts of iconic sculptures, one gained a basic familiarity and intimacy with classic art traditions. That is, the exercise doubled as a lesson in art history. I think that we can learn something from these traditions in terms of how we teach interaction and experience design.

It seems to me that this model works, is proven, and can be applied to interaction design. Furthermore, in the process, we can follow another established tradition and develop technique by reproducing the classics. In planning a curriculum, I would suggest that every example that I have included here should be evaluated as a candidate for an exercise in technique for the student.

Copying the classics is not only a good way to help gain an appreciation for the contributions of the past, it is also an excellent means of exercising the development of technique, without the compounding effect of trying to solve a new design problem at the same time. As a potential employer, if the student did not possess the technique to repeat any of the examples that I have discussed, I would question his or her fitness as an interaction designer.

The flip-side of the coin in all of this, of course, is that with it comes a need for the school or the company to provide the facilities or infrastructure that enables these types of activities to take place.

Moving Away from the Cult of the Individual

In any endeavour, the rule of thumb is that you get back that which you reward. So let's look at what we reward in our educational system and in our culture, in general. Based on sports, music, business, science, and the arts, I would suggest that the biases of our culture are all toward the cult of the individual. Our popular media are all about the superstar or the genius. But having had the privilege of working with world-class performers in sports, science, academia, the arts, business, and politics, I will tell you what I think. This whole cult of the individual is a superficial sham, and that following it blindly is causing us to throw away precious resources.

The myth of the individual superstar can be peeled back and revealed to be every much the fraud as Toto's pulling back of the curtain on the Wizard. What is inevitably found behind the facade is a group, a team, a community on which that individual's performance is founded. Even in the renaissance there was not a "Renaissance Man." Donatello would never have emerged if the Medici had not set up (in contemporary parlance) a "farm team" system of studios from which he could be "discovered." And it was not just the existence of the studio, where he began as an apprentice, or his talent, that led to his fame. It was also the patronage of the Medici, that is, the essentially idiosyncratic nature of their aesthetic taste, who just as easily could have chosen to support someone else with similar talent, but a different style.

To the extent that I am smart, I am smart because other smart people shared with me what they knew. The way that I pay them back is sharing what I know. This is the underpinning of academic life: the community of scholars.

The world works through mutual exploitation by consenting adults.

Though it may sound like it, this is not just me spouting off with an unsubstantiated personal opinion. There is good science that supports this viewpoint. Perhaps the most accessible is Ed Hutchin's (1995) classic study, *Cognition in the Wild*. In it, he showed how cognition was distributed among the group and embedded in the physical artifacts used by the subjects. In his case, the study was around ship navigation, but it could just as easily have been the design studio. What it suggests is that we simply cannot look at individual performance without considering the social, cultural and physical ecology that provides the context for that action. And if that ecology is so important, then it seems common sense that we begin to spend more time considering it as part of the design brief.

It is not a renaissance man or woman that we need to be cultivating, but the "renaissance team." In today's world of specialization, the problems are such as to require a great deal of depth in each of a range of disciplines. We have already mentioned a few: business, design, engineering, marketing, manufacturing, and science. No individual can possess all these skills at the level that is required to execute in a competitive way.

Yet, by the same token, those with depth in the requisite disciplines have been educated in a cultural silo, isolated by school, department, faculty, funding agency, tenure committee, and so on—all to the detriment of constructing the type of strong heterogeneous social networks (a.k.a. "renaissance teams") that are required to competently address today's problems.

On hearing me rant on this topic once upon a time, my friend Kelly Booth of the University of British Columbia came up with this:

> It's like this. Back in kindergarten, little Billy would come home from school with a glowing note from the teacher, with five gold stars, saying "Little Billy plays well with others." Loving such praise, Billy goes on the next year searching for more of the same. But from grade one on, through the rest of his formal schooling, do you know what that same behaviour was called: cheating.

Collective problem solving is not a significant part of our education. Virtually all rewards and examinations are about individual performance. And, in the few exceptions where group problem solving is involved, it is a homogenous group of peers, all from the same silo, that constitute the team.

Here is how I think about this:

> **How can you do experience design without a rich body of personal experience as an individual and as a group?**

Through a heterogeneous group, you inherently extend the range of experience that you can draw on. And, as a group, a prime mandate is to push the base of experience ever further—certainly further than could exist in any individual.

In the case study of the iPod or the mountain bike, we saw clearly the critical nature, and mutual dependence, of the different disciplines involved. Any company attempting to excel in developing new products must understand not only this dependence, but also the fact that the employees that they hire are coming from an educational system and culture that biases away from the trans-disciplinary group, and toward the individual or silo. Likewise, every school or department must acknowledge and compensate for the realities of today, and the inevitable fact that they exist as an island, populated by only a small subspecies of the larger community on which successful product design depends. The challenge is to work within the system today, and make the appropriate accommodations.

And Finally ...

Like the word "mathematics," I think the word "future" should be pluralized, as in "futures." As long as it is singular, there is a bias toward thinking that there is only one future. That takes the emphasis, and attendant responsibilities, away from the reality that there are many possible futures, and that it is our own decisions that will determine which one we end up with.

Design has a significant role to play in navigating through this space. Our focus in this book has been on experience design, and how that affects our approach to designing products. Throughout I have tried to ground the discussion and my arguments in terms that would relate to, and resonate with, those who have the most power to bring about change, namely those in business, research, and academia.

But at the same time, I can't deny my roots as both a musician and child of the 60s, and want to close with an attempt to raise the bar.

I spent a reasonable amount of time in 2003 and 2004 working with my friend, the Toronto designer Bruce Mau, and his students at the *Institute Without Boundaries*, on the creation of an exhibition called *Massive Change: The Future of Global Design* (Mau & Leonard 2003). One of the key notions underlying the show was this:

> It's not about the world of design; it's about the design of the world.

I have argued that we are at one crossroad in the history of design, namely one where products are assuming complex dynamic behaviours that require new skills and approaches to their design. Likewise, *Massive Change* argued that we are at another crossroad, one where the combined state of economics, technology, and design enables us to seriously address questions that we have never previously been capable of considering. This is reflected in yet another tag line/question/challenge of the exhibition:

> Now that we can do anything, what will we do?

I am by nature what I call a skeptimist, that is, someone who tries to find a workable balance between their skepticism and optimism. In my darker moments, I am prone to feel like John Ruskin, who lamented about the creative faculty of his day:

> We live in an age of base conceit and baser servility—an age whose intellect is chiefly formed by pillage, and occupied in desecration; one day mimicking, the next destroying, the works of all the noble persons who made its intellectual or art life possible to it: an age without honest confidence enough in itself to carve a cherry-stone with an original fancy, but with insolence enough to abolish the solar system, if it were allowed to meddle with it. (Ruskin 1870)

In the 1901 edition of his book that I have, there is a footnote to this that he added in 1887, where he says:

> Every day these bitter words become more sorrowfully true. (p. 234)

I suspect that he would say the same thing today, especially if he was speaking of our approach to technology or attitude to the environment.

Design and the attendant ability to change things is the only way that I can hold my inner skeptic at bay. I believe that what we have discussed in this book can be applied to design at many different levels, and if we can improve our overall practice and our ability to realize the potential of the resources available to us today, that these skills may have an impact beyond short-term commercial benefit.

Massive Change was largely prompted by a quote by historian Arnold Toynbee that Bruce noted in the 1957 Nobel Peace Prize lecture by Canada's Prime Minister, Lester B. Pearson:

> The 20th century will be chiefly remembered by future generations not as an era of political conflicts or technical inventions, but as an age in which human society dared to think of the welfare of the whole human race as a practical objective.

Now, rather than just daring to think about it, in the twenty-first century it is time to do something about it. At the risk of repeating myself, design—as we have discussed it in this book—has a fundamental role to play in this work.

Throughout this book I have scattered quotes that held meaning for me. Some reinforced what I was saying, and others challenged and questioned the whole endeavour. Since they have been such an important part of the process, I want to end with one of my favourites. It is from the introductory chapter of T.E. Lawrence's, *Seven Pillars of Wisdom* (1935):

> All men dream: but not equally. Those who dream by night in the dusty recesses of their minds wake in the day to find that it was vanity: but the dreamers of the day are dangerous men, for they may act their dreams with open eyes, to make it possible.

Let my last word be a wish:

May you dream in the day.

Matisse A Second Life
ROMAN SIGNER
ARTIST OF WONDERLAND
A New Testament
Turner's Britain
TURNER
CHARLOTTE PERRIAND
Type Selector
BILL WARD
RICHARD

References and Bibliography

References and Bibliography

A

Adamson, G. (2003). *Industrial strength design: How Brook Stevens shaped your world.* Cambridge: MIT Press.

Anthony, K.H. (1991). *Design juries on trial: The renaissance of the design studio.* New York: Van Nostrand.

Antonelli, P. (2003). *Objects of design from the Museum of Modern Art.* New York: Museum of Modern Art.

Apperley, M.D. & Spence, R. (1980). *Focus on information: The office of the professional (video).* London: Imperial College Television Studio, Production Number 1003.

Apperley, M.D., Tzavaras, I. & Spence, R. (1982). *A bifocal display technique for data presentation. Proceedings of Eurographics '82 Conference of the European Association for Computer Graphics,* 27–43.

Armitage, J. (2004). *Are agile methods good for design? Interactions* 11(1), 14–23.

Attfield, J. (2000). *Wild things: The material culture of everyday life.* Oxford: Berg Publishers.

Arvola, M. & Artman, H. (2006). *Enactments in interaction design: How designers make sketches behave. Artifact 1,* 70-83.

Avrahami, D. & Hudson, S. E. (2002). *Forming interactivity: A tool for rapid prototyping of physical interactive products. Proceedings of the Conference on Designing Interactive Systems: Processes, Practices, Methods, and Techniques (DIS'02),* 141–146.

B

Bacon, E. (1974). *Design of cities (revised edition).* New York: Penguin.

Baecker, R.M. (1969). *Picture-driven animation. Proceedings of the AFIPS Spring Joint Computer Conference,* 273–288.

Baecker, R.M., Grudin, J., Buxton, W. & Greenberg, S. (Eds.) (1995). *Readings in human computer interaction: Toward the year 2000.* San Francisco: Morgan Kaufmann Publishers.

Baecker, R.M. & Small, I.S. (1990). Animation at the interface. In B. Laurel, ed., The art of human-computer interface design. Reading, MA: Addison-Wesley, 251–267.

Bahamón, A. (2005). *Sketch • Plan • Build: World class architects show how it's done.* New York: Harper Design.

Bailey, B., Konstan, J. & Carlis, J. (2001). DEMAIS: Designing multimedia applications with interactive storyboards. *Proceedings of ACM Multimedia 2001,* 241–250.

Bailey, W., Knox, S. & Lynch, E. (1988). Effects of interface design upon user productivity. *Proceedings of the ACM-SIGCHI Conference on Human Factors in Computing Systems (CHI'88),* 207–212.

Balakrishnan, R., Fitzmaurice, G., Kurtenbach, G. & Buxton, W. (1999). Digital tape drawing. *Proceedings of the ACM Symposium on User Interface Software and Technology (UIST'99),* 161–169.

Balakrishnan, R., Fitzmaurice, G. & Kurtenbach, G (2001). User interfaces for volumetric displays. *IEEE Computer,* 34(3), 37–45.

Ballagas, R., Ringel, M., Stone, M. & Borchers, J. (2003). Between u and i: iStuff: A physical user interface toolkit for ubiquitous computing environments. *Proceedings of the ACM-SIGCHI Conference on Human Factors in Computing Systems (CHI'03),* 537–544.

Bardini, T. (2000). *Bootstrapping: Douglas Engelbart, coevolution, and the origins of personal computing.* Stanford: Stanford University Press.

Baum, L. F. (1900). *The wonderful wizard of Oz.* Chicago and New York: Geo. M. Hill Co. Text: http://etext.virginia.edu/toc/modeng/public/BauWiza.html.

Baxter, M. (1995). *Product design: Practical methods for the systematic development of new products.* London: Chapman & Hall.

Bayazit, N. (2004). *Investigating design: A review of forty years of design research.* Design Issues, 20(1), 16–29.

Bayles, D. & Orland, T. (2001). *Art & Fear: Observations on the perils (and rewards) of artmaking.* Santa Cruz, CA: Image ContinuumPress.

Beck, K. (1999). *Extreme programming explained: Embrace change.* Boston: Addison-Wesley.

Bekker, M. (1992). Representational issues related to communication in design teams. Posters and Short Talks: *Proceedings of the ACM-SIGCHI Conference on Human Factors in Computing Systems (CHI'92),* 151–152.

Belady, L.A. & Lehman, M.M. (1971). *Programming systems dynamics or the metadynamics of systems in maintenance and growth. IBM Research Report RC 3546.* Yorktown Heights: T.J. Watson Research Center.

Bel Geddes, N. (1932). Horizons. Boston, Little, Brown & Com

Benyon, D., Turner, P. & Turner, S. (2005). *Designing interactive systems: People, activities, contexts, technologies.* Edinburgh: Pearson Education Ltd.

Bergman, E. (Ed.) (2000). *Information appliances and beyond: Interaction design for consumer products.* San Francisco: Morgan Kaufmann Publishers.

Bergman, E. & Haitani, R. (2000). Designing the PalmPilot: A conversation with Rob Haitani. In Eric Bergman, ed., *Information appliances and beyond: Interaction design for consumer products.* San Francisco: Morgan Kaufmann Publishers, 81–102.

Bergman, E., Lund, A., Dubberly, H., Tognazzini, B. & Intille, S. (2004). Video visions of the future: A critical review. *Proceedings of the ACM-SIGCHI Conference on Human Factors in Computing Systems (CHI'04): Extended Abstracts,* 1584–1585.

Beyer, H. & Holtzblatt, K. (1999). *Contextual design. Interactions,* 6(1), 32–42.

Bicycle Magazine (Eds.) (2003). *The noblest invention: An illustrated history of the bicycle.* Emmaus, PA: Rodale Inc.

Black, A. (1990). Visible planning on paper and on screen: The impact of working medium on decision-making by novice graphic designers. Behavior and Information Technology, 9(4), 283–296.

Blomberg, J. & Henderson, D. A. (1990). Reflections on participatory design: Lessons from the trillium experience. Proceedings of the ACM-SIGCHI Conference on Human Factors in Computing Systems (CHI'90) 353–359.

Bly, S., Harrison, S. & Irwin, S. (1993). Media spaces: Bringing people together in a video, audio and computing environment. Communications of the ACM (CACM), 36(1), 28–47.

Bolter, J. & Gromala, D (2003). *Windows and mirrors: Interaction design, digital art, and the myth of transparency.* Cambridge, MA: MIT Press.

Boling, E. & Frick, T. (1997). Holistic rapid prototyping for web design: Early usability testing is essential. In B. Khan, ed., *Web-based instruction.* Englewood Cliffs, N.J.: Educational Technology Publications, 319–328.

Borgmann, A. (1984). *Technology and the character of contemporary life: A philosophical inquiry.* Chicago: University of Chicago Press.

Bowers, J., Pycock, J. (1994) Talking through design: Requirements and resistance in cooperative prototyping. *Proceedings of the ACM-SIGCHI Conference on Human Factors in Computing Systems (CHI'94),* 229–305.

Brand, S. (1987). *The Media Lab – Inventing the Future at MIT.* New York: Viking Penguin.

Brandt, E. & Grunnet, C. (2000). *Evoking the future: Drama and props in user centred design.* Proceedings of the Participatory Design Conference, 11–20.

Brooks, Frederick P. Jr. (1975). *The mythical man-month.* Reading, MA: Addison-Wesley.

Brouwer, A. & Wright, T.L. (1991). *Working in Hollywood.* NY: Avon Books.

Brown, G.S. (1969). The Laws of Form. London: George Allen and Unwin.

Brown, J.S., Denning, S., Groh, K. & Prusak, L. (2005). *Storytelling in organizations: Why storytelling is transforming 21st century organizations and management.* Burlington, MA: Elsevier Butterworth-Heinemann.

Buchenau, M. & Suri, J.F. (2000). Experience prototyping. *Proceedings of the Conference on Designing Interactive Systems: Processes, Practices, Methods, and Techniques,* 424–433.

Burns C., Dishman E., Verplank B. & Lassiter B. (1994). Actors, hair-dos and videotape – Informance design: Using performance techniques in multi-disciplinary, observation based design. *ACM-SIGCHI Conference on Human Factors in Computing Systems (CHI'94) Conference Companion,* 119–120.

Burns, C., Dishman, E., Johnson, B. & Verplank, B. (1995). "Informance": Min(d)ing future contexts for scenario based interaction design. Abstract for Aug. 8 BayCHI Talk. (www.Baychi.org/calendar/19950808/)

Burrows, M. (2000). *Bicycle design: Towards the perfect machine.* York: Open Road.

Butter, A. & Pogue, D. (2002). *Piloting Palm: The inside story of Palm, Handspring, and the birth of the billion-dollar hand-held industry.* Hoboken, NJ: John Wiley & Sons.

Buxton, W. (1986). Chunking and phrasing and the design of human-computer dialogues, *Proceedings of the IFIP World Computer Congress,* Dublin, Ireland, 475–480.

Buxton, W. (1990). The natural language of interaction: A perspective on non-verbal dialogues. In Laurel, B. (Ed.). *The art of human-computer interface design,* Reading, MA: Addison-Wesley. 405–416.

Buxton, W. (1997). Living in augmented reality: Ubiquitous media and reactive environments. In K. Finn, A. Sellen & S. Wilber, eds., *Video mediated communication.* Hillsdale, NJ: Erlbaum, 363–384.

Buxton, W. (2001). Less is more (more or less). In P. Denning, ed., *The invisible future: The seamless integration of technology in everyday life.* New York: McGraw Hill, 145–179.

Buxton. W. (2003). Performance by design: The role of design in software product development. *Proceedings of the Second International Conference on Usage-Centered Design.* Portsmouth, NH, 26–29 October 2003, 1–15.

Buxton, W. (2005). Innovation vs. invention. *Rotman Magazine, The Alumni Magazine of the Rotman School of Management,* Fall 2005, 52–53.

Buxton, W. (2006). What if Leopold didn't have a piano? *Rotman Magazine, The Alumni Magazine of the Rotman School of Management,* Spring/Summer 2006, 78–79.

Buxton, W., Fitzmaurice, G. Balakrishnan, R. & Kurtenbach, G. (2000). Large displays in automotive design. *IEEE Computer Graphics and Applications,* 20(4), 68–75.

Buxton, W., Lamb, M. R., Sherman, D. & Smith, K. C. (1983). Towards a comprehensive user interface management system. *Computer Graphics* 17(3), 31–38.

Buxton, W. & Moran, T. (1990). EuroPARC's integrated interactive intermedia facility (iiif): Early experience. In S. Gibbs & A.A. Verrijn-Stuart, eds., *Multi-user interfaces and applications, Proceedings of the IFIP WG 8.4 Conference on Multi-user Interfaces and Applications,* Heraklion, Crete. Amsterdam: Elsevier Science Publishers B.V. (North-Holland), 11–34.

Buxton, W. & Myers, B. (1986). A study in two-handed input. *Proceedings of the ACM-SIGCHI Conference on Human Factors in Computing Systems (CHI'86),* 321–326.

Buxton, W. & Sniderman, R. (1980). Iteration in the design of the human-computer interface. *Proceedings of the 13th Annual Meeting, Human Factors Association of Canada,* 72–81.

Buxton, W., Sniderman, R., Reeves, W., Patel, S. & Baecker, R. (1979). The evolution of the SSSP score editing tools. *Computer Music Journal* 3(4), 14–25.

C

Caputo, Tony (2003). *Visual story telling: The art and technique.* New York: Watson-Guptill Publications.

Card, S., Moran, T. & Newell, A. (1983), *The psychology of human-computer interaction,* Hillsdale, NJ: Lawrence Erlbaum Associates.

Card, S.K., Robertson, G.G., and Mackinlay, J.D. (1991). The information visualizer, an information workspace. *Proceedings of the ACM-SIGCHI Conference on Human Factors in Computing Systems (CHI'91),* 181–186.

Carpendale, M.S.T. (1999). *A framework for elastic presentation space.* Ph.D. Thesis, School of Computing Science at Simon Fraser University.

Carpendale, M.S.T., Cowperthwaite, D.J. & Fracchia, F.D. (1995). Three-dimensional pliable surfaces: For effective presentation of visual information. *Proceedings of the ACM Symposium on User Interface Software and Technology (UIST'95),* 217–226.

Carroll, J. (Ed.) (1995). *Scenario-based design: Envisioning work and technology in system development.* New York: Wiley & Sons.

Carroll, J. (2000). *Making use: Scenario-based design of human-computer interaction.* Cambridge: MIT Press.

Carter, J. (1992). Managing: To succeed with rapid prototyping. *Proceedings of the Human Factors Society 36th Annual Meeting,* 404–408.

Catani, M. B., & Biers, D.W. (1998). Usability evaluation and prototype fidelity: users and usability professionals. *Proceedings of the Human Factors and Ergonomics Society 42nd. Annual Meeting,* 1331–1335.

Chandler, G. (2004). *Cut by cut: Editing your film or video.* Studio City CA: Michael Wiese Productions.

Clark, A. (1997). *Being there: Putting brain, body, and world together again.* Cambridge: MIT Press.

Clark, A. (2001). *Mindware: An introduction to the philosophy of cognitive science.* Oxford: Oxford University Press.

Clement, A. & Van den Besselaar, P. (1993). A retrospective look at PD projects. *Communications of the ACM,* 36(6), 29–37.

Computer Science and Telecommunications Board of the National Research Council (2003). *Innovation in information technology.* Washington DC: The National Academies Press.

Cooper, A. (1999). *The inmates are running the asylum.* Indianapolis: SAMS.

Cooper, A. & Reimann, R. (2003). *About face 2.0: The essentials of interaction design.* Indianapolis: Wiley.

Cooper, R.G. (2001). *Winning at new products: Accelerating the process from Idea to Launch, 3e* (1e, 1993). New York: Basic Books.

Cullen, C.D. & Haller, L. (2004). *Design secrets—Products 2: 50 real-life projects uncovered.* Gloucester, MA: Rockport Publishers.

D

Dann, J. (2005). Spalding: An idea with bounce. *Technology Review.* 108(4), 36–37.

DeFanti, T.A.; Brown, M.D. & McCormick, B.H. (1989). Visualization: Expanding scientific and engineering research opportunities. *IEEE Computer,* 22(8), 12–16, 22–5.

Della Vigna, S. & Pollet, J.M. (2005). Attention, demographics, and the stock market. *NBER Working Paper No. W11211.*

Denning, P.J. (Ed.) (2001). *The invisible future: The seamless integration of technology in everyday life.* New York: McGraw Hill.

Denning, P.J. & Metcalfe, R.M. (Eds.) (1997). *Beyond calculation: The next fifty years of computing.* New York: Springer-Verlag.

Denning, S. (2001). *The springboard: How storytelling ignites action in knowledge-era organizations.* Boston: Butterworth Heinemann.

Deutschman, A. (2001). *The second coming of Steve Jobs.* New York: Broadway Books.

Dewey, J. (1929). *The quest for certainty: A study of the relation of knowledge and action.* New York: Minton, Balch & Company.

Dourish, P. (2001). *Where the action is: The foundations of embodied interaction.* Cambridge: MIT Press.

Dreyfuss, H. (1955). *Designing for people.* NY: Simon & Shuster.

Droste, M. (2006). *Bauhaus 1919–1933.* Köln: Taschen.

Dubberly, H. & Mitch, D. (1987). *The knowledge navigator.* Apple Computer, video tape.

te Duits, T. (Ed.) (2003). *The origin of things: Sketches, models, prototypes.* Rotterdam: Museum Boijmans Van Beuningen / NAI Publishers.

Dunne, A. (1999). *Hertzian tales: Electronic products, aesthetic experience and critical design.* London: Royal College of Art.

Dunne, A. & Raby, F. (2001). *Design noir: The secret life of electronic objects.* Basel: Birkhäuser.

E

Edwards, B. (1989). *Drawing on the right side of the brain: A course in enhancing creativity and artistic confidence* (revised edition). Los Angeles: Jeremy P Tarcher, Inc.

Ehn, P. (1988). *Work oriented design of computer artifacts.* Stockholm: Aretslivscentrum.

Ehn, P., Linde, P. (2004). Embodied interaction: Designing beyond the physical-digital divide. *Futureground, Design Research Society Int. Conf. 2004.* Melbourne: Monash University.

Erdmann, R.L. & Neal, A.S. (1971). Laboratory vs. field experimentation in human factors – An evaluation of an experimental self-service airline ticket vendor. *Human Factors* 13, 521–531.

Erickson, T. (1995). Notes on design practice: Stories and prototypes as catalysts for communication. In J. Carroll, ed., *Scenario-based design: Envisioning work and technology in system development.* New York: Wiley & Sons, 37–58.

Erickson, T. (1996). Design as storytelling. *Interactions* 3(4), 31–35.

Ericson, E. & Pihl, J. (2003). *Design for impact: Fifty years of airline safety cards.* NY: Princeton Architectural Press.

F

Fällman, D. (2003a). Design-oriented human-computer interaction. *Proceedings of the ACM-SIGCHI Conference on Human Factors in Computing Systems (CHI'03),* 225–232.

Fällman, D. (2003b). *In romance with the materials of mobile interaction.* PhD Thesis, Department of Informatics, Umeå University.

Ferguson, E. S. (1992). *Engineering and the mind's eye.* Cambridge MA: MIT Press.

Fiell, C. & Fiell, P. (2006). *Industrial design A-Z.* Köln: Taschen.

Fishkin, K. P. (In press). *A taxonomy for and analysis of tangible interfaces.* Journal of Personal and Ubiquitous Computing.

Fitzmaurice, G.W. (1993). Situated information spaces and spatially aware palmtop computers. *Communications of the ACM (CACM),* 36(7), 39–49.

Fitzmaurice, G.W. (1996). *Graspable user interfaces.* Ph.D. Thesis, Department of Computer Science, University of Toronto.

Fitzmaurice, G.W., Ishii, H. & Buxton, W. (1995). Bricks: Laying the foundations for graspable user interfaces. *Proceedings of the ACM-SIGCHI Conference on Human Factors in Computing Systems (CHI'95),* 442–449.

Forty, A. (1986). *Objects of desire design & society from Wedgwood to IBM.* NY: Pantheon Books.

Fraser, I. & Henmi, R. (1994). *Envisioning architecture: An analysis of drawing.* New York: John Wiley & Sons, Inc.

Frens, J. (2006). *Designing for rich interaction: Integrating form, interaction, and function.* PhD. Thesis, Technical University Eindhoven.

G

Gaver, W. (1991). Technology affordances. *Proceedings of the ACM-SIGCHI Conference on Human Factors in Computing Systems (CHI'91),* 79–84.

Gaver, W., Beaver, J. & Benford, S. (2003). Ambiguity as a resource for design. *Proceedings of the ACM-SIGCHI Conference on Human Factors in Computing Systems (CHI'03),* 233–240.

Gedenryd, H. (1998). *How designers work: Making sense of authentic cognitive activities.* PH.D. Dissertation, Lund University Cognitive Science. (http://asip.lucs.lu.se/People/Henrik. Gedenryd/HowDesignersWork/)

Gelernter, D. (1998). *Machine beauty: Elegance and the heart of technology.* New York: Basic Books Inc.

Gershon, N. & Page, W. (2001). What Storytelling can do for information visualization. *Communications of the ACM (CACM),* 44(8), 31–37.

Gibbons, M. & Hopkins, D. (1980). How experiential is your experience-based program? *The Journal of Experiential Education,* 3(1), 32–37.

Gibson, J.J. (1979). *The ecological approach to visual perception.* New York: Houghton Mifflin.

Gladwell, M. (2000). *The tipping point: How little things can make a big difference.* Boston: Little, Brown & Co.

Goel, V. (1995). *Sketches of thought.* Cambridge, MA: MIT Press.

Goldschmidt, G. (2003). *The backtalk of self-generated sketches. Design Issues,* 19(1), 72-88.

Goldschmidt, G. (1991). *The dialectics of sketching. Creativity Research Journal,* 4(2), 123–143.

Gould, J. (1988). How to design usable systems. In M. Helander, ed., *Handbook of human-computer interaction.* Amsterdam: North-Holland Elsevier, 757–789. Reprinted in Baecker et al. (1995), op. cit., 93–121.

Gould, J., Conti, J. & Hovanvecz, T. (1983). Composing letters with a simulated listening typewriter. *Communications of the ACM (CACM),* 26(4), 295–308.

Greenbaum, J. & Kyng, M. (Eds.) (1991). *Cooperative design of computer systems.* Hillsdale, NJ: Lawrence Erlbaum Associates.

Greene, L. M. (2001). *Inventorship: The art of innovation.* New York: John Wiley & Sons.

Greene, S. (1990). Prototyping: An integral component of application development. *Proceedings of the Human Factors Society 34th Annual Meeting,* 266.

Greenberg, S. (2005). Collaborative physical user interfaces. In K. Okada, T. Hoshi & T. Inoue, eds., *Communication and Collaboration Support Systems.* Amsterdam: IOS Press.

Greenberg, S. & Boyle, M. (2002). Customizable physical interfaces for interacting with conventional applications. *Proceedings of the 15th Annual ACM Symposium on User Interface Software and Technology (UIST'02),* 31–40.

Greenberg, S. & Fitchett, C. (2001). Phidgets: Easy development of physical interfaces through physical widgets. *Proceedings of the 14th Annual ACM Symposium on User Interface Software and Technology (UIST'01),* 209–218.

Griffin, J.H. (1961). *Black like me.* Boston: Houghton Mifflin.

Gross, M.D. (1998). The proverbial back of the envelope. *IEEE Intelligent Systems,* 13(3), 10–13.

Gross, M. D. & Do, E.Y. (1996). Ambiguous intentions. *Proceedings of the ACM-SIGCHI Conference on Human Factors in Computing Systems (CHI'96),* 183–192.

H

Hanks, K. & Belliston, L. (1990). *Rapid Viz: A new method for the rapid visualization of ideas.* Menlo Park, CA: Crisp Publications.

Hannah, G.G. (2002). *Elements of design: Rowenda Reed Kostellow and the structure of visual relationships.* New York: Princeton Architectural Press.

Hart, J. (1999). *The art of the storyboard: Storyboarding for film, TV, and animation.* Woburn MA: Focal Press.

Hatch, P. & McDonagh, D. (Eds.) (2006). *Realize design means business.* Dulles, VA: Industrial Designers Society of America.

Hawn, C. (2004). Steve Jobs, Apple and the limits of innovation. *Fast Company,* 78 (January), 68–74.

Henderson, D. Austin (1986). The trillium user interface design environment. *Proceedings of the ACM-SIGCHI Conference on Human Factors in Computing Systems (CHI'86),* 221–227.

Henderson, K. (1999) on line and on paper: *Visual representations, visual culture, and computer graphics in design engineering.* Cambridge, MA: MIT Press.

Herbert, D.M. (1993). *Architectural Study Drawings.* New York: Van Nostand Reinhold.

Highsmith, J. (2002). *Agile software development ecosystems.* Boston: Addison-Wesley.

Hirshberg, J. (1998). *The creative priority: Putting innovation to work in your business.* New York: HarperBusinss.

Hix, D. & Hartson, R. (1993). *Developing user interfaces: Ensuring usability through product & process.* New York: John Wiley & Sons.

Holmquist, L. E. (2005). Prototyping: Generating ideas or cargo cult designs? *Interactions,* 12(2), 48–54.

Holmquist, L. E., Gellersen, H-W, Kortuem, G., Schmidt, A., Strohbach, M., Antifakos, S. et al. (2004). Building intelligent environments with Smart-Its. *IEEE Computer Graphics and Applications,* 56–64.

Holmquist, L. E., Mazé, R. & Ljungblad, S. (2003). Designing tomorrow's smart products – Experience with the Smart-Its platform. *Proceedings of DUX 2003,* 1–4.Holzner, S. (2005). *How Dell does it: Using speed and innovation to achieve extraordinary results.* New York: McGraw-Hill, Inc.

Hopps, W. & Davidson, S. (1997). *Robert Rauschenberg: A retrospective* (exh. cat.). New York: Guggeneheim Museum.

Hummels, C. (2000). *Gestural design tools: Prototypes, experiments and scenarios.* PhD Thesis, Technical University Delft.

Hummels, C., Djajadiningrapt, T. & Overbeeke, K. (2001). Knowing, doing and feeling: Communication with your digital products. *Interdisziplinäres Kolleg Kognitions und Neurowissenschaften, Günne am Möhnesee,* March 2–9 2001, 289–308.

Houde, S., Hill, C. (1997). What do prototypes prototype? In M. Helander, T. Landauer, and P. Prabhu, eds., *Handbook of Human-Computer Interaction, 2e.* Amsterdam: Elsevier Science B.V., 367–381.

Hutchins, E. (1995). *Cognition in the wild.* Cambridge, MA: MIT Press.

Huurdeman, A. (2003). *The worldwide history of telecommunications.* Hoboken NJ: John Wiley & Sons.

I

Iacucci, G., Iacucci, C. & Kuutti, K. (2002). Imagining and experiencing in design, the role of performances. *Proceedings of NordiCHI, The Nordic Conference on Human-Computer Interaction,* 167–176.

Iacucci, G. & Kuutti, K. (2002). Everyday life as a stage in creating and performing scenarios for wireless devices. *Personal and Ubiquitous Computing* 6(4): 299–306.

Iacucci, G., Kuutti, K. & Ranta, M. (2000). On the move with a magic thing: Role playing in concept design of mobile services and devices. *Proceedings of the Conference on Designing Interactive Systems: Processes, Practices, Methods, and Techniques (DIS'00),* 193–202.

Illich, I. (1971). *Deschooling society.* New York: Harper and Row.

Illich, I. (1973). *Tools for conviviality.* New York: Harper & Row Publishers.

Industrial Designers Association of America (IDSA) (2001). *Design secrets: Products.* Gloucester, MA: Rockport Publishers.

Isaacs, E. & Walendowski, A. (2002). *Designing from both sides of the screen: How designers and engineers can collaborate to build cooperative technology.* Indianapolis: New Riders.

Ishii, H. & Kobayashi, M. (1992). ClearBoard: A seamless medium for shared drawing and conversation with eye contact. *Proceedings of the ACM-SIGCHI Conference on Human Factors in Computing Systems (CHI'92),* 525–532, 705–706.

Isensee, S. & James, R. (1996). *The art of rapid prototyping: User interface design for Windows and OS/2.* London: International Thomson Computer Press.

J

Jenson, S. (2002). *The simplicity shift.* Cambridge: Cambridge University Press.

Johnson, J., Roberts, T.L., Verplank, W., Smith, D.C., Irby, C.H., Beard, M., Mackey, K. (1989). The Xerox Star: A retrospective, *IEEE Computer,* 22(9), 11– 26, 28–29.

Jones, J.C. (1991). *Designing designing.* London: Architecture Design and Technology Press.

Jones, J.C. (1992). *Design methods, 2e.* New York: Van Nostrand Reinhold.

K

Kaptelinin, V. & Czerwinskim M. (Eds)(2007). *Beyond the desktop metaphor: Designing integrated digital work environments.* Cambridge: MIT Press.

Katz, S.D. (1991). *Film directing: Shot by shot: Visualizing from concept to screen.* NY: Focal Point Press.

Keller, A.I. (2005). *For inspiration only: Designer interaction with informal collections of visual material.* PhD Thesis, Technical University Delft.

Kelley,T. & Littman J. (2001). *The art of innovation: Lessons in creativity from ideo, America's leading design Firm.* Doubleday/Currency Books.

Kicherer, S. (1990). *Olivetti: A study of the corporate management of design.* New York: Rizzoli Inc.

Klemmer, S., Li, J., Lin, J. & Landay, J. (2004). Paper-mâché: Toolkit support for tangible input. *Proceedings of the ACM-SIGCHI Conference on Human Factors in Computing Systems (CHI'04),* 399–406.

Kolb, D. (1984). Experiential learning: Experience as the source of learning and development. Englewood Cliffs, NJ: Prentice-Hall.

Koskinen, I., Battarbee, K. & Mattelmäki, T. (2003). *Empathic design: User experience in product design.* Edita: IT Press.

Kranzberg, M. (1986). Technology and history: "Kranzberg's laws." *In Technology and Culture,* 27(3): 544–560.

Kubo, M. & Prat, R. (Eds.)(2005) *Seattle Public Library: OMA / LMN.* Barcelona: Actar.

Kuhn, T. (1962). *The structure of scientific revolutions.* Chicago: University of Chicago Press.

Kuniavsky, M. (2003). *Observing the user experience: A practitioner's guide.* San Francisco: Morgan Kaufmann.

L

Laing, G. (2004). *Digital retro: The evolution and design of the personal computer.* Alameda CA: Ilex Press.

Landay, J.A. (1996a). *Interactive sketching for the early stages of user interface design.* PhD Thesis, Carnegie Mellon University.

Landay, J.A. (1996b). SILK: Sketching interfaces like krazy. Conference Companion. *Proceedings of the ACM-SIGCHI Conference on Human Factors in Computing Systems (CHI'96),* 398–399.

Landay, J.A. & Myers, B.A. (1995). Interactive sketching for the early stages of user interface design. *Proceedings of the ACM-SIGCHI Conference on Human Factors in Computing Systems (CHI'95),* 43–50.

Landay, J.A. & Myers, B.A. (2001). Sketching interfaces: Toward more human interface design. *IEEE Computer,* 34(3), 56–64.

LaSalle, D. & Britton, T. (2003). Priceless: *Turning ordinary products into extraordinary experiences.* Boston: Harvard Business School Press.

Laseau, P. (1980). *Graphic thinking for architects and designers.* New York: Van Nostrand Reinhold Company.

Latour, B. (1996). *Aramis or the love of technology.* Cambridge, MA: Harvard University Press.

Laurel, B. (Ed.) (2003). *Design research: Methods and perspectives.* Cambridge, MA: MIT Press.

Lawrence, T.E. (1935). *Seven pillars of wisdom: A triumph.* London: Jonathan Cape.

Lawson, B. (1997). *How designers think: The design process demystified,* 3e. Amsterdam: Elsevier.

Lawson, B. (2004). *What designers know.* Amsterdam: Elsevier.

Lefèvre, W. (Ed.) (2004). *Picturing machines 1400-1700.* Cambridge, MA: The MIT Press.

Leppälä, K., Kerttula, M. & Tuikka, T. (2003). *Virtual design of smart products.* Edita: IT Press.

Levy, J. (2002). *Really useful: The origins of everyday things.* Willowdale, Canada: Firefly Books.

Levy, S. (2006). *The perfect thing: How the iPod shuffles commerce, culture and coolness.* New York: Simon & Shuster.

Lidwell, W., Holden, K. & Butler, J. (2003). *Universal principles of design.* Gloucester, MA: Rockport Publishers.

Li, Y. & Landay, J.A. (2005). Informal prototyping of continuous graphical interactions by demonstration. *Proceedings of the ACM Symposium on User Interface Software and Technology (UIST'05),* 221–230.

Lin, J., Newman, M.W., Hong, J.I. & Landay, J.A. (2000). DENIM: Finding a tighter fit between tools and practice for web site design. *Proceedings of the ACM-SIGCHI Conference on Human Factors in Computing Systems (CHI '00),* 510–517.

Lin, J., Thomsen, M. & Landay, J.A. (2002). A visual language for sketching large and complex interactive designs. *Proceedings of the ACM-SIGCHI Conference on Human Factors in Computing Systems (CHI'02),* 307–314.

Linzmayer, O. (2004). *Apple confidential 2.0.* San Francisco: No Starch Press.

Ljungblad, S., Skog, T. & Gaye, L. (2003). Are designers ready for ubiquitous computing? A formative study. *Proceedings of the ACM-SIGCHI Conference on Human Factors in Computing Systems (CHI'03)* - Extended Abstracts, 992–993.

Logan, J. & Molotch, H. (1987). *Urban fortunes: The political economy of place.* Berkeley: University of California Press.

Löwgren, J. (1995). Applying design methodology to software development. *Proceedings of the Conference on Designing Interactive Systems: Processes, Practices, Methods, & Techniques,* 87–95.

Löwgren, J. & Stolterman, E. (1999). Design methodology and design practice. *Interactions,* 6(1),13–20.

Löwgren, J. & Stolterman, E. (2004). *Thoughtful interaction design.* Cambridge, MA: The MIT Press.

Löwgren, J. (2005). Inspirational patterns for embodied interaction. *Proceedings of the Nordic Design Research Conference (Nordes),* Copenhagen.

M

Macbeth, S., Moroney, W. & Biers, D. (2000). Development and evaluation of symbols and icons: A comparison of the production and focus group methods. *Proceedings of the IEA 200/HFS 200 Congress,* 327–329.

MacLean, A., Young, R. & Moran, T. (1989). Design rationale: The argument behind the artifact. *Proceedings of the ACM-SIGCHI Conference on Human Factors in Computing Systems (CHI'89),* 247–252.

Maclean, N. (1976). *A river runs through it and other stories.* Chicago: University of Chicago Press.

Maeda, J. (2006). *The laws of simplicity.* Cambridge, MA: MIT Press.

Martin, R. (2004). The Design of Business. *Rotman Magazine,* Winter 2004, 6–10.

Massironi, M. (2002). *The psychology of graphic images: Seeing, drawing, communicating.* Mahwah, NJ: Lawrence Erlbaum.

Mattioda, M., Norlen, L. & Tabor, P. (Eds.) (2004). *Interaction design almanac 2004.* Ivrea, Italy: Interaction Design Institute.

Mau, B. & Leonard, J. (2003). *Massive change: A manifesto for the future of global design.* London: Phaidon.

Maulsby, D., Greenberg, S. & Mander, R. (1993). *Prototyping an intelligent agent through Wizard of Oz. Proceedings of the ACM-SIGCHI Conference on Human Factors in Computing Systems (CHI'93),* 277-284.

McCabe, B. (1999). *Dark knights and holy fools: The art and films of terry gilliam.* London: Orion.

McCarthy, J. & Wright, P. (2004). *Technology as experience.* Cambridge, MA: MIT Press.

McCloud, S. (1993). *Understanding comics: The invisible art.* Northampton, MA: Tundra.

McCloud, S. (2000). *Reinventing comics.* New York: Paradox Press.

McCormick, B.H., DeFanti, T.A. & Brown, M.D. (Eds.) (1987). Visualization in scientific computing. *Computer Graphics* 21(6).

McCullough, M. (1996). *Abstracting craft: The practiced digital hand.* Cambridge, MA: MIT Press.

McCullough, M. (2004). *Digital ground: Architecture, pervasive computing, and environmental knowing.* Cambridge, MA: MIT Press.

McDonough, W. & Braungart, M. (2002). *Cradle to cradle: Remaking the way we make things.* New York: North Point Press.

McGee, D. (2004). *The origins of early modern machine design. In Wolfgang Lefèvre, ed., Picturing machines 1400-1700.* Cambridge, MA: The MIT Press, 53–84.

Messner, R. (1979). *Everest: Expedition to the ultimate.* London: Kaye & Ward.

Meyer, J. (1996). EtchaPad – Disposable sketch based interfaces. *Proceedings of the ACM-SIGCHI Conference on Human Factors in Computing Systems (CHI'91)* – Short Papers, 195–196.

Mitchell, C.T. (1993). *Redefining designing. From form to experience.* New York: Van Nostrand Reinhold.

Moll, S. (2004). Framing carbon fiber. *I.D. Magazine,* May 2004, 64–68.

Molotch, H. (2003). *Where stuff comes from: How toasters, toilets, cars, computers, and many other things come to be as they are.* New York: Routledge.

Moggridge, B. (1993). Design by story-telling. *Applied Ergonomics,* 24(1), 15–18.

Moggridge, B. (2006). *Designing Interaction.* Cambridge, MA: The MIT Press.

Moore, P. & Conn, C.P. (1985). *Disguised: A true story.* Waco, Texas: Word Books.

Mountford, S.J. ,Buxton, W., Krueger, M., Laurel, B. & Vertelney, L. (1989). Drama and personality in user interface design. *Proceedings of the ACM-SIGCHI Conference on Human Factors in Computing Systems (CHI'89),* 105–108.

Muller, M. J. 1991. *PICTIVE—an exploration in participatory design. Proceedings of the ACM-SIGCHI Conference on Human Factors in Computing Systems (CHI'91),* 225-231.

Muller, M.J. (2003). Participatory design: The third space in HCI. In J.A. Jacko & A. Sears, eds., *The human-computer interaction handbook: Fundamentals, evolving technologies, and emerging applications,* Lawrence Erlbaum Associates, Chapter 54, 1051–1068.

Mumford, L. (1966). *The myth of the machine: Technics and human development.* New York: Harcourt Brace Jovanovich Inc.

Mumford, L. (1970). *The myth of the machine: The pentagon of power.* New York: Harcourt Brace Jovanovich Inc.

Myers, B. A. (1998). A brief history of human-computer interaction technology. *Interactions* 5(1), 44–54.

N

Nelson, H. G. & Stolterman, E. (2003). *The design way: Intentional change in an unpredictable world. Fundamentals of design competence.* Englewood Cliffs: Educational Technology Publications.

Nielsen, J. (1990). Paper versus computer implementation as mockup scenarios for heuristic evaluation. *Proceedings of IFIP INTERACT '90: Human-Computer Interaction,* 315–320..

Nielsen, J. (1993). *Usability engineering.* Boston: Academic Press.

Noguchi-san (2000). *100 years of bicycle component and accessory design: Authentic reprint edition of the data book.* San Francisco: Van der Plas Publications.

Nokes, S., Major, I., Greenwood, A., Allen, D. & Goodman, M. (2003). *The definitive guide to project mangement: The fast track to getting the job done on time and on budget.* London: FT Prentice Hall.

Norman, D.A. (1988). *The psychology of everyday things.* New York: Basic Books Inc.

Norman, D. (1993). *Things that make us smart: Defending human attributes in the age of the machine.* Reading, MA: Addison-Wesley.

Norman, D. (2004). *Emotional design: Why we love (or hate) everyday things.* New York. Basic Books.

Norman, D. A. & Draper, S. W. (Eds.) (1986), *User centered system design: New perspectives on human-computer interaction.* Hillsdale, NJ: Lawrence Erlbaum Associates.

Novick, D. (2000). Testing documentation with "low-tech" simulation. *Proceedings of IEEE professional communication society international professional communication conference and Proceedings of the 18th annual ACM international conference on Computer documentation: Technology & teamwork,* 55–68.

NTT, (1994). Seamless media design. *SIGGRAPH Video Review, CSCW '94 Technical Video Program,* Issue 106, Item 10, ACM: New York.

O

Olsen, D. R., Buxton, W., Ehrich, R., Kasik, D., Rhyne, J. & Sibert, J. (1984). A context for user interface management. *IEEE Computer Graphics and Applications* 4(12), 33–42.

Osborn, A.F. (1953). *Applied imagination: Principles and procedures of creative thinking.* New York: Charles Scribner's.

Oulasvirta, A., Kurvinen, E. & Kankainen, T. (2003). Understanding contexts by being there: Case studies in bodystorming. *Personal and Ubiquitous Computing,* 7(2): 125–134.

P

Papanek, V. (1995). *The green imperative: Natural design for the real world.* New York: Thames and Hudson.

Pering, C. (2002). Interaction design prototyping of communicator devices: Towards meeting the hardware-software challenge. *Interactions* 9(6), 36–46.

Pfiffner, P. (2003). *Inside the publishing revolution: The Adobe story.* Berkeley: Peachpit Press.

Piedmont-Palladino, S.C. (Ed.)(2007). *Tools of the imagination: Drawing tools and technologies from the eighteenth century to the present.* New York: Princeton Architectural Press.

Pine, B.J. & Gilmore, J.H. (1999). *The experience economy: Work is theatre and every business a stage.* Boston: Harvard Business School Press.

Plimmer, B. & Apperley, M. (2003a). Software to sketch interface designs. *Proceedings of INTERACT'03,* 73–80.

Plimmer, B. & Apperley, M. (2003b). Interacting with sketched interface designs: An evaluation study. *Proceedings of the ACM-SIGCHI Conference on Human Factors in Computing Systems (CHI'03),* 1337–1340.

Porter, M. E. (1987). From competitive advantage to corporate strategy. *Harvard Business Review.* May-June 1987.

Postman, N. (1985). *Amusing ourselves to death: Public discourse in the age of show business.* New York: Viking Penguin.

Powell, D. (2002). *Presentation techniques: A guide to drawing and presenting design ideas.* Boston: Little, Brown and Company.

Preece, J., Rogers, Y. and Sharp, H. (2002). *Interaction design.* New York: John Wiley & Sons.

Pruitt, J. & Adin, T. (2006). *The persona lifecycle: Keeping people in mind throughout product design.* San Francisco: Morgan Kaufmann Publishers.

Pugh, S. (1990). *Total design: Integrated methods for successful product engineering.* Reading MA: Addison-Wesley.

R

Ragnetti, A, Stuardi, L., Marzano, S., Verbrucken, M.,Rigot, B., Philips, I., Baxter, A., Garlick, N. & Van Kuyck. (2006). *The simplicity event 2006.* Eindhoven: Koninklijke Philips Electronics.

Reinertsen, D. (1997). *Managing the design factory: A product developer's toolkit.* New York: The Free Press.

Rescher, N. (1980). *Unpopular essays on technological progress.* Pittsburgh: University of Pittsburgh Press.

Rettig, M. (1994). Prototyping for tiny fingers. *Communications of the ACM (CACM),* 37(4), 21–27.

Rudd, J., Stern, K. & Isensee, S. (1996). Low vs. high-fidelity prototyping debate. *Interactions,* 76–85.

Robbins, E. (1994). *Why architects draw.* Cambridge: MIT Press.

Robertson, G. G., Card, S. K. & Mackinlay, J. D. (1993). Information visualization using 3d interactive animation. *Communications of the ACM (CACM),* 36(4), 57–71.

Rowe, P. (1987). *Design thinking.* Cambridge: MIT Press.

Ruskin, J. (1870). *Lectures on art: Delivered before the University of Oxford in Hilary Term, 1870.* Oxford: Clarendon Press.

S

Schrage, M. (2000). *Serious play – How the world's best companies simulate to innovate.* Boston: Harvard Business School Press.

Schön, D.A. (1967). *Technology and change: The new Heraclitus.* New York: Delacorte Press.

Schön, D. A. (1983). *The reflective practitioner: How professionals think in action.* New York: Basic Books.

Schön, D. A. (1987). *Educating the reflective practitioner.* San Francisco: Jossey-Bass.

Schön, D. A. & Wiggins, G. (1992). Kind of seeing and their functions in designing. *Design Studies,* 13(2), 135–156.

Sefelin, R. (2002). Comparison of paper- and computer based low-fidelity prototypes. *CURE Technical Report 03.* Vienna: Centre for Usability Research & Engineering.

Sefelin, R., Tscheligi, M. & Giller, V. (2003). Paper prototyping—What is it good for? A comparison of paper- and computer-based low-fidelity prototyping. *Proceedings of the ACM-SIGCHI Conference on Human Factors in Computing Systems (CHI'03),* 778–779.

Sharp, H., Rogers, Y. & Preece, J. (2007). *Interaction design: Beyond human-computer interaction. 2nd Ed.* New York: John Wiley & Sons.

Shaw, C. & Ivens, J. (2002). *Building Great Customer Experiences.* NY: Palgrave Macmillan.

Siegel, D. & Dray, S. (2005). Avoiding the next schism: Ethnography and USABILITY. *Interactions,* 12(2), 58–61.

Slade, Giles (2006). *Made to break: Technology and obsolescence in America.* Cambridge MA: Harvard University Press.

Snodgrass, A. & Coyne, R. (2006). *Interpretation in architecture: Design as a way of thinking.* London: Routledge.

Snyder, C. (2003). *Paper prototyping: The fast and simple techniques for designing and refining the user interface.* San Francisco: Morgan Kaufmann Publishers.

Sparke, P. (1987). *Design in context.* London: Guild Publishing.

Spence, R. (2007). *Information visualization: Design for Interaction.* Harlow: Pearson Prentice Hall.

Spence, R. & Apperly, M. (1982). Data base navigation: An office environment for the professional. *Behaviour and Information Technology,* 1(1): 43–54.

Stahovich, T.F. (1998). The engineering sketch. *IEEE Intelligent Systems,* 13(3), 17–19.

Stappers, P.J. (2006). Creative connections: User, designer, context, and tools. *Personal Ubiquitous Computing.* 10(2), 95–100.

Startt, J. (2000). *Tour de France/Tour de Force: A visual history of the world's greatest bicycle race.* San Francisco: Chronicle Books.

Stolterman, E. (1999). The design of information systems – Parti, formants and sketching. *Information Systems Journal,* 9(1), 3–20.

Strömberg, H., Pirttilä, V. & Ikonen, V. (2004). Interactive scenarios – Building ubiquitous computing concepts in the spirit of participatory design. *Personal and Ubiquitous Computing,* 8, 200–207.

Suh, Nam P. (1990). *The principles of design.* New York: Oxford University Press.

Sutherland, I. (1963). *Sketchpad: A man-machine graphical communication system.* PhD Thesis, MIT. http://www.cl.cam.ac.uk/techreports/UCAM-CL-TR-574.html

Suwa, M. and Tversky, B. (1996). What architects see in their sketches: Implications for design tools. *Proceedings of the ACM-SIGCHI Conference on Human Factors in Computing Systems (CHI'96) Conference Companion,* 191–192.

Suwa, M. and Tversky, B. (2002). External representations contribute to the dynamic construction of ideas. In M. Hegarty, B. Meyer, and N. H. Narayanan, eds., *Diagrams 2002.* NY: Springer-Verlag, 341–343.

Svanæs, D. & Seland, G. (2004). Putting the users center stage: Role playing and low-fi prototyping enable end users to design mobile systems. *Proceedings of the ACM-SIGCHI Conference on Human Factors in Computing Systems (CHI'04),* 479–486.

T

Tang, J.C. & Minneman, S.L. (1991a). Videowhiteboard: Video shadows to support remote collaboration. *Proceedings of the ACM-SIGCHI Conference on Human Factors in Computing Systems (CHI'91),* 315–322.

Thurlow, L. (1992). *Head to head: The coming economic battle among Japan, Europe, and America.* New York: William Morrow.

Tohidi, M., Buxton, W., Baecker, R. & Sellen, A. (2006a). Getting the right design and the design right: Testing many is better than one. *Proceedings of the ACM-SIGCHI Conference on Human Factors in Computing Systems (CHI'06),* 1243–1252.

Tohidi, M., Buxton, W., Baecker, R. & Sellen, A. (2006b). User sketches: A quick, inexpensive, and effective way to elicit more reflective user feedback. Proceedings of NordiCHI, The Nordic Conference on Human-Computer Interaction, 105–114.

Tretiack, P. (1999). *Raymond Loewy and streamlined design.* New York: Universe.

Tsang, M., Fitzmaurice, G., Kurtenbach, G., Khan, A. & Buxton, W. (2002). Boom chameleon: Simultaneous capture of 3D viewpoint, voice and gesture annotations on a spatially-aware display. *Proceedings of the 2002 ACM Symposium on User Interface Software and Technology (UIST'02),* 111–120.

Tullis, T. (1990). High-fidelity prototyping throughout the design process. *Proceedings of the Human Factors Society 34th Annual Meeting,* 266.

Turner, P. & Davenport, E. (Eds.) (2005). *Spaces, spatiality and technology.* Dordrecht: Springer.

Tversky, B. (1999). What does drawing reveal about thinking? In J. S. Gero & B. Tversky, eds., *Visual and spatial reasoning in design.* Sydney, Australia: Key Centre of Design Computing and Cognition, 93–101.

Tversky, B. (2002). What do sketches say about thinking? *Proceedings of AAAI Spring Symposium on Sketch Understanding.* 148–151.

U

Ullman, D.G., Wood, S. & Craig, D. (1990). The Importance of drawing in the mechanical design process. *Computer & Graphics,* 14(2), 263–274.

Ulrich, K. & Eppinger, S. (1995). *Product Design and Development.* New York: McGraw-Hill, Inc.

Umbach, S., Musgrave, K. & Gluskoter, S. (2006). Dell design means business. In P. Hatch & D. McDonagh. (Eds.). *Realize design means business.* Dulles, VA: Industrial Designers Society of America.

V

Valéry, P. (1927). De la simulation. NRF, T. XXVIII.

Van der Lugt, R. (2001). *Sketching in Design Idea Generation Meetings.* PhD. Thesis. Technical University Delft.

Van der Lugt, R. (2002). Functions of sketching in design idea generation meetings. *Proceedings of the Fourth Conference on Creativity & Cognition,* 72–79.

Van der Lugt, R. (2005). How sketching can affect the idea generation process in design group meetings. *Design Studies,* 26 (2), 101–122.

Van der Lugt, R. & Stappers, P.J. (Eds.)(2006). *Design and the growth of knowledge: Best practices and ingredients for successful design research.* Delft: StudioLab Press.

Van Gundy, A.B. (1981). *Techniques of structured problem solving.* New York: Van Nostrand Reinhod Co.

Van Rijn, H., Bahk, Y.N., Stappers, P.J. & Lee, K.P. (2006). Three factors for contextmapping in East Asia: Trust, control, and Nunchi. *Codesign: International Journal of CoCreation in Design and Arts,* 2(3), 157–177.

Vertelney, L. (1989). Using video to prototype user interfaces. *SIGCHI Bulletin* 21(2), 57–61. Reprinted in Baecker et al. (1995), op. cit.,142–146.

Vicente, K. (2004). *The human factor: Revolutionizing the way people live with technology.* New York: Routledge.

Villar, N. (2005). *The Friday afternoon project: A two-hour pin & play prototyping exercise.* University of Lancaster: Unpublished manuscript.

Villar, N., Lindsay, A. & Gellersen, H. (2005). Pin & play & perform: A rearrangeable interface for musical composition and performance. *Proceedings of the Conference on New Instruments for Musical Expression (NIME) '05,* 188–191.

Virzi, R. A., Sokolov, J., L. & Karis, D. (1996). Usability problem identification using both low- and high-fidelity prototypes. *Proceedings of the ACM-SIGCHI Conference on Human Factors in Computing Systems (CHI'96),* 236–243.

Vitruvius (1960). *The ten books on architecture.* New York: Dover Publications.

Vogel, C., Cagan, J. & Boatwright, P. (2005). *The design of things to come: How ordinary people create extraordinary products.* Upper Saddle River, NJ: Wharton School Publishing.

W

Waldrop, M. (2001). *The dream machine: J.C.R. Licklider and the revolution that made computing personal.* New York: Viking Penguin.

Walker, M., Takayama, L. & Landay, J. (2002). High-fidelity or low-fidelity, paper or computer medium? *Proceedings of the Human Factors and Ergonomics Society 46th Annual Meeting: HFES2002.* 661–665.

Walsh, V., Roy, R., Bruce, M. & Potter, S. (1992). *Winning by design – Technology, product design and international competitiveness.* Oxford: Blackwell Business.

Weisberg, R. (1993). *Creativity: Beyond the myth of genius.* New York: W.H. Freeman.

Weiser, M. (1991). *The computer for the 21st century. Scientific American,* 265(3), 94–104.

Wensveen, S. (2005). *A tangibility approach to affective interaction.* PhD Thesis, Technical University Delft.

Whisler, R. H. (1943). Inertia controlled shock absorber. United States Patent 2,329,803. Filed Oct. 6, 1941. Issued Sept. 21, 1943.

Whiteway, M. (2001). *Christopher Dresser 1834–1904.* Milan: Skira Edinore.

Whitford, F. (1984). *Bauhaus.* London: Thames & Hudson.

Wiklund, M., Thurrott, C. & Dumas, J. (1992). Does the fidelity of software prototypes affect the perception of usability? *Proceedings of the Human Factors Society 36th Annual Meeting,* 399–403.

Williams, Richard (2001). *The animator's survival kit: A manual of methods, principles and formulas for classical, computer, games, stop motion, and internet animators.* London: Faber & Faber.

Winograd, T. (Ed.) (1996). *Bringing design to software.* New York: ACM Press.

Wong, Y.Y. (1992). Rough and ready prototypes: Lessons from graphic design. Posters and Short Talks: *Proceedings of the ACM-SIGCHI Conference on Human Factors in Computing Systems (CHI'92),* 83–84.

Wong, Y.Y. (1993). Layer tool: Support for progressive design. *Proceedings of the ACM-SIGCHI Conference on Human Factors in Computing Systems (INTERACT '93 and CHI '93),* 127–128.

Woodward, D. & Lewis, G.M. (Eds.)(1998). *The history of cartography, Volume two, book three: Cartography in the traditional African, American, Arctic, Australian, and Pacific societies.* Chicago: University of Chicago Press.

Index